METHODS IN MOLECULAR BIOLOGY

For further volumes:
http://www.springer.com/series/7651

For over 35 years, biological scientists have come to rely on the research protocols and methodologies in the critically acclaimed *Methods in Molecular Biology* series. The series was the first to introduce the step-by-step protocols approach that has become the standard in all biomedical protocol publishing. Each protocol is provided in readily-reproducible step-by-step fashion, opening with an introductory overview, a list of the materials and reagents needed to complete the experiment, and followed by a detailed procedure that is supported with a helpful notes section offering tips and tricks of the trade as well as troubleshooting advice. These hallmark features were introduced by series editor Dr. John Walker and constitute the key ingredient in each and every volume of the *Methods in Molecular Biology* series. Tested and trusted, comprehensive and reliable, all protocols from the series are indexed in PubMed.

NF-κB Transcription Factors

Methods and Protocols

Edited by

Guido Franzoso

Centre for Molecular Immunology and Inflammation, Department of Immunology and Inflammation, Imperial College London, London, UK

Francesca Zazzeroni

Department of Biotechnological and Applied Clinical Sciences, University of L'Aquila, L'Aquila, Italy

Humana Press

Editors
Guido Franzoso
Centre for Molecular Immunology and Inflammation, Department of Immunology and Inflammation
Imperial College London
London, UK

Francesca Zazzeroni
Department of Biotechnological and Applied Clinical Sciences
University of L'Aquila
L'Aquila, Italy

ISSN 1064-3745 ISSN 1940-6029 (electronic)
Methods in Molecular Biology
ISBN 978-1-0716-1668-0 ISBN 978-1-0716-1669-7 (eBook)
https://doi.org/10.1007/978-1-0716-1669-7

Cover Illustration Caption: See Chapter 3 for more details.

This Humana imprint is published by the registered company Springer Science+Business Media, LLC, part of Springer Nature.
The registered company address is: 1 New York Plaza, New York, NY 10004, U.S.A.

Dedication

In memory of Uli

Preface

NF-κB Transcription Factors are pleiotropic and ubiquitous proteins which are extensively studied by researchers worldwide because of their key roles in the regulation of some of the most fundamental aspects of metazoan life and their frequent dysregulation in human diseases. This systematic examination of the analytical methods to study NF-κB in physiology and disease aims to provide an up-to-date guidebook to navigate both conventional and highly specialized methods to detect and analyze the different signaling pathways of NF-κB activation and contextualize them within organismal physiology and disease pathogenesis, using genetic and biochemical techniques and some of the most advanced computational and systems biology methods. This volume also provides several examples of approaches utilized by leading experts in the NF-κB field to analyze and modulate NF-κB signaling in specific physiological and disease contexts, along with examples of some of the most promising approaches to pharmacologically target the NF-κB pathway in human disease.

Considering the recent exponential acceleration of technological evolution, this new volume of *Methods in Molecular Biology* on the NF-κB pathway provides an up-to-date guide intended for both basic and translational scientists who are working in the NF-κB field.

London, UK — ***Guido Franzoso***
L'Aquila, Italy — ***Francesca Zazzeroni***

Contents

Contributors

YOUSEF ABU-AMER • *Department of Orthopedic Surgery, Washington University School of Medicine and Shriners Hospital for Children, Saint Louis, MO, USA; Department of Cell Biology & Physiology, Washington University School of Medicine and Shriners Hospital for Children, Saint Louis, MO, USA; Division of Biology and Biomedical Sciences, Washington University School of Medicine, Saint Louis, MO, USA*

EMMA J. ADAMS • *Division of Haematology & Immunology, Leeds Institute of Medical Research at St. James's University Hospital, University of Leeds, Leeds, UK*

SURESH ADAPALA • *Department of Orthopedic Surgery, Washington University School of Medicine and Shriners Hospital for Children, Saint Louis, MO, USA; Department of Cell Biology & Physiology, Washington University School of Medicine and Shriners Hospital for Children, Saint Louis, MO, USA*

PABLO J. AGUIRRE CARRIÓN • *Department of Biology, Boston University, Boston, MA, USA*

EDOARDO ALESSE • *Department of Biotechnological and Applied Clinical Sciences (DISCAB), University of L'Aquila, L'Aquila, Italy*

ANNAPAOLA ANGRISANI • *Department of Molecular Medicine, Sapienza University of Rome, Rome, Italy*

MANOJ ARRA • *Department of Orthopedic Surgery, Washington University School of Medicine and Shriners Hospital for Children, Saint Louis, MO, USA; Division of Biology and Biomedical Sciences, Washington University School of Medicine, Saint Louis, MO, USA*

BALAJI BANOTH • *Systems Immunology Laboratory, National Institute of Immunology, Aruna Asaf Ali Marg, New Delhi, India; Department of Immunology, St. Jude Children's Research Hospital, Memphis, TN, USA*

SOUMEN BASAK • *Systems Immunology Laboratory, National Institute of Immunology, Aruna Asaf Ali Marg, New Delhi, India*

MICHAEL BATIE • *Department of Biochemistry, Institute of Integrative Biology, University of Liverpool, Liverpool, UK*

FEDERICA BEGALLI • *Centre for Molecular Immunology and Inflammation, Department of Immunology and Inflammation, Imperial College London, London, UK; Centre for Endocrinology, William Harvey Research Institute, Barts and the London School of Medicine and Dentistry, Queen Mary University of London, London, UK*

JASON BENNETT • *Centre for Molecular Immunology and Inflammation, Department of Immunology and Inflammation, Imperial College London, London, UK*

CORBETT T. BERRY • *Department of Pathobiology, The University of Pennsylvania School of Veterinary Medicine, Philadelphia, PA, USA*

DAVID BRAY • *Department of Biology, Boston University, Boston, MA, USA; Biological Design Center, Boston University, Boston, MA, USA; Bioinformatics Program, Boston University, Boston, MA, USA*

DINIS P. CALADO • *The Francis Crick Institute, London, UK; Peter Gorer Department of Immunobiology, School of Immunology & Microbial Sciences, King's College London, London, UK*

Daria Capece • *Department of Biotechnological and Applied Clinical Sciences (DISCAB), University of L'Aquila, L'Aquila, Italy; Centre for Molecular Immunology and Inflammation, Department of Immunology and Inflammation, Imperial College London, London, UK*

Carmela Casale • *Institute of Genetics and Biophysics 'Adriano Buzzati-Traverso', Consiglio Nazionale delle Ricerche (CNR), Naples, Italy*

Alain Chariot • *Interdisciplinary Cluster for Applied Genoproteomics (GIGA), University of Liege, Liège, Belgium; Laboratory of Medical Chemistry, GIGA Stem Cells, University of Liège, Liège, Belgium; Walloon Excellence in Life Sciences and Biotechnology (WELBIO), Wavre, Belgium*

Lin-Feng Chen • *Department of Biochemistry, University of Illinois at Urbana-Champaign, Urbana, IL, USA*

Jessica Cornice • *Department of Biotechnological and Applied Clinical Sciences (DISCAB), University of L'Aquila, L'Aquila, Italy*

Elvira Crescenzi • *Istituto di Endocrinologia ed Oncologia Sperimentale, Consiglio Nazionale delle Ricerche (CNR), Naples, Italy*

Daniel D'Andrea • *Centre for Molecular Immunology and Inflammation, Department of Immunology and Inflammation, Imperial College London, London, UK; MRC Centre for Neuropsychiatric Genetics and Genomics, Cardiff University, Cardiff, UK*

Alvina Deka • *Systems Immunology Laboratory, National Institute of Immunology, Aruna Asaf Ali Marg, New Delhi, India*

Nilushi S. De Silva • *Institut Curie, PSL Research University, INSERM U932, Paris, France*

Enrico De Smaele • *Department of Experimental Medicine, Sapienza University of Rome, Rome, Italy*

Barbara Di Francesco • *Department of Biotechnological and Applied Clinical Sciences (DISCAB), University of L'Aquila, L'Aquila, Italy*

Jude Fitzgibbon • *Haemato-Oncology, Barts Cancer Institute, Queen Mary University of London, London, UK*

Guido Franzoso • *Centre for Molecular Immunology and Inflammation, Department of Immunology and Inflammation, Imperial College London, London, UK*

Bruce D. Freedman • *Department of Pathobiology, The University of Pennsylvania School of Veterinary Medicine, Philadelphia, PA, USA*

Francesca Fusco • *Institute of Genetics and Biophysics 'Adriano Buzzati-Traverso', Consiglio Nazionale delle Ricerche (CNR), Naples, Italy*

Andreas Gewies • *Research Unit Cellular Signal Integration, Institute of Molecular Toxicology and Pharmacology, Helmholtz-Zentrum München – German Research Center for Environmental Health, Neuherberg, Germany*

Thomas D. Gilmore • *Department of Biology, Boston University, Boston, MA, USA*

Tatiana Goncharov • *Department of Early Discovery Biochemistry, Genentech, South San Francisco, CA, USA*

Carina Graß • *Research Unit Cellular Signal Integration, Institute of Molecular Toxicology and Pharmacology, Helmholtz-Zentrum München – German Research Center for Environmental Health, Neuherberg, Germany*

Heather Hook • *Department of Biology, Boston University, Boston, MA, USA; Biological Design Center, Boston University, Boston, MA, USA*

Jessica L. Keenan • *Department of Biology, Boston University, Boston, MA, USA; Biological Design Center, Boston University, Boston, MA, USA; Bioinformatics Program, Boston University, Boston, MA, USA*

Ulf Klein • *Division of Haematology & Immunology, Leeds Institute of Medical Research at St. James's University Hospital, University of Leeds, Leeds, UK*

Daniel Krappmann • *Research Unit Cellular Signal Integration, Institute of Molecular Toxicology and Pharmacology, Helmholtz-Zentrum München – German Research Center for Environmental Health, Neuherberg, Germany*

Antonio Leonardi • *Dipartimento di Medicina Molecolare e Biotecnologie Mediche, Federico II University of Naples, Naples, Italy*

Michael J. May • *Department of Biomedical Sciences, The University of Pennsylvania School of Veterinary Medicine, Philadelphia, PA, USA*

Bernard D. Maybury • *The Francis Crick Institute, London, UK; Haemato-Oncology, Barts Cancer Institute, Queen Mary University of London, London, UK*

Shigeki Miyamoto • *McArdle Laboratory for Cancer Research, Department of Oncology, University of Wisconsin-Madison, Madison, WI, USA*

Nikita Tushar Modi • *Department of Biochemistry, University of Illinois at Urbana-Champaign, Urbana, IL, USA*

Marta Moretti • *Department of Experimental Medicine, Sapienza University of Rome, Rome, Italy*

Tapas Mukherjee • *Systems Immunology Laboratory, National Institute of Immunology, Aruna Asaf Ali Marg, New Delhi, India; Department of Immunology, University of Toronto, Toronto, ON, Canada; Department of Laboratory Medicine and Pathobiology, University of Toronto, Toronto, ON, Canada*

Mauro Di Vito Nolfi • *Department of Biotechnological and Applied Clinical Sciences (DISCAB), University of L'Aquila, L'Aquila, Italy*

Francesco Pacifico • *Istituto di Endocrinologia ed Oncologia Sperimentale, Consiglio Nazionale delle Ricerche (CNR), Naples, Italy*

Alessandra Pescatore • *Institute of Genetics and Biophysics 'Adriano Buzzati-Traverso', Consiglio Nazionale delle Ricerche (CNR), Naples, Italy*

Smarajit Polley • *Department of Biophysics, Bose Institute, Kolkata, India*

Zhaoxia Qu • *UPMC Hillman Cancer Center, Pittsburgh, PA, USA; Department of Microbiology and Molecular Genetics, University of Pittsburgh School of Medicine, Pittsburgh, PA, USA*

Yashika Ratra • *Systems Immunology Laboratory, National Institute of Immunology, Aruna Asaf Ali Marg, New Delhi, India*

Sonia Rocha • *Department of Biochemistry, Institute of Integrative Biology, University of Liverpool, Liverpool, UK*

Menotti Ruvo • *Istituto di Biostrutture e Bioimmagini, Consiglio Nazionale delle Ricerche (CNR), Naples, Italy*

Yolanda Saavedra-Torres • *The Francis Crick Institute, London, UK*

Annamaria Sandomenico • *Istituto di Biostrutture e Bioimmagini, Consiglio Nazionale delle Ricerche (CNR), Naples, Italy*

Dilem Shakir • *Department of Biochemistry, Institute of Integrative Biology, University of Liverpool, Liverpool, UK*

Kateryna Shostak • *Interdisciplinary Cluster for Applied Genoproteomics (GIGA), University of Liege, Liège, Belgium*

TREVOR SIGGERS • *Department of Biology, Boston University, Boston, MA, USA; Biological Design Center, Boston University, Boston, MA, USA; Bioinformatics Program, Boston University, Boston, MA, USA*

THOMAS J. A. SNOEKS • *The Francis Crick Institute, London, UK*

FAN SUN • *UPMC Hillman Cancer Center, Pittsburgh, PA, USA; Department of Microbiology and Molecular Genetics, University of Pittsburgh School of Medicine, Pittsburgh, PA, USA*

GAURAV SWARNKAR • *Department of Orthopedic Surgery, Washington University School of Medicine and Shriners Hospital for Children, Saint Louis, MO, USA; Department of Cell Biology & Physiology, Washington University School of Medicine and Shriners Hospital for Children, Saint Louis, MO, USA*

SYLVIA TIELENS • *Interdisciplinary Cluster for Applied Genoproteomics (GIGA), University of Liege, Liège, Belgium*

LAURA TORNATORE • *Centre for Molecular Immunology and Inflammation, Department of Immunology and Inflammation, Imperial College London, London, UK*

MATILDE VALERIA URSINI • *Institute of Genetics and Biophysics 'Adriano Buzzati-Traverso', Consiglio Nazionale delle Ricerche (CNR), Naples, Italy*

EUGENE VARFOLOMEEV • *Department of Early Discovery Biochemistry, Genentech, South San Francisco, CA, USA*

DAVIDE VECCHIOTTI • *Department of Biotechnological and Applied Clinical Sciences (DISCAB), University of L'Aquila, L'Aquila, Italy*

DANIELA VERZELLA • *Department of Biotechnological and Applied Clinical Sciences (DISCAB), University of L'Aquila, L'Aquila, Italy*

DOMAGOJ VUCIC • *Department of Early Discovery Biochemistry, Genentech, South San Francisco, CA, USA*

CAROLINE WATHIEU • *Interdisciplinary Cluster for Applied Genoproteomics (GIGA), University of Liege, Liège, Belgium*

LEAH M. WILLIAMS • *Department of Biology, Boston University, Boston, MA, USA*

ZHAO-HUI WU • *Department of Radiation Oncology, University of Tennessee Health Science Center, Memphis, TN, USA; Department of Pathology and Laboratory Medicine, University of Tennessee Health Science Center, Memphis, TN, USA*

GUTIAN XIAO • *UPMC Hillman Cancer Center, Pittsburgh, PA, USA; Department of Microbiology and Molecular Genetics, University of Pittsburgh School of Medicine, Pittsburgh, PA, USA*

FRANCESCA ZAZZERONI • *Department of Biotechnological and Applied Clinical Sciences (DISCAB), University of L'Aquila, L'Aquila, Italy*

ROSE W. ZHAO • *Department of Biology, Boston University, Boston, MA, USA; Biological Design Center, Boston University, Boston, MA, USA*

Part I

General Methods for Analyzing NF-κB Activation

Chapter 1

Measuring NF-κB Phosphorylation and Acetylation

Nikita Tushar Modi and Lin-Feng Chen

Abstract

Posttranslational modifications of NF-κB, including phosphorylation, acetylation, and methylation, have emerged as important regulatory mechanisms to control the transcriptional outcomes of this important transcription factor. These modifications work independently, sequentially or in combination to modulate the diverse biological functions of NF-κB in cancer and inflammatory response. Here, we describe some experimental methods to detect the in vitro and in vivo phosphorylation and acetylation of NF-κB, specifically focusing on the RelA subunit of NF-κB. These methods include labeling the phospho- or acetyl- groups with radioisotopes in vitro and immunoblotting with site-specific anti-phospho-serine or acetyl-lysine antibodies in culture cells and tissue samples.

Key words NF-κB, Phosphorylation, Acetylation

1 Introduction

Transcription factor NF-κB plays a critical role in the innate and adaptive immune response [1]. NF-κB is activated when the cells respond to a variety of environmental challenges, including pro-inflammatory cytokines, bacterial and viral infections, and DNA damage [1]. Activated NF-κB stimulates the expression of many inflammatory factors to fight against pathogens or cell damage. After clearance of infection or damaged cells, NF-κB needs to be inactivated to prevent prolonged inflammatory response since sustained NF-κB activation leads to many human diseases (e.g., chronic inflammation and cancer). Cells utilize many different mechanisms to activate and terminate NF-κB and posttranslational modifications serve as important mechanisms to turn on or off NF-κB signaling. These modifications target various cytoplasmic or nuclear proteins in the NF-κB signaling pathways and modulate the transcriptional and biological outcomes in response to various extracellular signals.

It has been well documented in literature that RelA subunit of NF-κB is subject to a variety of posttranslational modifications

Guido Franzoso and Francesca Zazzeroni (eds.), *NF-κB Transcription Factors: Methods and Protocols*, Methods in Molecular Biology, vol. 2366, https://doi.org/10.1007/978-1-0716-1669-7_1,

(PTMs), including phosphorylation, acetylation, methylation, ubiquitination, nitrosylation and glycosylation in the cytosol or in the nucleus [2–4]. These modifications, alone or in combination, regulate the activation and termination of NF-κB and fine-tune the expression of selective NF-κB target genes, adding a different layer of complexity to the transcriptional regulation of NF-κB [4–6].

Phosphorylation of RelA is the first identified and most abundant posttranslational modification of RelA, it occurs in both cytosol and nucleus, mediated by various kinases, depending on the extracellular stimuli [7–9]. Many phosphorylation residues, including eight serines and three threonines, have been identified within the N-terminal Rel homologous domain and the C-terminal transactivation domain [9]. Phosphorylation of RelA at these different residues regulates diverse activity of NF-κB, including transcriptional activity, protein stability, protein–protein interaction, and its cross-reaction with other modifications [4, 9]. For example, phosphorylation of RelA at serine 276 (S276) by the catalytic domain of protein kinase A prevents the association of RelA with its inhibitor IκBα and promotes its interaction with acetyltransferase p300, resulting in the enhanced acetylation of lysine 310 (K310) of RelA and the transcriptional activation of NF-κB [10, 11]. Phosphorylation of threonine 254 (T254) by an unknown kinase alters the conformation of RelA, increasing its stability, nuclear accumulation and transcriptional activity [12]. In addition, the phosphorylation of S536 of RelA by IκB kinase 2 (IKK2) enhances the transcriptional activity of NF-κB by increasing its interaction with p300 and the acetylation of K310 [8].

Acetylation of RelA mediated by histone acetyltransferases (HAT) is another important PTM that regulates diverse functions of NF-κB, including DNA binding activity, transcriptional activity, and its ability to associate with other proteins. Acetylation plays an important role in the NF-κB-mediated inflammatory response and cancer [4, 5]. Several acetylated lysines have been identified within RelA, including K122, K123, K218, K221, K310, K314, and K315 [13–15]. Acetylation of these different lysines controls different properties of NF-κB. For example, acetylation of K221 increases the DNA binding of NF-κB and, together with acetylation of K218, prevents RelA's interaction with IκBα [14]. Acetylation of K314 and K315 by p300 differentially regulates the expression of a subset of NF-κB target genes [15, 16]. On the other hand, acetylation of K122 and K123 by p300/CBP or PCAF negatively regulates NF-κB-mediated transcription by reducing RelA binding to the κB enhancer [13]. K310 is the best-characterized acetylation site of RelA due to its requirement for the full transcriptional activation of NF-κB [14, 17]. Acetylation of K310 also regulates the stability of RelA by preventing its methylation at K314/315 and maintaining the sustained NF-κB activity in cancer cells [18]. The importance of acetylation of K310 is highlighted by the presence of this modification in cancer cells and in human cancer patient samples [19–21].

Due to the essential role of phosphorylation and lysine acetylation in NF-κB target gene expression and its biological functions, it is important to monitor these modifications in cultured cells and tissues. Many assays have been used to successfully detect the phosphorylation and acetylation of RelA, including the initial in vitro isotopic labeling of the phospho- or acetyl-groups with radioactive ATP or acetyl-CoA, immunoblotting with site-specific phospho-RelA or acetyl-RelA antibodies [14, 15, 22–24]. Mass spectrometry has also been used to identify the phosphorylation and acetylation of RelA [25]. In vitro kinase or acetylation assay allows the rapid identification and validation of RelA kinases or HATs, respectively. In these assays, active enzymes transfer radiolabeled phospho- or acetyl- groups from ATP or acetyl-CoA to recombinant RelA [26]. With the availability of site-specific antibodies against phosphorylated or acetylated RelA, immunoblotting has become a more common and powerful approach to detect these modifications in various form of RelA, including recombinant RelA, overexpressed RelA or endogenous RelA in cultured cells and in human samples [10, 23, 24, 27, 28].

2 Materials

2.1 Cell Lines, Primary Cells and Tissues

1. HEK293T cells co-transfected with HA-tagged IKK2 and T7-tagged RelA.
2. HEK293T cells co-transfected with T7-tagged RelA and p300.
3. Bone marrow-derived macrophages (BMDMs): prepare BMDMs from bone marrow of mice [29]. Isolate bone marrow from tibia and femur using aseptic technique. To differentiate bone marrow cells into macrophages culture them in BMDM medium for 7 days in sterile plastic Petri dishes.
4. Formalin fixed paraffin embedded tumor xenografts sections (4μm) from A549 cells.
5. Dulbecco's Modified Eagle Medium (DMEM) supplemented with 10% fetal bovine serum (FBS), 100 U/ml penicillin, 100μg/ml streptomycin and 2 mM L-Glutamine.
6. BMDM medium is DMEM/F12 supplemented with 10% FBS, 2 mM L-glutamine, 100 U/ml penicillin, 100μg/ml streptomycin, 10 mM HEPES buffer, and 20% conditioned medium from mouse-L929 fibroblasts cells.
7. 100 ng/ml LPS (*E. coli*, Sigma).

2.2 Antibodies

1. Site-specific anti-phosphorylated serine-536 RelA antibody (Cell signaling Technology).

2. Anti-pan acetylated lysine antibody (Cell Signalling Technology).
3. Site-specific anti-acetylated lysine-310 antibody (Cell Signaling Technology).
4. Anti-NF-κB/RelA antibody (F-6) (Santa Cruz Biotechnologies).
5. HRP-linked secondary antibodies.
6. Biotinylated secondary antibodies.
7. FITC conjugated-secondary antibodies.
8. Enhanced chemiluminescence solution (ECL).

2.3 Recombinant Proteins

1. Recombinant RelA protein (Active Motif).
2. Active recombinant IKK2 (Abcam).
3. Recombinant catalytic domain of p300 proteins (Active Motif).

2.4 Radioisotope

[γ-^{32}P] ATP and [^{14}C]-Acetyl-CoA (PerkinElmer).

2.5 Buffers and Reagents

1. SDS-PAGE gels.
2. Nitrocellulose membrane.
3. PBS-T: Phosphate buffer saline (PBS) and 0.1% Tween-20.
4. Blocking solution: 5% milk in PBS-T.
5. Blocking solution 2: 0.5% BSA in PBS.
6. 5× kinase buffer (Thermo Fisher Scientific).
7. 5× acetylation buffer (Upstate Biotechnology).
8. 2× sample buffer (BioRad).
9. Lysis or immunoprecipitation (IP) buffer: 50 mM HEPES, pH 7.4, 250 mM NaCl, 0.1% NP-40, 1 mM EDTA, pH 8.0. Add 1 mM phenylmethylsulfonyl fluoride (PMSF) (Sigma), and 1 tab Protease inhibitor cocktail (Roche)/50 ml buffer at the time of use.
10. Lysis buffer 1: 50 mM HEPES-KOH, pH 7.5, 140 mM NaCl, 1 mM EDTA, 10% glycerol and 0.5% Triton X-100.
11. Lysis buffer 2: 10 mM Tris–HCl, pH 8.0, 200 mM NaCl and 1 mM EDTA.
12. Lysis buffer 3: 10 mM Tris–HCl, pH 8.0, 100 mM NaCl, 1 mM EDTA, 0.1% sodium deoxycholate and 20% N-lauroylsarcosine.
13. HAT assay buffer: 250 mM Tris base, pH 8.0, 50% glycerol, 0.5 mM EDTA, 5 mM DTT.
14. Low salt wash buffer: 0.1% SDS, 1% TritonX-100, 2 mM EDTA, 20 mM Tris–HCl (pH 8.0), 150 mM NaCl.

15. High salt wash buffer: 0.1% SDS, 1% Triton X-100, 2 mM EDTA, 20 mM Tris–HCl (pH 8.0), 500 mM NaCl.
16. LiCl wash buffer: 0.25 M LiCl, 1% NP-40, 1% deoxycholate, 1 mM EDTA, 10 mM Tris–HCl (pH 8.0).
17. TE/NaCl buffer: 1× TE buffer, 50 mM NaCl.
18. Elution buffer: 10 mM Tris–HCl, pH 8.5, 10 mM EDTA, 1% SDS.
19. Triton X-100.
20. Distilled Water.
21. Salmon Sperm DNA/Protein A agarose beads.
22. 50% slurry of anti-T7 antibody conjugated agarose beads.
23. Protein-G agarose beads.
24. 5 M NaCl.
25. Proteinase K (20 mg/ml).
26. QIAquick Purification kit (Qiagen).
27. Fixing solution: isopropanol: water: acetic acid (25:65:10, v/v).
28. Amplify Fluorographic Reagent (Amersham Biosciences).
29. X-ray films.
30. Xylene.
31. Graded series of ethanol:100%, 95%, 70% and 50%.
32. Sodium citrate buffer: 10 mM Sodium Citrate, 0.05% Tween 20, pH 6.0.
33. Autoblocker (R&D Systems).
34. Avidin/biotin solution (Vector Laboratories).
35. Avidin-biotin complex (Vector Laboratories).
36. Diaminobenzidine substrate.
37. Hematoxylin.
38. Vectashield Antifade mounting media (Vector laboratories).
39. 4% paraformaldehyde.
40. 2.5 M Glycine.
41. 1% BSA solution.
42. DAPI.

3 Methods

3.1 Phosphorylation of RelA

Phosphorylation of RelA can be assessed by in vitro kinase assay using recombinant RelA proteins and in cultured cells using site-specific anti-phosphorylated antibodies. [^{32}P]-labeled ATP is often used in the in vitro phosphorylation assay with recombinant RelA and recombinant kinases or kinase complexes immunoprecipitated from stimulated cells. In vitro kinase assay using [^{32}P] allows for the identification of novel kinases and new phosphorylation residues. Several RelA phosphorylation sites have been identified and the corresponding site-specific antibodies are commercially available, allowing the rapid detection of in vitro or in vivo phosphorylated RelA. These site-specific phosphorylated RelA antibodies are useful in the immunoblotting or chromatin immunoprecipitation (ChIP) assays to detect the stimulus-coupled phosphorylation of endogenous RelA in cells and on the promoter of certain genes [18, 24, 27, 28]. Phosphorylated RelA can also be detected by immunohistochemistry (IHC) in primary tumors and in xenografts [30, 31]. Since phosphorylation of S536 is the most abundant phosphorylation within RelA and has often been used as a marker for activated RelA, we will demonstrate how to detect phosphorylated S536 of RelA. Similar approaches could be used to detect other phosphorylated residues within RelA.

3.1.1 *In Vitro Phosphorylation Assay*

1. Mix 1μg of recombinant RelA proteins (His-tagged full-length RelA or GST-tagged RelA C-terminal regions) with 4μl of 5× kinase buffer (*see* **Note 1**).
2. Add 0.5μg of recombinant IKK2 or 10μl IKK2 immunoprecipitates agarose beads washed with 1× kinase buffer (*see* **Note 2**).
3. Add 1μCi of [γ-^{32}P] ATP.
4. Add water to make a total volume to 20μl.
5. Mix well by tapping the bottom of the tube.
6. Spin the tube and incubate at 30 °C for 30 min.
7. Stop the reaction by adding 20μl 2× sample buffer and boil for 5 min.
8. Analyze the products by SDS-PAGE followed by electrophoretic transfer to nitrocellulose membrane and exposure to Hyperfilm MP (Amersham Biosciences).

3.1.2 *In Vivo Phosphorylation of RelA S536*

Detection of Phosphorylated Overexpressed RelA (Co-transfection with IKK2)

1. Lyse HEK293T cells co-transfected with T7-tagged RelA and HA-tagged IKK2 36 h after transfection with 200μl 1× sample buffer (*see* **Note 3**).
2. Resolve the proteins on a SDS-PAGE gel and transfer to a nitrocellulose membrane.
3. Incubate the membrane in a blocking solution for 1 h at room temperature.
4. Rinse the membrane in PBS-T three times for 5 min each.

5. Incubate the membrane overnight at 4 °C in 1:200–1:500 diluted anti-phosphorylated S536 RelA antibody on a shaker.
6. Rinse the membrane with PBS-T three times for 10 min each.
7. Incubate the membrane with Donkey anti-rabbit IgG, HRP-linked secondary antibodies for 30 min.
8. Rinse the membrane with PBS-T two times for 15 min each.
9. Incubate the membrane with ECL following manufacturer instructions and image using a ChemiDoc imaging system or with an X-ray film in a dark room.

Detection of Phosphorylation of Endogenous RelA in BMDMs

For detection of endogenous RelA phosphorylation, it is critical to stimulate the cells with TNF-α or LPS to activate the kinases. Due to the abundance of phosphorylated S536 RelA, the phosphorylation of RelA can be detected by directly immunoblotting the whole cell extracts with anti-phosphorylated S536 RelA antibodies [10].

1. Seed 2 ml BMDMs (1.2×10^6) in each well of a six-well plate and culture overnight at 37 °C.
2. Treat the cells with 100 ng/ml LPS for 5, 15, 30 and 60 min (*see* **Note 4**).
3. Lyse cells in 200μl of 1× sample buffer and boil for 5 min at 95 °C.
4. Resolve the samples on 8% SDS-PAGE gel and transfer the gel to a nitrocellulose membrane.
5. Incubate the membrane in a blocking solution for 1 h at room temperature with agitation.
6. Rinse the membrane with PBS-T three times for 5 min each.
7. Incubate the membrane with diluted antibodies for phosphorylated S536-RelA overnight at 4 °C with agitation. (*see* **Note 5**).
8. Rinse the blot the following day with PBS-T and incubate the membrane with the corresponding secondary antibody in blocking solution for 1 h at room temperature.
9. Rinse the membrane with PBS-T two times for 15 min each.
10. Incubate the membrane with ECL following manufacturer instructions and image using a ChemiDoc imaging system or with an X-ray film in a dark room.

Detection of Chromatin-Associated Phosphorylated RelA by Chromatin Immunoprecipitation (ChIP)

1. Seed BMDMs (2×10^7) in 150 mm dishes and culture overnight at 37 °C.
2. Stimulate cells with 100 ng/ml of LPS for 30 min.
3. Cross link proteins to DNA by adding 1% formaldehyde to the cells and incubate for 10 min at room temperature.

4. Stop the fixation with 1/20th volume of 2.5 M glycine for 5 min.
5. Use a cell scraper to collect the cells into a conical tube and pellet cells for 5 min, at 500 × *g* at 4 °C.
6. Wash the cells once with cold PBS and resuspend the cell pellets in lysis buffer 1.
7. Incubate the cells on ice for 10 min and collect the cell pellet by centrifugation at 200 × *g* for 5 min at 4 °C.
8. To remove any detergents, resuspend the pellet in lysis buffer 2. Agitate the samples at room temperature on a rotator for 10 min and collect the nuclei by centrifugation.
9. Resuspend the nuclei in 600μl of lysis buffer 3 and incubate for 30 min on ice.
10. Sonicate the samples under conditions optimized to generate DNA fragments with an average size of 200–1000 base pairs. Ensure that the samples remain cool at all times.
11. Centrifuge samples for 10 min at 13,000 × *g* at 4 °C. Collect the supernatant and add 1%. Triton X-100. Store 10μl of sample as input DNA at −80 °C. Transfer 200μl of the sonicated cell supernatant to a new Eppendorf tube. The remaining supernatants can be stored at −80 °C for future use.
12. Wash 60μl of Salmon Sperm DNA/Protein A agarose beads with 1 ml of blocking solution 2 for each sample three times.
13. Incubate the Protein A beads with 2μg of anti-phosphorylated S536-RelA antibodies in 1 ml of blocking solution overnight at 4 °C on a rotator.
14. On the following day, wash the Protein A beads three times with 1 ml blocking solution 2. Drain the blocking solution using aspiration.
15. Add 200μl of samples to the antibody/Protein A agarose beads and incubate overnight at 4 °C on a rotator.
16. Pellet the agarose beads by centrifugation at 100 × *g* for 1 min at 4 °C. Discard the supernatant.
17. Wash the agarose beads on a rotating platform for 5 min each with 1 ml of the following buffers: low salt wash buffer, high salt wash buffer and LiCl wash buffer.
18. Wash once with by TE/NaCl buffer. Collect the agarose beads by centrifugation and remove any residual TE buffer.
19. Add 250μl of the elution buffer to the agarose beads at 65 °C for 15 min. Resuspend the agarose beads by vortexing every few minutes. Centrifuge the samples at 100 × *g* for 1 min and collect 230μl of supernatant.

20. Incubate the samples overnight at 65 °C after addition of 20μl of 5 M NaCl to reverse crosslink the protein and DNA.
21. Digest the samples with 2μl of proteinase K for 1 h at 65 °C. Reverse crosslink the previously stored input DNA overnight at 65 °C by addition of 3 volumes of elution buffer.
22. Purify the DNA using Qiagen QuickSpin Purification kit and elute in 50μl DNA elution buffer.
23. Use quantitative real-time PCR with specific primers to detect the presence of different DNA fragments.

3.2 Acetylation of RelA

The first direct evidence of acetylation of RelA in vivo was obtained using radioactive labeling with [^{3}H]-acetate [22]. However, the weak signal of [^{3}H]-acetate limits its applications in studying the kinetics or stoichiometry of RelA acetylation. With the availability of pan or site-specific anti-acetylated RelA antibodies, immunoblotting has become a simple and common method for detection of acetylation. Both polyclonal and monoclonal anti-acetylated lysine antibodies from Cell Signaling (#9441 and #9681) recognize in vitro acetylated recombinant RelA by p300 or in vivo overexpressed RelA with co-transfected p300 plasmids. The site-specific anti-K310 antibodies also recognize the acetylation of endogenous RelA in cultured cells and cancer samples.

3.2.1 In Vitro Acetylation Assay

1. Mix 1μg of recombinant RelA protein, 0.5μg of recombinant p300 or p300 HAT domain in 4μl of 5× HAT assay buffer (*see* **Note 6**).
2. Add 2μl of [^{14}C]-acetyl-CoA or acetyl-CoA to the mixture.
3. Add distilled water to a total volume of 20μl.
4. Mix the contents of the tube and incubate the tube at 30 °C for 1 h with occasionally shaking.
5. Stop the reaction by adding 12μl of 2× sample buffer to the reaction and boil for 5 min (*see* **Note 7**).
6. Run the samples on a SDS-PAGE gel. When using [^{14}C]-acetyl-CoA as the acetyl-group donor, fix the gel using a fixing solution for approximately 30 min (*see* **Note 8**).
7. Soak the gel in Amplify Fluorographic Reagent (sufficient for the gel to be free floating) with agitation for 15–30 min.
8. Dry the gel and hold the gel in close contact with an appropriate X-ray film at −80 °C overnight.
9. Detect the radioactive signal by autoradiography.

3.2.2 In Vivo Acetylation Assay

In vivo acetylated RelA can be detected by radiolabeling using radiolabeled sodium acetate or by immunoblotting using anti-acetylated lysine antibodies. Before the antibodies for acetylated

RelA became available, radiolabeling using [^{3}H]-acetate was used to demonstrate the acetylated RelA in cultured cells [22]. In vivo labeling the acetylated RelA using radioisotope has been described previously [32]. Here, we focus on how to detect the acetylated RelA in vivo using anti-acetylated K310-RelA antibodies. Some antibodies against lysines other than K310 are also commercially available and could be similarly employed in the in vivo detection of acetylated RelA.

Detection of Acetylation of Overexpressed RelA by Immunoblotting

1. Lyse HEK293T cells co-transfected with T7-tagged RelA and p300 36 h after transfection using 350μl of IP buffer (*see* **Note 9**).
2. Immunoprecipitate T7-RelA by adding 20μl of 50% slurry of anti-T7 antibody conjugated agarose beads to the lysates. Incubate the samples for 2 h on a rotator at 4 °C.
3. Wash the beads three times with IP buffer and collect the beads.
4. Boil the beads at 95 °C in 50μl of 2× sample buffer for 5 min.
5. Resolve the proteins on a SDS-PAGE gel and transfer to a nitrocellulose membrane.
6. Immunoblot the membrane with anti-pan acetylated lysine antibodies or site-specific anti-acetylated K310 antibodies overnight.
7. Rinse the blot the following day with PBS-T and incubate the membrane with the corresponding secondary antibody in blocking solution for 1 h at room temperature.
8. Rinse the membrane with PBS-T two times for 15 min each.
9. After washing the membrane, incubate the membrane with ECL following manufacturer instructions and visualize the signal using a ChemiDoc imaging system or with an X-ray film in a dark room.

Detection of Acetylation of Endogenous RelA in BMDMs

Endogenous RelA can be detected using site-specific anti-acetylated RelA such as anti-acetylated K310 antibodies from Cell Signaling and Abcam. Stimulation of cells with TNF-α and LPS will allow the translocation of RelA into nucleus, where most HATs localize, and enhance the acetylation signal.

1. Seed 2 ml BMDMs (1.2 × 10^{6} cells) in each well of a six-well plate and culture overnight at 37 °C. Prepare two wells for each sample.
2. Treat the cells with 100 ng/ml of LPS for 0, 15, 30, and 60 min.
3. Wash the cells twice with ice-cold PBS and lyse the cells with 350μl of IP buffer.

4. Collect the cell lysates from the two wells using a cell scraper into 1.5-ml Eppendorf tubes.
5. Pre-clear the cell lysates using 20μl of protein-G agarose beads slurry and collect the supernatant in fresh Eppendorf tubes.
6. Add 5μl of anti-RelA antibodies to the supernatant and incubate for 3 h at 4 °C on a rotator.
7. Add 25μl of Protein-G agarose beads (50% slurry) to cell lysates and incubate over night at 4 °C on a rotator.
8. Collect the agarose beads by pulse centrifugation at 13,000 × *g* for 30 s.
9. Wash the agarose beads with 500μl of ice-cold IP buffer 3–5 times.
10. Resuspend the agarose beads in 30μl of 2× sample buffer.
11. Boil the samples at 95 °C for 5 min.
12. Resolve the samples on a SDS-PAGE gel and transfer the gel to a nitrocellulose membrane.
10. Immunoblot the membrane with anti-acetyl K310 RelA antibodies overnight at 4 °C.
11. Rinse the blot the following day with PBS-T and incubate the membrane with the corresponding secondary antibody in blocking solution for 1 h at room temperature.
12. Rinse the membrane with PBS-T two times for 15 min each.
13. After washing the membrane, incubate the membrane with ECL following manufacturer instructions and visualize the signal using a ChemiDoc imaging system or with an X-ray film in a dark room (*see* **Note 10**).

Detection of Acetylated Endogenous RelA by Immunohistochemistry (IHC)

IHC can be used to detect acetylation of RelA in mouse xenografts or tissue microarray of cancer samples.

1. Slice formalin fixed and paraffin embedded tumor xenografts from A549 cells to 4μm using a microtome and mount on slides.
2. For deparaffinization, place slides on a slide rack and wash twice with Xylene for 3 min each.
3. Wash sections in a 1:1 mixture of xylene and 100% ethanol for 3 min.
4. Wash sections with a graded series of ethanol—100%, 95%, 70%, and 50% for 3 min each time and rinse with distilled water to remove ethanol.
5. Treat sections for antigen retrieval with a sodium citrate buffer.

6. All sections are treated with Autoblocker to inhibit endogenous peroxidase and avidin/biotin solution to block endogenous biotin.
7. Incubate sections with rabbit anti-acetylated K310 polyclonal antibodies with 1:100 dilution at 4 °C overnight.
8. Wash sections with PBS-T for 2 min twice.
9. Incubate sections with 7.5μg/ml of biotinylated goat anti-rabbit secondary antibody for 1 h at room temperature.
10. Wash sections with PBS-T and incubate with avidin-biotin complex for 15 min.
11. The antigen-antibody reaction is shown using diaminobenzidine as substrates and the sections are counter-stained with hematoxylin.
12. Add a coverslip with Vectashield Antifade mounting media and visualize the acetylated RelA signals under an appropriate microscope.

Detection of Acetylated RelA by Immunocytochemsitry (ICC)

1. Plate BMDMs (3×10^5 cells) on 2-well chamber slides and culture overnight at 37 °C.
2. Treat cells with 100 ng/ml of LPS for 30–60 min (*see* **Note 11**).
3. Fix the cells with 4% paraformaldehyde for 15 min at room temperature.
4. Wash cells with 1 ml of 1× PBS three times.
5. Block cells with 1% BSA for 1 h at room temperature.
6. Incubate cells with diluted acetylated K310-RelA antibodies (1:200) in 0.1% BSA overnight at 4 °C.
7. Rinse cells three times for 5 min with 1× PBS at room temperature.
8. Incubate cells with goat anti-rabbit IgG secondary antibodies conjugated with FITC for 1 h at room temperature.
9. Rinse cells two times for 5 min with 1× PBS at room temperature. Counterstain the nuclei with DAPI for 1 min and add a coverslip with Vectashield Antifade mounting media.
10. Visualize the acetylated RelA by fluorescence microscopy.

4 Notes

1. RelA recombinant proteins with serine 536 mutated to alanine can be used as a negative control.
2. Recombinant IKK2 or endogenous IKK2 immunoprecipitates from stimulated cells can be used as active IKK2 for the in vitro

kinase assay. Other recombinant kinase proteins or kinase immunoprecipitates can be used similarly to test their abilities to phosphorylate RelA in vitro.

3. Transfection of HEK293T cells can be done using calcium phosphate method or Lipofectamine 2000 (Invitrogen) using 0.5μg of T7-tagged RelA and 1μg HA-tagged IKK2 per well of a six-well plate. RelA-S536A expression vector can be used as a control for the validation of S536A phosphorylation. Expression vectors for different kinases can be similarly used in the experiment.
4. Different stimuli might have different kinetics of kinase activation.
5. The levels of phosphorylated RelA may vary in different cell types with different phosphorylation residues. To increase the phosphorylation signal, phosphorylated RelA can be immunoprecipitated with site-specific anti-phosphorylated RelA antibodies followed by immunoblotting with anti-RelA antibodies.
6. Immunoprecipitated p300 from transfected HEK293T cells with p300 expression plasmids has high HAT activity and can be used in the in vitro acetylation assay [32].
7. When p300 immunoprecipitates are used as HAT, collect the supernatants and add 2× sample buffer to stop the reaction.
8. When nonradioactive acetyl-CoA is used as the acetyl-group donor, transfer the gel to a nitrocellulose membrane, followed by immunoblotting with anti-pan-acetylated lysine antibodies or site-specific anti-acetylated RelA antibodies.
9. Transfection of HEK293T cells can be done using calcium phosphate method or Lipofectamine 2000 (Invitrogen) using 1μg of expression vector for T7-tagged RelA and 2μg of expression vector for p300 per well of a six-well plate. K310R-RelA expressing vector can be used a control.
10. In addition to the nuclear RelA from stimulated cells, cancer cells have constitutive nuclear acetylated RelA. However, the amount of nuclear acetylated RelA might not be sufficient to be detected by direct immunoblotting of whole cell extracts. Immunoprecipitation of nuclear extracts of cancer cells with anti-acetylated RelA antibodies, followed by immunoblotting with anti-RelA antibodies would allow for the efficient detection of RelA acetylation.
11. Pretreatment of cells with Trichostatin A (0.5μM) for 2 h would enhance the acetylation signal.

Acknowledgments

The work described in this article was supported in part by funds from University of Illinois at Urbana-Champaign.

References

1. Ghosh S, Karin M (2002) Missing pieces in the NF-kappaB puzzle. Cell 109(Suppl):S81–S96
2. Yang WH, Park SY, Nam HW, Kim DH, Kang JG, Kang ES et al (2008) NFkappaB activation is associated with its O-GlcNAcylation state under hyperglycemic conditions. Proc Natl Acad Sci U S A 105(45):17345–17350
3. Kelleher ZT, Matsumoto A, Stamler JS, Marshall HE (2007) NOS2 regulation of NF-kappaB by S-nitrosylation of p65. J Biol Chem 282(42):30667–30672
4. Huang B, Yang XD, Lamb A, Chen LF (2010) Posttranslational modifications of NF-kappaB: another layer of regulation for NF-kappaB signaling pathway. Cell Signal 22(9):1282–1290
5. Perkins ND (2006) Post-translational modifications regulating the activity and function of the nuclear factor kappa B pathway. Oncogene 25(51):6717–6730
6. Chen LF, Greene WC (2004) Shaping the nuclear action of NF-kB. Nat Rev Mol Cell Biol 5(5):392–401
7. Zhong H, SuYang H, Erdjument-Bromage H, Tempst P, Ghosh S (1997) The transcriptional activity of NF-kappaB is regulated by the IkappaB-associated PKAc subunit through a cyclic AMP-independent mechanism. Cell 89 (3):413–424
8. Sakurai H, Chiba H, Miyoshi H, Sugita T, Toriumi W (1999) IkB kinases phosphorylate NF-kB p65 subunit on serine 536 in the transactivation domain. J Biol Chem 274 (43):30353–30356
9. Christian F, Smith EL, Carmody RJ (2016) The regulation of NF-kappaB subunits by phosphorylation. Cell 5(1):12
10. Chen LF, Williams SA, Mu Y, Nakano H, Duerr JM, Buckbinder L et al (2005) NF-kappaB RelA phosphorylation regulates RelA acetylation. Mol Cell Biol 25 (18):7966–7975
11. Zhong H, Voll RE, Ghosh S (1998) Phosphorylation of NF-kappa B p65 by PKA stimulates transcriptional activity by promoting a novel bivalent interaction with the coactivator CBP/p300. Mol Cell 1(5):661–671
12. Ryo A, Suizu F, Yoshida Y, Perrem K, Liou YC, Wulf G et al (2003) Regulation of NF-kappaB signaling by Pin1-dependent prolyl isomerization and ubiquitin-mediated proteolysis of p65/RelA. Mol Cell 12(6):1413–1426
13. Kiernan R, Bres V, Ng RW, Coudart MP, El Messaoudi S, Sardet C et al (2003) Post-activation turn-off of NF-kB-dependent transcription is regulated by acetylation of p65. J Biol Chem 278(4):2758–2766
14. Chen LF, Mu Y, Greene WC (2002) Acetylation of RelA at discrete sites regulates distinct nuclear functions of NF-kB. EMBO J 21 (23):6539–6548
15. Buerki C, Rothgiesser KM, Valovka T, Owen HR, Rehrauer H, Fey M et al (2008) Functional relevance of novel p300-mediated lysine 314 and 315 acetylation of RelA/p65. Nucleic Acids Res 36(5):1665–1680
16. Rothgiesser KM, Fey M, Hottiger MO (2010) Acetylation of p65 at lysine 314 is important for late NF-kappaB-dependent gene expression. BMC Genomics 11:22
17. Yeung F, Hoberg JE, Ramsey CS, Keller MD, Jones DR, Frye RA et al (2004) Modulation of NF-kappaB-dependent transcription and cell survival by the SIRT1 deacetylase. EMBO J 23(12):2369–2380
18. Yang XD, Tajkhorshid E, Chen LF (2010) Functional interplay between acetylation and methylation of the RelA subunit of NF-kappaB. Mol Cell Biol 30(9):2170–2180
19. Zou Z, Huang B, Wu X, Zhang H, Qi J, Bradner J et al (2014) Brd4 maintains constitutively active NF-kappaB in cancer cells by binding to acetylated RelA. Oncogene 33 (18):2395–2404
20. Wu X, Qi J, Bradner JE, Xiao G, Chen LF (2013) Bromodomain and extraterminal (BET) protein inhibition suppresses human T cell leukemia virus 1 (HTLV-1) Tax protein-mediated tumorigenesis by inhibiting nuclear factor kappaB (NF-kappaB) signaling. J Biol Chem 288(50):36094–36105
21. Lee H, Herrmann A, Deng JH, Kujawski M, Niu G, Li Z et al (2009) Persistently activated Stat3 maintains constitutive NF-kappaB activity in tumors. Cancer Cell 15(4):283–293
22. Chen LF, Fischle W, Verdin E, Greene WC (2001) Duration of nuclear NF-kB action

regulated by reversible acetylation. Science 293 (5535):1653–1657

23. Yang XD, Huang B, Li M, Lamb A, Kelleher NL, Chen LF (2009) Negative regulation of NF-kappaB action by Set9-mediated lysine methylation of the RelA subunit. EMBO J 28 (8):1055–1066
24. Lu T, Jackson MW, Wang B, Yang M, Chance MR, Miyagi M et al (2010) Regulation of NF-kappaB by NSD1/FBXL11-dependent reversible lysine methylation of p65. Proc Natl Acad Sci U S A 107(1):46–51
25. Savaryn JP, Skinner OS, Fornelli L, Fellers RT, Compton PD, Terhune SS et al (2016) Targeted analysis of recombinant NF kappa B (RelA/p65) by denaturing and native top down mass spectrometry. J Proteome 134:76–84
26. Benson LJ, Annunziato AT (2004) In vitro analysis of histone acetyltransferase activity. Methods 33(1):45–52
27. Ea CK, Baltimore D (2009) Regulation of NF-kappaB activity through lysine monomethylation of p65. Proc Natl Acad Sci U S A 106(45):18972–18977
28. Levy D, Kuo AJ, Chang Y, Schaefer U, Kitson C, Cheung P et al (2010) Lysine methylation of the NF-kappaB subunit RelA by SETD6 couples activity of the histone methyltransferase GLP at chromatin to tonic repression of NF-kappaB signaling. Nat Immunol 12 (1):29–36
29. Trouplin V, Boucherit N, Gorvel L, Conti F, Mottola G, Ghigo E (2013) Bone marrow-derived macrophage production. J Vis Exp 81:e50966
30. Wang J, Cai Y, Shao LJ, Siddiqui J, Palanisamy N, Li R et al (2011) Activation of NF-{kappa}B by TMPRSS2/ERG fusion isoforms through toll-like receptor-4. Cancer Res 71(4):1325–1333
31. Zhang L, Shao L, Creighton CJ, Zhang Y, Xin L, Ittmann M et al (2015) Function of phosphorylation of NF-kB p65 ser536 in prostate cancer oncogenesis. Oncotarget 6 (8):6281–6294
32. Chen LF, Greene WC (2005) Assessing acetylation of NF-kappaB. Methods 36(4):368–375

Chapter 2

Biochemical Methods to Analyze the Subcellular Localization of NF-κB Proteins Using Cell Fractionation

Davide Vecchiotti, Daniela Verzella, Daria Capece, Mauro Di Vito Nolfi, Barbara Di Francesco, Jessica Cornice, Guido Franzoso, Edoardo Alesse, and Francesca Zazzeroni

Abstract

Cell fractionation is a method used to study different cellular events like protein translocation and sequestration by disrupting cells and fractionating their contents, thus allowing an enrichment of the protein of interest. Using different concentrations of sucrose or detergent buffer formulations in combination with centrifugations, the cell fractions are separated based on their density and size.

Key words Centrifugation, Sucrose, Nuclear protein, Cytoplasmatic protein, NF-κB

1 Introduction

The localization and characterization of different cellular proteins within the cells is achievable by subcellular fractionation methods. In 1946, Albert Claude was the first person to describe this method, stating that "by means of differential centrifugation at various speeds, it has been possible to separate three main fractions which appear to be morphologically and biochemically distinct" [1, 2]. The use of sucrose density gradients followed by others advanced isolation methods or detergent solutions that combine purity with high yield have allowed to understand cellular architecture, composition and function of organelles and cellular compartments (e.g., nucleus and cytoplasm), as well as cellular proteins [3–7].

To date, there are several generic fractionation protocols, but depending of the sample type (cell or tissue) and protein localization and function, method optimization is required in order to obtain a good quality of the cellular components and protein enrichment. The most important methods rely on sucrose

Guido Franzoso and Francesca Zazzeroni (eds.), *NF-κB Transcription Factors: Methods and Protocols*, Methods in Molecular Biology, vol. 2366, https://doi.org/10.1007/978-1-0716-1669-7_2,

gradient-centrifugation as core procedure with the addition of detergent solutions. The cellular fraction will be separated based on size and density; for example, nuclei are collected at the bottom of the centrifuge tube after low speed centrifugation, while mitochondria are accumulated at higher speed. Lysis of cell membranes is needed to allow the release of proteins from each fraction, but at the same time, it is necessary to avoid protein denaturation. Of course, the homogenization medium is carefully chosen, as well as multiple protease inhibitors need to be added. The extracted nuclear and cytoplasmatic proteins are then determined by sodium dodecyl sulfate (SDS) polyacrylamide-gel electrophoresis (SDS-PAGE) [8].

The NF-κB shuttle between the nucleus and the cytoplasm is a common event in response to stress or following stimulation with several factors including LPS, growth factors and cytokines [9–12]. NF-κB proteins can be found in the cytoplasm and nucleus depending on their activation state. Consistent with this, while in its inactivated state NF-κB complexes are sequestered in the cytosol by inhibitory proteins (IκBs), when activated, free NF-κB dimers translocate to the nucleus exerting their transcriptional functions [9–12]. It is known that disease can be characterized by alterated localization and amount of proteins respect to physiological state. In fact, NF-κB is known to be overexpressed in most human diseases including cancer [10]. This chapter attempts to describe a fractionation method (Fig. 1) in order to characterize and localize NF-κB proteins within the cytoplasmatic and nuclear subcellular compartments (Fig. 2).

2 Materials

Prepare the solution using ultrapure water. It is recommended to use protease inhibitors in order to have reagents free of any contaminating proteases and nucleases. The ideal lysis buffer should release an adequate amount of protein of interest from the sample.

2.1 Lysis Buffer for Nuclear and Cytoplasmatic Extracts

Prepare the two extraction buffers as reported below and store them at −20 °C.

1. Subcellular fraction (SF) buffer: 350 mM Sucrose, 10 mM HEPES (4-(2-hydroxyethyl)-1-piperazineethanesulfonic acid), 10 mM Potassium Chloride (KCl), 1.5 mM Magnesium Chloride ($MgCl_2$), 1 mM EDTA (Ethylenediaminetetraacetic acid), 0.2% IGEPAL (*see* **Note 1**). Add H_2O to a volume of 50 ml. Immediately before starting, add the following protease inhibitors: 5 mM Sodium metavanadate ($NaVO_3$) (*see* **Note 2**), 10 mM Sodium Fluoride (NaF), 10 μg/ml Aprotinin and 0.1 mg/ml Phenylmethanesulfonyl fluoride (PMSF) (*see* **Note 3**).

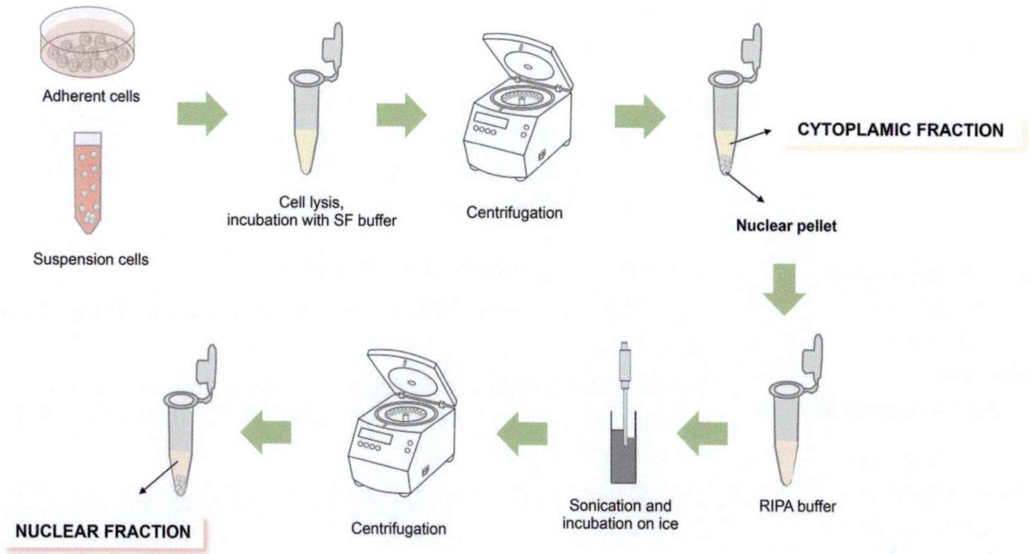

Fig. 1 Schematic diagram of cell fractionation protocol

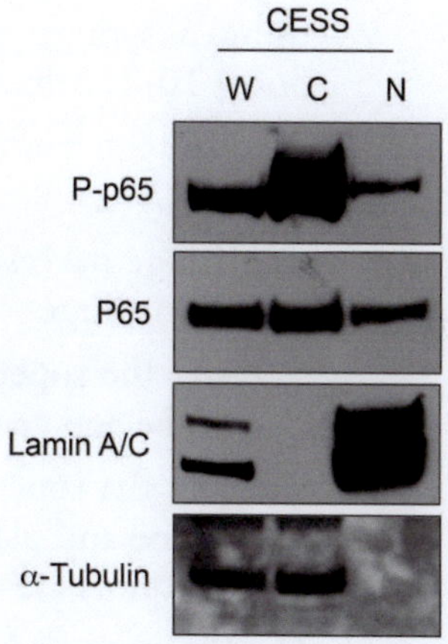

Fig. 2 Cytosolic and nuclear protein extracts isolated from acute myeloid leukemia (AML) cell line, CESS, at basal level, according to the protocol described in Methods. Whole cells (W), Cytosol (C) and nuclear (N) fractions were immunoblotted with α-NF-κB. α-Tubulin and Lamin A/C are shown as loading controls

2. RIPA buffer: 1× Phosphate buffered saline (PBS), 150 mM Sodium chloride (NaCl), 5% Sodium deoxycholate 0, 0.1% Sodium dodecyl sulfate (SDS) (*see* **Note 4**), 1% IGEPAL. Add H_2O to a volume of 10 ml. Immediately before starting, add the following protease inhibitors: 5 mM Sodium metavanadate ($NaVO_3$), 10 mM Sodium Fluoride (NaF), 10 μg/ml Aprotinin and 0.1 mg/ml Phenylmethanesulfonyl fluoride (PMSF).

3 Methods

Keep the preparation on ice where indicated. All buffers, centrifuge rotors and equipment should be maintained at 2–8 °C. This protocol is for processing around 5 × 10^6 cultured cells. Of note, it can be scaled up and down accordingly.

3.1 NF-κB Nuclear and Cytoplasmic Extracts from Adherent and Suspension Cells

3.1.1 Adherent Cell Culture Preparation and Cytoplasmic Proteins Extraction

1. Remove the culture medium from the dish.
2. Place the cell culture dish on ice and wash the cells gently with ice-cold 1× PBS.
3. Remove and discard PBS from the plate and add a suitable amount of ice-cold SF buffer (400 μl for 100 mm^2 dish; 600 μl for 150 mm^2 dish).
4. Using a cold cell scraper, detach the cells off the surface of the dish.
5. Transfer gently the mixture of lysed cells into a pre-cooled conical microcentrifuge tube.
6. [*Optional*] Gently vortex (for 10 s) to suspend the cells or using a syringe needle gently pull the suspension up and down 10–15 times on ice (*see* **Note 5**).
7. Incubate the lysate for 30 min at 4 °C on a rocker (35 rpm) (*see* **Note 5**).
8. Centrifuge harvested cells at 700 × *g* for 10 min at 4 °C in a microcentrifuge.
9. Transfer the supernatant containing cytoplasmic proteins to a clean tube and centrifuge at 10,000 × *g* for 10 min at 4 °C.
10. Transfer the supernatant containing cytoplasmic proteins to a clean tube and place immediately this tube on ice until use or store aliquots at −80 °C for future use (*see* **Note 6**).
11. Gently resuspend and wash the nuclear pellet with a suitable amount of SF buffer (900 μl for 150 mm^2 dish; 500 μl for 100 mm^2 dish) by pipetting up and down (*see* **Note 7**).
12. Centrifuge at 700 × *g* for 10 min at 4 °C.
13. Aspirate and discard the supernatant.

3.1.2 Suspension Cell Culture Preparation and Cytoplasmic Proteins Extraction

1. Centrifuge harvested cells at 700 × *g* for 5 min.
2. Remove and discard the culture medium keeping cell pellets.
3. Wash the cells gently with ice-cold 1× PBS.
4. Transfer the cells to a pre-cooled microcentrifuge tube and centrifuge at 700 × *g* for 5 min. Repeat this step twice.
5. Remove and discard PBS and add a suitable amount of ice-cold SF buffer (around 250–500 μl).

6. Gently vortex to suspend the cells or using a syringe needle gently pull the suspension up and down 10–15 times on ice (*see* **Note 5**).
7. Incubate the lysate for 30 min at 4 °C on a rocker (35 rpm) (*see* **Note 5**).
8. Centrifuge harvested cells at 700 × *g* for 10 min at 4 °C in a microcentrifuge.
9. Transfer the supernatant containing cytoplasmic proteins to a clean tube and centrifuge at 10,000 × *g* for 10 min at 4 °C.
10. Transfer the supernatant containing cytoplasmic proteins to a clean tube and place immediately this tube on ice until use or store aliquots at −80 °C for future use (*see* **Note 6**).
11. Gently resuspend and wash the nuclear pellet with a suitable amount of SF buffer (500–900 μl) by pipetting up and down (*see* **Note 7**).
12. Centrifuge at 700 × *g* for 10 min at 4 °C.
13. Aspirate and discard the supernatant.

3.1.3 Nuclear Protein Extraction from Adherent and Suspension Cells

1. Resuspend the nuclear pellet with ice-cold RIPA buffer (100–300 μl) by pipetting up and down (*see* **Note 8**).
2. Sonicate the samples for 10 s and incubate on ice for 10 s. Repeat this step three times (*see* **Note 9**).
3. Once sonicated the sample, incubate it on ice for 15 min.
4. Centrifuge sonicated cells for 10–30 min at 4 °C at 14,000 × *g*.
5. Carefully transfer the nuclear protein extract to a pre-cooled clean microcentrifuge tube.
6. Place microcentrifuge tube on ice until use or store nuclear extract aliquots at −80 °C (*see* **Note 10**).
7. *See* **Note 11**.

4 Notes

1. IGEPAL is a nonionic, non-denaturing detergent suitable for isolation and purification of membrane protein complexes. It selectively disrupts plasma membrane and it is not able to solubilize nuclear membrane. Normally it is used as neutral detergent in homogenizing buffer.
2. $NaVO_3$ is a commonly used protease inhibitor.
3. PMSF is a serine proteases inhibitor. It works in the range of 0.1–1 mM and is half-life is around 1 h at pH 7.5. For this reason, it is recommended to add it just prior to use.

4. SDS is an anionic detergent used as a protein solubilization reagent. SDS precipitates at 4 °C. Before use warm it to remove any precipitates.
5. The broken cell membranes are visible under microscope after staining with trypan blue. The nuclei will appear blue. Mechanical force needed to lyse cells depending on cell types; low force will not lyse all the cells, leading to low cytoplasmic proteins and contaminated nuclear fraction, while excess force can damage nuclei, leading to a contaminated cytoplasmic fraction with nuclear proteins.
6. Make sure not to remove the nuclei pellet. You can leave a low aliquot of supernatant and proceed with the next step (see on the protocol). If you have low cytoplasmic proteins, due to incomplete lysis of the cells or insufficient homogenization of the tissues, you can increase the amount of SF buffer and optimize the homogenization. You can also vortex thoroughly.
7. This step is performed in order to reduce contamination between fractions.
8. The amount of RIPA buffer depends on the size of the pellet. For example, for 150 mm^2 dish add 250 μl, while add 100 μl for 100 mm^2 dish. By using a sharp pipette tip, remove the sticky lump formed by dead cells and some lysed nuclei, if any.
9. The sonication protocol should be optimized based on sample type and sonicator.
10. In case you have low nuclear protein yield probably due to incomplete nuclei isolation or cell pellet not dissolved sufficiently you can increase the time of centrifugation or vortex thoroughly.
11. Before using cytoplasmic and nuclear proteins, the determination of protein concentration is necessary. The assay used for determining the protein concentration depends on the buffers used for protein extraction. For this protocol, BCA assay is needed, and the samples must be diluted at least 1:2 before running the assay.

Acknowledgments

The work was supported by the MIUR PRIN grant n° 2017WLKYAM_1 to F.Z.

References

1. Claude A (1946) Fractionation of mammalian liver cells by differential centrifugation: I. Problems, methods, and preparation of extract. J Exp Med 84:51–59
2. Claude A (1946) Fractionation of mammalian liver cells by differential centrifugation: II. Experimental procedures and results. J Exp Med 84:61–89
3. de Duve C, Pressman BC, Gianetto R et al (1955) Tissue fractionation studies. 6. Intracellular distribution patterns of enzymes in rat-liver tissue. Biochem J 60:604–617
4. Blobel G, Potter VR (1966) Nuclei from rat liver: isolation method that combines purity with high yield. Science 154:1662–1665
5. Bronfman M, Loyola G, Koenig CS (1998) Isolation of intact organelles by differential centrifugation of digitonin-treated hepatocytes using a table Eppendorf centrifuge. Anal Biochem 255:252–256
6. Srinivas KS, Chandrasekar G, Srivastava R et al (2004) A novel protocol for the subcellular fractionation of C3A hepatoma cells using sucrose density gradient centrifugation. J Biochem Biophys Methods 60(1):23–27
7. Suzuki K, Bose P, Leong-Quong RY et al (2010) REAP: a two minute cell fractionation method. BMC Res Notes 3:294
8. Graham JM, Rickwood (1997) Subcellular fractionation a practical approach. IRL Press Oxford University Press, Oxford
9. Capece D, Verzella D, Di Francesco B et al (2020) NF-κB and mitochondria cross paths in cancer: mitochondrial metabolism and beyond. Semin Cell Dev Biol 98:118–128
10. Begalli F, Bennett J, Capece D et al (2017) Unlocking the NF-κB conundrum: embracing complexity to achieve specificity. Biomedicine 5 (3):50
11. Bennett J, Capece D, Begalli F et al (2018) NF-κB in the crosshairs: rethinking an old riddle. Int J Biochem Cell Biol 95:108–112
12. Baud V, Karin M (2009) Is NF-κB a good target for cancer therapy? Hopes and pitfalls. Nat Rev Drug Discov 8(1):33–40

Chapter 3

Immunohistochemical Analysis of Expression, Phosphorylation, and Nuclear Translocation of NF-κB Proteins in Human Tissues

Davide Vecchiotti, Daniela Verzella, Daria Capece, Jessica Cornice, Mauro Di Vito Nolfi, Barbara Di Francesco, Guido Franzoso, Edoardo Alesse, and Francesca Zazzeroni

Abstract

Immunohistochemistry (IHC) is a technique aimed at detecting specific antigens on tissue sections by the use of targeting reagents labeled with reporter molecules. This technique allows a snapshot of the structure of tissue and determines the cellular and subcellular localization of a target antigen. This chapter describes how to identify and localize NF-κB proteins in human tissue using immunohistochemical staining on formalin-fixed paraffin-embedded and frozen tissue.

Key words Fixation, Antigen retrieval, Antibodies, Detection, Microscope, Identification (localization), NF-κB

1 Introduction

Immunohistochemistry (IHC) is a method used to detect the expression and localization of specific proteins within intact tissue. By using labeled target-specific antibodies, IHC is able to visualize localization, translocation, processing, trafficking, and targeting of specific cellular components in both physiological and pathological events, suggesting that this technique is widely recognized as a useful method for clinical diagnosis and research [1–3]. Developed by Coons and colleagues in 1941, IHC was used to localized pneumococcal antigens in infected tissues using antibodies labeled with fluorescein isothiocyanate (FITC) and then visualized by fluorescent microscopy [4]. The introduction of immunoperoxidase method used for detecting antigens in formalin fixed paraffin

* Davide Vecchiotti and Daniela Verzella contributed equally to the work as joint first author

Guido Franzoso and Francesca Zazzeroni (eds.), *NF-κB Transcription Factors: Methods and Protocols*, Methods in Molecular Biology, vol. 2366, https://doi.org/10.1007/978-1-0716-1669-7_3,

embedded (FFPE) tissues in 1974, increased the use of this assay in laboratory, thus improving the poor reproducibility [5–9]. The growing use of IHC in diagnostic and research laboratories was due to the discovery of monoclonal antibodies, new sensitive detection systems and new automated instruments, that all together increased stability and sensitivity of this technique.

IHC is a semi-quantitative method and a balance between sensitivity and low background signals is necessary to have a good staining [10–12]. Accordingly, several factors should be considered in order to have the best immunostaining. Among these factors, antigen localization, fixation, antigen retrieval and antibodies need to be carefully reviewed. Fixation is the first important step to consider during the IHC protocol. It aims to make a snapshot of the tissue at a particular moment in time, maintaining antigenicity and morphology [13–15]. There are three type of fixation (e.g., air drying, snap freezing, and chemical fixation), but the commonly used method is the immersion of the tissue in (or in some cases perfusion via vascular system with) a suitable chemical fixative including formaldehyde, alcohols, and acetone. It is important to know that the fixation method mainly depends on both epitope sensitivity and antibody specificity while tissue type and size are crucial for fixation time. There are some antigens that do not resist to standard fixation, for this reason tissues are snap freezing with or without a cryo-agent (e.g., optimal cutting temperature (OCT)) using liquid nitrogen. Following the fixation, the specimen is dehydrated and embedded in a paraffin block to make it easier for sectioning and mounting on microscopic slides [13, 14]. Other important steps for immunostaining are the epitope retrieval, used to reverse the antigen-masking effects of fixative, and the detection of the specific antigen. Since the core of the IHC is the antigen-antibody binding, before incubation with the antibody of interest, the fixed tissue requires antigen retrieval step to unmask the epitope, thus allowing the antibody to bind the selected antigen. The two widely recognized method to breaking the protein-cross-linking due to chemical fixation are: heat induced epitope retrieval (HIER) and proteolytic induced epitope retrieval (PIER) [13, 14]. The first method utilizes heat carried out by microwave, high pressure or water bath, while the second one uses digestion enzymes such as proteases, pepsin and trypsin. The choice between these two methodologies relies on the antigen nature. However, HIER is the most commonly used antigen retrieval technique.

Direct or indirect detection method is another key step in IHC [13, 14, 16]. The direct detection, used especially for high-density antigens, involves a labeled primary antibody reacting directly with the antigen in the tissue (Fig. 1a). On the other hand, for localize weakly expressed antigens, two-step staining so-called indirect method is more appropriate. In this case, the primary antibody, bound to antigen, is recognized by a conjugated secondary

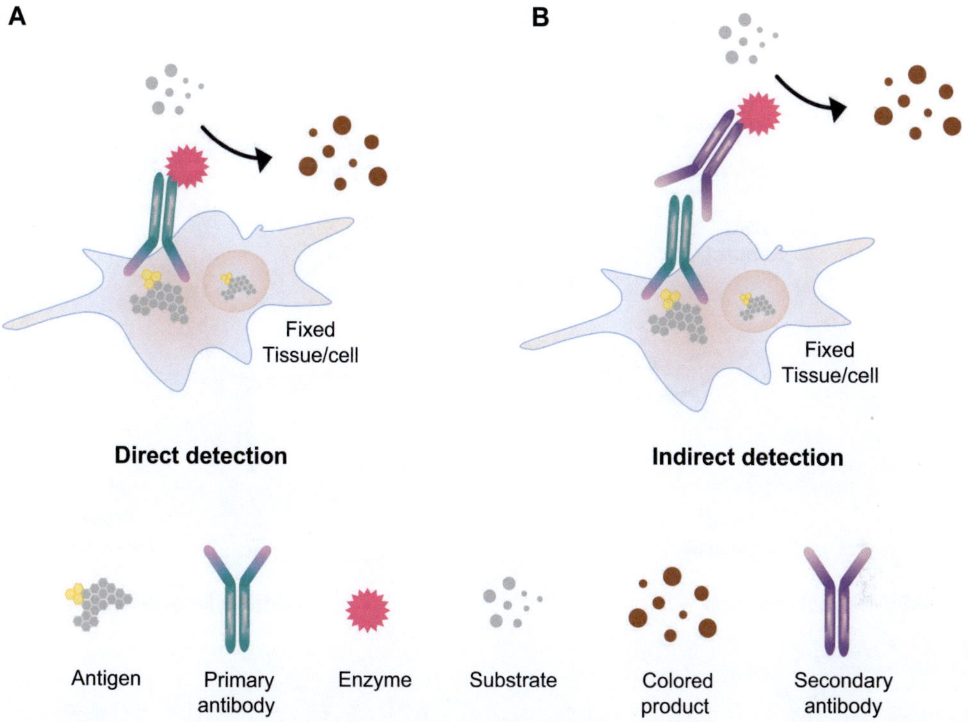

Fig. 1 IHC detection methods. (**a**) Direct method. (**b**) Indirect method

antibody (Fig. 1b). As more than two labeled secondary antibodies can bind different antigenic sites on the same primary antibody, the final signal results amplificated indicating that this method is more sensitive than direct detection. In both systems, the antigen-antibody complex is identified using chemically linked reporter molecules like enzymes (e.g., alkaline phosphatase (AP) or horseradish peroxidase (HRP)), fluorochromes or colloidal gold. The final product, which can be either a colored precipitated or fluorescent end product of an enzyme/substrate reaction or fluorescent, is then visualized by light, fluorescence or electron microscopy.

In order to enhance the signal and thus the sensitivity, amplification systems like Biotin-avidin/streptavidin-based methods (e.g., avidin-biotin complex (ABC), labeled avidin/streptavidin-biotin (LAB/LSAB)) (Fig. 2a, b), soluble enzyme-anti-enzyme immune complex (e.g., peroxidase-anti-peroxidase complex (PAP) and alkaline phosphatase-anti-alkaline phosphatase complex (APAAP)) have been developed [16, 17]. To overcome the limitations due to tissue endogenous biotin as well as the duration of the staining, other next-generation amplification systems have been elaborated. The most relevant systems are: chain polymer-based IHC such as enhanced polymer one step staining system (EPOS) and polymer-enhanced two-step IHC detection system (e.g., EnVision, EV), that utilize a conjugated antibodies or enzymes-polymer backbone;

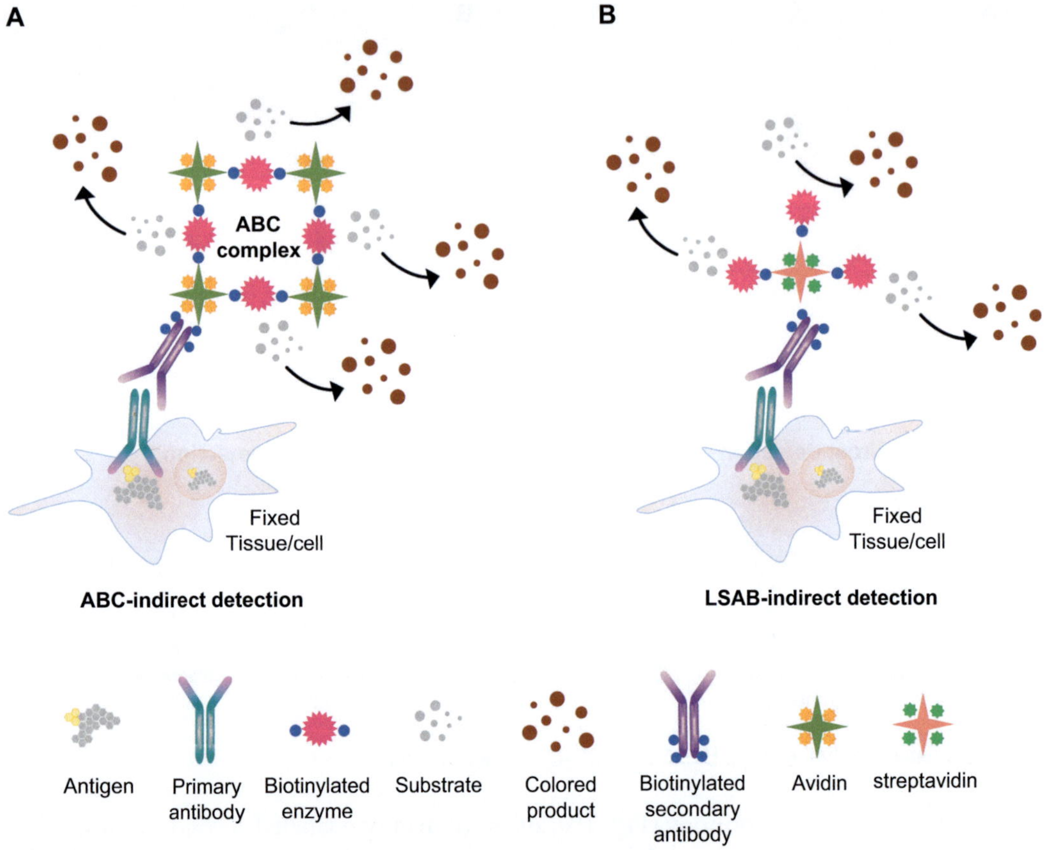

Fig. 2 IHC amplification detection methods. (**a**) Avidin-biotin complex (ABC) method. (**b**) Labeled streptavidin-biotin (LSAB) method

biotinylated tyramide-based signal amplification (e.g., tyramine signal amplification (TSA), catalyzed signal amplification (CSA)) both based on an analyte-dependent reporter enzyme (ADRE) and finally rolling circle amplification (RCA) in which a short DNA/RNA primer is amplified by a RNA/DNA polymerases using a circular DNA template [16, 18, 19].

Despite the numerous advantages of these new technologies, the ABC system remains widely popular for signal amplification in research laboratory. This system takes advantage of the high affinity that streptavidin/avidin have for biotin as well as for the number of biotin-binding sites on streptavidin/avidin. The streptavidin/avidin bound to the biotinylated-secondary antibody represents a link between antigen-primary antibody complex and the biotinylated-reporter.

Of course, the choice of an IHC detection method depends on many variables including sensitivity, costs and, last but not least, time.

To further defining the localization of the antigen of interest, it is recommended to use a nuclear counterstain (like Hematoxylin, DAPI, HOECHST), allowing the visualization of the cells within the tissue, without obscuring positive staining and then helping the interpretation of the immunostaining.

NF-κB belongs to a family of transcription factors that involve five proteins known as RelA/p65, RelB, c-Rel, NF-κB1 (p105/p50), and NF-κB2 (p100/p52) and regulate several biological processes such as inflammation, cell survival, immunity, apoptosis, and differentiation. NF-κB consists of homo- or heterodimers of different subunits and the heterodimer p65/p50 represent the most abundant active complex in mammalian cells [20–23]. Under normal condition, NF-κB is sequestered in the cytoplasm by NF-κB inhibitory proteins (IκBs) in an inactive state [20–23]. The activation of NF-κB occurs through two signaling pathways known as the classical (canonical) and the alternative (non-canonical) pathway. During canonical pathways, stimuli like proinflammatory cytokines, lipopolysaccharide (LPS) and growth factors, activate the IκB kinase (IKK) complex, that in turn promotes the phosphorylation of IκB, thus triggering IκB polyubiquitination and proteasomal degradation. Activated NF-κB is now able to translocate to the nucleus and activate target gene expression. On the other hand, during the alternative pathways, the activation of p100/RelB complex relies on NF-κB inducing kinase (NIK), that phosphorylates and activates IKKα complex, leading to the processing of p100 into p52 and subsequent nuclear translocation of the p52/RelB active heterodimer [20, 24].

This chapter gives an overview of NF-κB immunostaining, highlighting the key steps of the IHC on FFPE (Fig. 3) and frozen-OCT embedded (Fig. 4) tissue sections as well as the principal factors that can influence the results of a *bona fide* NF-κB staining.

2 Materials

2.1 Reagents

In order to obtain a successful IHC, all reagents should be freshly prepared before starting the assay.

1. 0.1 M phosphate buffered saline (PBS) 10×, pH 7.4: 77 mM Na_2HPO_4 (anhydrous), 23 mM NaH_2PO_4 (anhydrous), 1.5 M NaCl. Add distilled water to a volume of 1 L. Mix to dissolve and adjust pH to 7.4. This solution can be stored at room temperature (RT) for up 3 months. Before use, dilute 1:10 in H_2O.
2. 10% neutral buffered formalin (NBF): 100 mL of (37–40%) Formaldehyde, 6.5 g of Na_2HPO_4 (dibasic/anhydrous), 4.0 g of NaH_2PO_4 (monobasic). Add distilled water to a volume of 1 L. Mix to dissolve and store at RT.

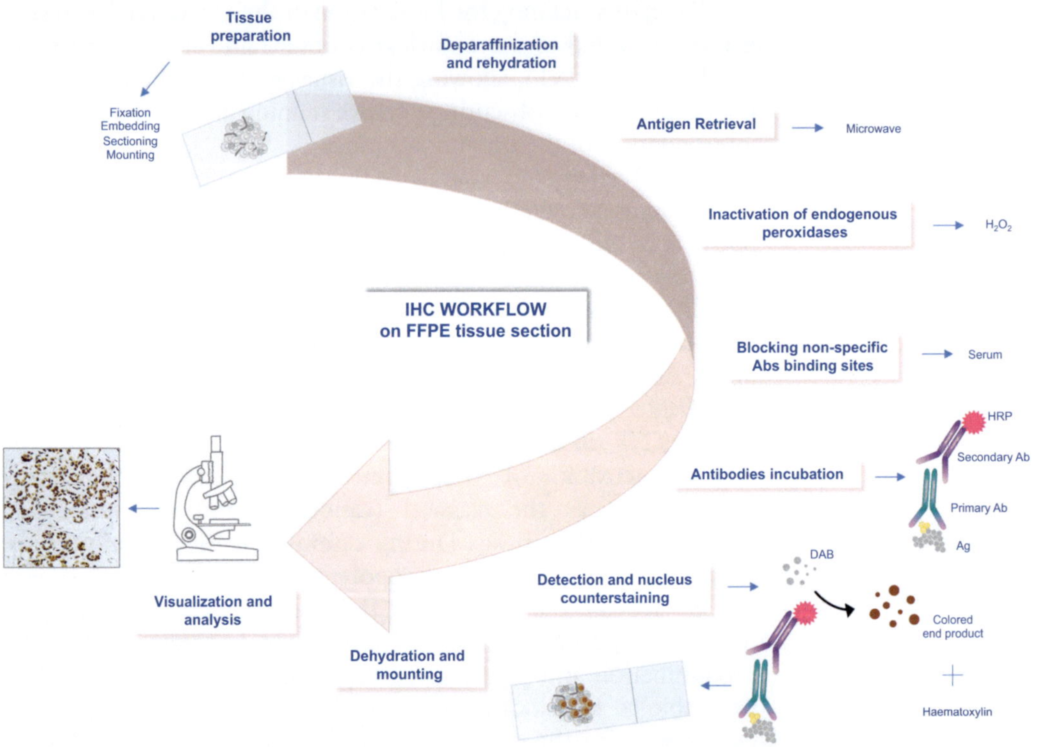

Fig. 3 IHC workflow on formalin-fixed paraffin-embedded tissue sections. Schematic representation of key events characterizing a typical FFPE immunostaining

3. 10 mM Citrate buffer, pH 6.0: 2.94 g of citric acid (anhydrous). Add distilled water to a volume of 1 L. Mix to dissolve and adjust pH to 6.0 with 1 N NaOH. Store solution at 4 °C.
4. Phosphate Buffer Saline (1×) (PBS).
5. 10% Neutral Buffer Formalin, 1 L: 4 g of sodium dihydrogen orthophosphate (monohydrate), 6.5 g of disodiumhydrogen orthophosphate (anhydrous), 100 mL of formaline (40% aqueous solution of formaldehyde). Add distilled water to a volume of 1 L.
6. Paraffin.
7. Xylene.
8. 100% ethanol.
9. 95% ethanol.
10. 80% ethanol.
11. 70% ethanol.
12. 50% ethanol.
13. 3% and 0.3% hydrogen peroxide solution (H_2O_2).
14. Antibody diluent (Lab Vision).
15. Serum from the species of secondary antibody diluted in 1× PBS.

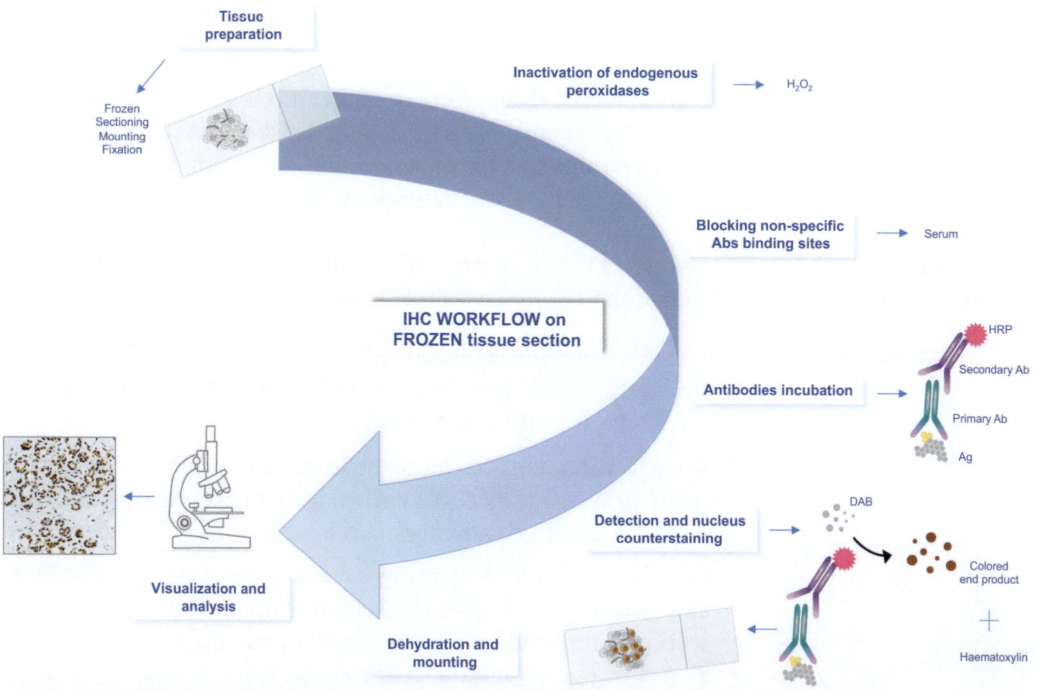

Fig. 4 IHC workflow on frozen tissue sections. Schematic representation of key events characterizing a typical immunostaining on frozen-OCT embedded tissues

16. Anti NF-κB primary antibody.
17. Biotinylated-secondary antibody.
18. Strept(avidin) complex conjugated with HRP (Vector Laboratories).
19. 3,3′-Diaminobenzidine (DAB) substrate solution (Vector Laboratories).
20. Hematoxylin solution.
21. Slide mounting solution (Vector Laboratories).
22. Optimal Cutting Temperature compound (OCT).
23. 4% Paraformaldehyde.

2.2 Equipment

1. Tweezers.
2. Brushes.
3. Hydrophobic wax barrier pen.
4. Microtome.
5. Cryotome cryostat.
6. Glass slides.
7. Oven.
8. Optical Microscope.

3 Methods

Since no universal IHC protocol for NF-κB is available for all tissue types, the following protocol should be considered as a starting point for researcher. Optimization and standardization of procedures are needed to ensure reproducible results (*see* **Notes 1** and **2**).

3.1 NF-κB IHC on FFPE Tissue Section

Keep in mind that a successful IHC is due to a tissue preparation, with particular regard to sample fixation.

3.1.1 Tissue Preparation

1. Harvest the fresh tissue and wash it with PBS to remove blood.
2. Fix freshly tissue immediately with 10% NBF or other fixative for 12–24 h at RT (*see* **Notc 3**).
3. After fixation, embed the tissue in paraffin using an automated tissue processing machine known as tissue processor (*see* **Note 4**). The paraffin tissue block can be stored at RT for years.
4. Cut the FFPE block by using a microtome at 3–5μm thickness and transfer it onto glass slide suitable for IHC (normally tissue section is mounted on positively charged coated slide).
5. Dry the slide overnight at 37 °C or RT in order to remove water trapped under the tissue section during the sectioning. The slide is now ready to use.

3.1.2 Deparaffinization and Rehydration

1. Incubate the slide at 60 °C for 1 h in oven in order to remove excess wax.

   Dewax the slide in xylene 3–5 min each. Repeat this step twice (*see* **Note 5**).
2. Rehydrate by immersing the slide in a series of decreasing alcohols as following:

   2× 100% ethanol 3–5 min each.

   1× 95% ethanol 3–5 min.

   1× 80% ethanol 3–5 min.

   1× 70% ethanol 3–5 min.

   1× 50% ethanol 3–5 min.
3. Rinse the slide in water for 5 min. Repeat this step twice.

3.1.3 Antigen Retrieval: HIER

1. Preheat the citrate buffer (antigen retrieval solution) to 95–100 °C (*see* **Note 6**).
2. Immerse the FFPE sections into coplin jar filled with preheated citrate buffer using a sufficient volume to cover the slide for about 10–20 min (*see* **Notes** 7 and **8**).
3. Leave the slide to cool at RT for about 15–30 min before proceeding with the next step.

4. Rinse slide in 1× PBS at RT for 5 min. Repeat this step twice.
5. *See* **Note 9**.

3.1.4 Inactivation of Endogenous Peroxidases

1. Block endogenous peroxidase activity by incubating sections in 3% H_2O_2 solution at room temperature for 10–20 min (*see* **Note 10**).
2. Rinse slide in 1× distilled H_2O for 5 min at RT.
3. Wash slide in PBS at RT for 5 min. Repeat this step twice.
4. *See* **Note 11**.

3.1.5 Blocking Non-specific Antibody Binding Sites

1. Add 100μL of serum from the species of secondary antibody diluted in 1× PBS onto sections and incubate in a humidified chamber for 30–60 min at RT (*see* **Notes 12** and **13**).
2. Discard the blocking solution before proceeding with antibody incubation.
3. [*Optional*] Wash the slide in 1× PBS for 5 min at RT.

3.1.6 Antibodies Incubation

1. Dilute primary antibody with antibody diluent to the optimal concentration.
2. Incubate the section with 100μL of diluted primary antibody (α-NF-κB) in a humidified chamber for 1 h at RT (*see* **Notes 14** and **15**).
3. Wash the slide with 1× PBS at RT for 5–10 min. Repeat this step three times.
4. Dilute biotinylated secondary antibody with antibody diluent to the optimal concentration.
5. Apply approximately 100μL of diluted secondary antibody to the section on the slide and incubate in a humidified chamber for 1 h at RT.
6. Rinse slide in 1× PBS at RT for 5–10 min. Repeat this step three times.

3.1.7 Detection and Nucleus Counterstain

1. Incubate section with approximately 100μL of strept(avidin) complex conjugated with HRP for 30 min in a humidified chamber protected from light at RT accordingly to the manufacturer's instruction.
2. Wash the slide with 1× PBS at RT for 5–10 min. Repeat this step three times.
3. Prepare DAB substrate solution and apply approximately 100μL of DAB substrate solution to the section for 1 min at RT in the dark to reveal the color of antibody staining (*see* **Note 16**).
4. Remove the DAB washing the slide with distilled H_2O for 5 min. Repeat this step twice.

5. Counterstain nuclei by immersing slide in hematoxylin for 1–2 min.
6. Rinse the slide in running tap water for at least 10 min.
7. Wash slide in distilled water for 10 min. Repeat this step twice.

3.1.8 Dehydration and Mounting

1. Dehydrate the tissue slide by immersing the slide in a series of increasing alcohols as following:
   1× 50% ethanol 3–5 min.
   1× 70% ethanol 3–5 min.
   1× 80% ethanol 3–5 min.
   1× 95% ethanol 3–5 min.
   2× 100% ethanol 3–5 min each.
2. Clear slide in xylene 5 min. Repeat this step twice.
3. Coverslip the slide using mounting solution and leaves the mounted slide to dry at RT for at least 1 h.

3.1.9 Visualization

1. Observe NF-κB immunostaining under the microscope (Fig. 5).

3.2 NF-κB IHC on Frozen-OCT Embedded Tissue Section

Keep in mind that all instruments including tweezers, brushes and pencil (to circle the sections in order to reduce and better identify the area of interest) should be placed in the chamber to equilibrate to temperature (−20 °C).

3.2.1 Tissue Preparation

1. Harvest small fresh tissue and wash it with cold 1× PBS to remove blood.
2. Place tissue immediately with the freeze-embedding medium OCT into a pre-labeled tissue base mold. Frozen the mold into liquid nitrogen till the complete freezing and store at −80 °C until use (*see* **Note 17**).
3. Allocate the OCT block to a cryostat chamber prior to sectioning for 1 h in order to let the temperature of the OCT block to equilibrate to the temperature of the cryotome cryostat (*see* **Note 18**).
4. Cut the OCT block using a cryotome cryostat at 3–10μm thickness and transfer it onto glass slide suitable for IHC (normally tissue section is mounted on positively charged coated slide to decrease the section dissociating from the slide during the staining) (*see* **Notes 19** and **20**).
5. Air-dry the slide overnight at RT or at least 1 h. The slide is now ready to use or store it at −80 °C for later use (*see* **Note 21**).
6. Collect the slide from −80 °C and rehydrate in 1× PBS for about 10 min.
7. Fix tissue slide into 4% paraformaldehyde for 10 min at RT.

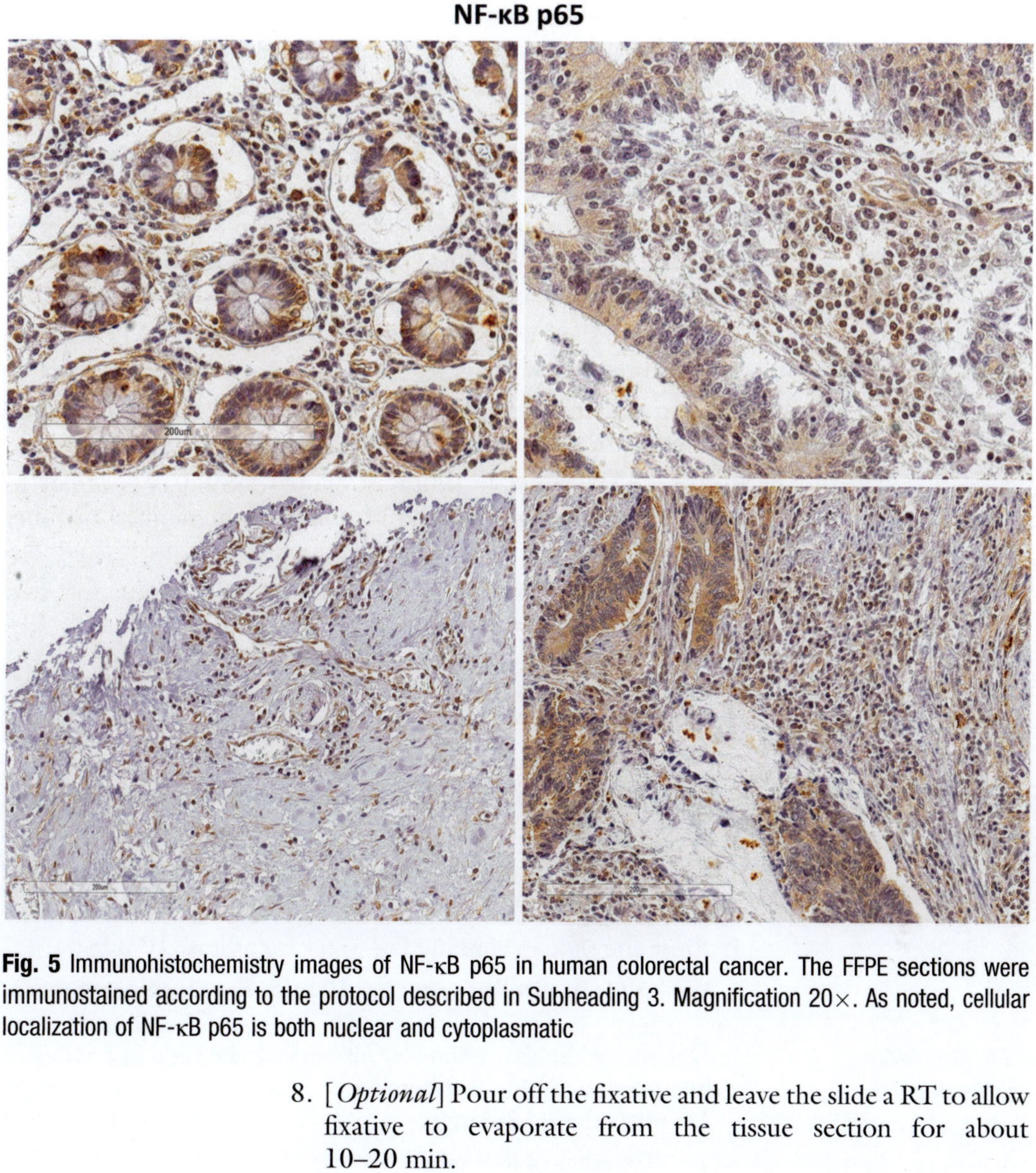

Fig. 5 Immunohistochemistry images of NF-κB p65 in human colorectal cancer. The FFPE sections were immunostained according to the protocol described in Subheading 3. Magnification 20×. As noted, cellular localization of NF-κB p65 is both nuclear and cytoplasmatic

8. [*Optional*] Pour off the fixative and leave the slide a RT to allow fixative to evaporate from the tissue section for about 10–20 min.
9. Rinse slide in 1× PBS for 5 min. Repeat this step three times.

3.2.2 Inactivation of Endogenous Peroxidases

1. Block endogenous peroxidase activity by incubating sections in 0.3% H_2O_2 at RT for 10–20 min (*see* **Note 8**).
2. Wash section by dipping it on 1× distilled H_2O at RT.
3. Rinse slide in 1× PBS at RT for 5 min. Repeat this step twice.

3.2.3 Blocking Nonspecific Antibody Binding Sites

1. Add 100μL commercially blocking solution or serum from the species of secondary antibody diluted in 1× PBS onto section and incubate in a humidified chamber for 30–60 min at RT (*see* **Notes 12** and **13**).

2. Discard the blocking solution before proceeding with antibody incubation.
3. [*Optional*] Wash the slide in 1× PBS for 5 min at RT.

3.2.4 Antibodies Incubation

1. Dilute primary antibody with antibody diluent to the optimal concentration.
2. Incubate the section with 100μL of diluted primary antibody (α-NF-κB) in a humidified chamber for 1 h at RT (*see* **Notes 14** and **15**).
3. Wash the slide with 1× PBS at RT for 5–10 min. Repeat this step three times.
4. Dilute secondary antibody-HRP conjugated with antibody diluent to the optimal concentration.
5. Apply approximately 100μL of diluted secondary antibody to the section on the slide and incubate in a humidified chamber for 1 h at RT.
6. Rinse slide in 1× PBS at RT for 5–10 min. Repeat this step three times.

*3.2.5 Detection and Nucleus Counterstain (See **Note 22**)*

1. Prepare DAB substrate solution and apply approximately 100μL of DAB substrate solution to the section for 1 min at RT in the dark to reveal the color of antibody staining (*see* **Note 16**).
2. Remove the DAB washing the slide with distilled H_2O for 5 min. Repeat this step twice.
3. Counterstain nuclei by immersing slide in Hematoxylin for 1–2 min.
4. Rinse the slide in running tap water for at least 10 min.
5. Wash slide in distilled water for 10 min. Repeat this step twice.

3.2.6 Dehydration and Mounting

1. Dehydrate the tissue slide by immersing the slide in a series of increasing alcohols as following:
   1× 50% ethanol 3–5 min.
   1× 70% ethanol 3–5 min.
   1× 80% ethanol 3–5 min.
   1× 95% ethanol 3–5 min.
   2× 100% ethanol 3–5 min each.
2. Clear slide in xylene 5 min. Repeat this step twice.
3. Coverslip the slide using mounting solution and leaves the mounted slide to dry at RT for at least 1 h.

3.2.7 Visualization

1. Observe NF-κB immunostaining under microscope.

4 Notes

1. It is important to run controls during any immunostainings in order to confirm the accuracy, authenticity and reliability of results as well as to exclude experimental artifacts. As a general rule, antigen controls and reagent controls are in IHC. Regarding the first control type, a slide containing a tissue section known to express the antigen of interest, named positive control, are used as validation of the staining, since a positive result is expected. Note that a negative result from the positive control indicates that the procedure is not working and needs to be optimized. Contrariwise, negative control is a tissue section in which the protein of interest is not expressed (e.g., knockout tissue), allowing to discriminate non-specific antibody binding and false positive results. Instead reagent control ensures that the staining is produced from binding of the primary antibody to the target antigen and not from the detection system or the tissue. The best way to enrich this issue, is to carry out the IHC without primary antibody and/or isotype controls.
2. The choice and optimization of the primary antibody is a key step for a successful immunostaining. The most important features that should be considered are the specificity of the antibody, the clonality (monoclonal vs polyclonal), and the host species that preferably should be different to the species of the samples in order to avoid cross-reactivity with endogenous immunoglobulins. To achieve specific staining with low background concentrations, incubation time, and temperature of antibody need to be optimized. Note that monoclonal antibody is more sensitive to fixation-induced alterations, pH, and salt concentrations than polyclonal antibody.
3. The ideal fixation time will depend on the size and type of the tissue. Under-fixation can result in a strong signal on the edges and no signal in the middle of the section while over-fixation can mask the epitope and no signal might be seen at the end of the staining. It is widely accepted that 1:10 ratio volume tissue to fixative and 12–24 h is a good starting point.
4. The automated processing allows a better quality of the specimen. Is possible to perform the procedure manually as following: (a) due to hydrophobic properties of paraffin, dehydrate the tissue by immersing specimen in a series of increasing ethanol solution until pure (e.g., 50%, 70%, 80%, 95%, and 100%) for 45–60 min each. A typical sequence would be: 50% ethanol 45 min; 70% ethanol 45 min; 80% ethanol 45 min; 95% ethanol 45 min; 3× 100% ethanol 1 h each. (b) Clear the tissue using a clearing agent such as xylene prior to infiltrate the tissue with paraffin. It is important to clearing the tissue two times,

1 h each (this step is necessary because paraffin and ethanol are immiscible; furthermore, it is important to remove a substantial amount of fat from the tissue and to impart an optical transparency to the tissue). (c) Immerse the tissue in paraffin-based histological wax, 3 × 1 h each. (d) Embed the tissue in a paraffin block using an embedding station (the paraffin is liquid at 56–60 °C and it solidifies at 20 °C).

5. It is better to carry out the technique under the hood since the clearing solution is toxic and RT.
6. It is possible to achieved this by placing the coplin jar filled with retrieval solution into a water bath at 95 °C.
7. The optimal incubation time and pH should be determined experimentally by the researcher. A control experiment is recommended beforehand.
8. Three parameters are responsible for the successful of HIER step: the methylene bridge formation within tissue, the susceptibility of the antigen, and the conditions of the antigen retrieval process. Bear in mind that over-retrieved of antigens can lead to dissociation of the tissue, thus destructing the antigen of interest; while under-retrieved can led to weak staining.
9. The antigen retrieval can be performed by proteolytic enzymes such as trypsin, proteinase K, pepsin, and others. In this scenario, depending on type of tissue and fixation, the slide will be incubated with 0.05–0.1% enzyme for 10–30 min at 37 °C. The incubation time should be determined previously.
10. Inactivation of endogenous peroxidases is fundamental step to avoid non-specific background and needs to be done before the incubation with primary antibody. Dilute peroxidase blocking solution in methanol or PBS following the manufacturer's instruction. The blocking solutions depend on the enzyme conjugated to antibody used. In case of AP-conjugated antibody, the AP inhibitors available include levamisole hydrochloride or tetramisole hydrochloride.
11. When using ABC detection method, in order to avoid non-specific background, an avidin/biotin blocking step can be done if biotin is normally express in the organs such as kidney, liver and spleen. This step can be overcome if working with paraffin tissue sections, where the endogenous biotin is loss by tissue pretreatment.
12. The amount of serum solution depends on tissue size. Commercially available blocking solution can be used as well.
13. To avoid higher background, do not block with normal serum from the host species of primary antibody.

14. The incubation with primary antibody could be done overnight at 4 °C in a humidified chamber. The amount of antibody depends on tissue size.
15. In case the incubation is performed overnight, the day after, leaves the slides for at least 1 h at RT.
16. If the antigen of interest is poorly expressed, it is recommended to incubate until the desired color intensity is reached by monitoring the reaction under a bright-field microscope. The amount of blocking solution depends on tissue size.
17. Be careful not submerge completely the OCT-based mold with liquid nitrogen. The OCT should change the color to white.
18. This step will ensure that the OCT retains its plasticity.
19. Cut the OCT block when it is too cold will result in a brittle sample and wrinkling.
20. Tissue sectioning aims at obtaining a thin slice of a tissue sample, a one-plane view of the complex 3D structure of the tissue. It is important to cut the section as thinly as possible for two main reasons: the IHC process will stain only the superficial layer of the section and therefore a thick section will not improve/increase the signal detected; a thick slice will also carry multiple focal points that will confuse the images during acquisition. Additionally, if sections are cut thin, two consecutive sections can be considered as a single plane and be directly comparable (imagine having two sections stained with different markers to compare relative distribution/localisation).
21. This step will ensure good adhesion to the slide.
22. If using amplification signal method (e.g., ABC), the secondary antibody should be biotinylated. Proceed with the following two steps before incubating with DAB:
    (a) Incubate section with approximately 100μL of strept(avidin) complex conjugated with HRP for 30 min in a humidified chamber protected from light at RT accordingly to the manufacturer's instruction.
    (b) Wash the slide with 1× PBS at RT for 5–10 min. Repeat this step three times.

Acknowledgments

The work was supported by the MIUR PRIN grant n° 2017WLKYAM_1 to F.Z.

References

1. Tuffaha MSA, Guski H, Kristiansen G (2018) Immunohistochemistry in tumor diagnostic. https://doi.org/10.1007/978-3-319-53577-7
2. Garcia CF, Swerdlow SH (2009) Best practices in contemporary diagnostic immunohistochemistry: panel approach to hematolymphoid proliferation. Arch Pathol Lab Med 133(5):756–765
3. Leong ASY, Wright J (1987) The contribution of immunohistochemical staining in tumour diagnosis. Histopathology 11(12). https://doi.org/10.1111/j.1365-2559
4. Coons AH, Creech HJ, Jones RN (1941) Immunological properties of an antibody containing a fluorescent group. Exp Biol Med 47:200–202
5. Coons AH, Kaplan MH (1950) Localization of antigen in tissue cells ii. improvements in a method for the detection of antigen by means of fluorescent antibody. J Exp Med 91(1):1–13
6. Nakane PK, Pierce GB (1967) Enzyme-labeled antibodies for the light and electron microscopic localization of tissue antigens. J Cell Biol 33:307–318
7. Taylor CR, Burns J (1974) The demonstration of plasma cells and other immunoglobulin-containing cells in formalin-fixed, paraffin-embedded tissues using peroxidase-labelled antibody. J Clin Pathol 27:14–20
8. Taylor CR, Mason DY (1974) The immunohistological detection of intracellular immunoglobulin in formalin-paraffin sections from multiple myeloma and related conditions using the immunoperoxidase technique. Clin Exp Immunol 18:417–429
9. Mason DY, Sammons R (1978) Alkaline phosphatase and peroxidase for double immunoenzymatic labelling of cellular constituents. J Clin Pathol 31:454–460
10. Jensen K, Krusenstjerna-Hafstrøm R, Lohse J et al (2017) A novel quantitative immunohistochemistry method for precise protein measurements directly in formalin-fixed, paraffin-embedded specimens: analytical performance measuring HER2. Mod Pathol 30:180–193
11. Laurinaviciene A, Plancoulaine B, Baltrusaityte I et al (2014) Digital immunohistochemistry platform for the staining variation monitoring based on integration of image and statistical analyses with laboratory information system. Diagn Pathol 9(Suppl 1):S10
12. Taylor CR, Shi SR, Barr NJ (2013) Techniques of immunohistochemistry: principles, pitfalls, and standardization. In: Dabbs DJ (ed) Diagnostic immunohistochemistry, 3rd edn. W.B. Saunders, Philadelphia, PA, pp 1–42
13. Buchwalow IB, Böcker W (2010) Immunohistochemistry: basics and Methods. Springer, Heidelberg Dordrecht London New York
14. Renshaw S (2017) Immunohistochemistry and immunocytochemistry essential methods, 2nd edn. Wiley & Sons, ltd., New York, NY
15. Kalynzhny AE (2016) Immunohistochemistry: essential elements and beyond, 1st edn. Springer, Cham, pp 49–61
16. Shojaeian S, Maslehat Lay N, Zarnani AH (2018) Detection systems in immunohistochemistry. In: Streckfus CF (ed) Immunohistochemistry – the ageless biotechnology. InTechOpen, London
17. Hsu SM, Raine L, Fanger H (1981) Use of avidin-biotin-peroxidase complex (ABC) in immunoperoxidase techniques: a comparison between ABC and unlabeled antibody (PAP) procedures. J Histochem Cytochem 29(4):577–580
18. Dixon AR, Bathany C, Tsuei M et al (2015) Recent developments in multiplexing techniques for immunohistochemistry. Expert Rev Mol Diagn 15(9):1171–1186
19. Warford A, Akbar H, Riberio D (2014) Antigen retrieval, blocking, detection and visualisation systems in immunohistochemistry: a review and practical evaluation of tyramide and rolling circle amplification systems. Methods 70(1):28–33
20. Capece D, Verzella D, Di Francesco B et al (2020) NF-κB and mitochondria cross paths in cancer: mitochondrial metabolism and beyond. Semin Cell Dev Biol 98:118–128
21. Begalli F, Bennett J, Capece D et al (2017) Unlocking the NF-κB conundrum: embracing complexity to achieve specificity. Biomedicine 5(3):50
22. Bennett J, Capece D, Begalli F et al (2018) NF-κB in the crosshairs: rethinking an old riddle. Int J Biochem Cell Biol 95:108–112
23. Baud V, Karin M (2009) Is NF-κB a good target for cancer therapy? Hopes and pitfalls. Nat Rev Drug Discov 8(1):33–40
24. Oeckinghaus A, Ghosh S (2009) The NF-kappaB family of transcription factors and its regulation. Cold Spring Harb Perspect Biol 1(4):a000034

Chapter 4

High-Throughput Analysis of the Cell and DNA Site-Specific Binding of Native NF-κB Dimers Using Nuclear Extract Protein-Binding Microarrays (NextPBMs)

Heather Hook, Rose W. Zhao, David Bray, Jessica L. Keenan, and Trevor Siggers

Abstract

Nuclear factor-kappa B (NF-κB) transcription factors coordinate gene expression in response to a broad array of cellular signals. In vertebrates, there are five NF-κB proteins (c-Rel, RelA/p65, RelB, p50, and p52) that can form various dimeric combinations exhibiting both common and dimer-specific DNA-binding specificity. In this chapter, we describe the use of the nuclear extract protein-binding microarray (nextPBM), a high-throughput method to characterize the DNA binding of transcription factors present in cell nuclear extracts. NextPBMs allow for sensitive analysis of the DNA binding of NF-κB dimers and their interactions with cell-specific cofactors.

Key words NF-kB, DNA binding, Protein-binding microarrays, Nuclear extract

1 Introduction

Nuclear factor-kappa B (NF-κB) proteins constitute a family of transcription factors (TFs) that regulate diverse biological processes [1]. In vertebrate organisms there are five NF-κB proteins (c-Rel, RelA/p65, RelB, p50, and p52) that can form various homodimeric and heterodimeric combinations [2, 3]. Homologs of vertebrate NF-κB proteins have also been identified in diverse phyla, including Arthropods (e.g., insects), Cnidaria (e.g., corals) and even single-celled protists [4]. NF-κB dimers are dynamic TFs that transmit cell signal events into gene expression changes. For example, in response to a range of stimuli, NF-κB dimers (e.g., p65:p50, RelB:p52) can translocate into the nucleus and bind regulatory elements to regulate expression of target genes

Heather Hook and Rose Zhao contributed equally to this work.

Guido Franzoso and Francesca Zazzeroni (eds.), *NF-κB Transcription Factors: Methods and Protocols*, Methods in Molecular Biology, vol. 2366, https://doi.org/10.1007/978-1-0716-1669-7_4,

[3, 5]. Different NF-κB proteins, such as p65 and RelB, have even been observed to dynamically replace one another at binding sites within the genome [3]. NF-κB dimers can, therefore, distinguish themselves via their dynamic response to cell signals; however, they also exhibit differences in DNA binding. Studies examining the DNA-binding specificity of NF-κB dimers have demonstrated conserved yet distinct sequence preferences between dimers [6, 7]. These studies highlight the need for methods to assay the dynamic and dimer-specific DNA binding of NF-κB in a stimulus-dependent manner to advance our understanding of NF-κB regulatory specificity.

Perhaps the most widely used approach for detecting and examining protein-DNA binding is the electrophoretic mobility shift assay (EMSA) [8, 9]. EMSAs remain a useful experimental method but are generally considered a "low-throughput" approach, as they allow on the order of tens of distinct DNA sequences to be examined per experiment. In contrast, the last 15 years have seen the development of a number of "high-throughput" approaches that utilize DNA microarrays or next-generation sequencing to assay thousands to millions of distinct DNA sequences in a single experiment [10]. One such approach—the protein-binding microarray (PBM) [11–13]—has been used to study the DNA-binding specificity of NF-κB dimers from a wide range of species, allowing for detailed comparisons of the DNA binding of NF-κB dimer homologs and orthologs [6, 7, 14–17]. These high-throughput approaches provide a much more detailed description of protein-DNA binding specificity and have led to new insights into the complexities of TF-DNA binding specificity [10, 18, 19].

A common feature of most high-throughput approaches for studying protein-DNA binding is that they use purified or *in vitro*-produced protein samples [11, 20, 21], or epitope-tagged proteins overexpressed in cells (e.g., HEK293) [22, 23]. As a consequence, these approaches do not capture the impact of posttranslational modifications (PTMs) or interactions with other nuclear proteins on binding, both of which can occur in a cell- or stimulus-dependent manner. NF-κB proteins can be regulated by both PTMs [24] and interactions with other proteins [25–29]; however, the impact of these regulatory mechanisms on NF-κB DNA-binding specificity has not been well studied. To address these complexities in TF-DNA binding, our lab developed the nuclear extract PBM (nextPBM) as a high-throughput approach to assay the DNA binding of proteins directly from cell nuclear extracts, where native PTMs and interacting proteins are present. Here, we describe the nextPBM methodology and its application to the study of NF-κB binding.

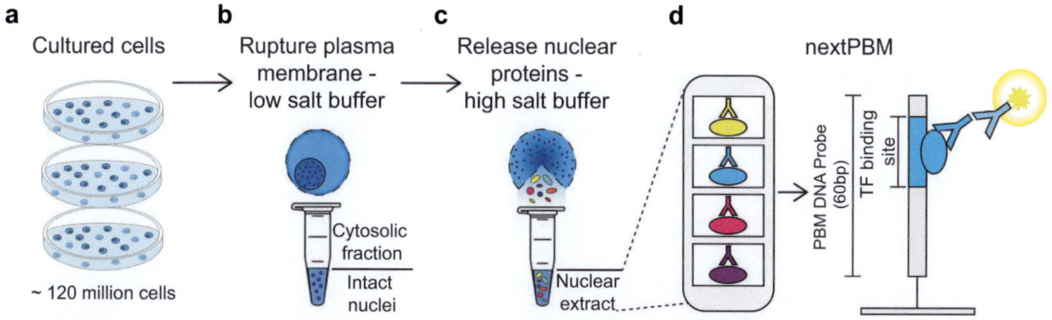

Fig. 1 nextPBM workflow. (**a**) Cells are cultured as the source of the protein sample for the nextPBM experiment. Using the Agilent 4 × 180K microarray format, ~120 million cells are required to perform a single four-chambered experiment (i.e., 30 million cells per independent chamber). (**b**) To harvest the nuclear extract from cultured cells, the plasma membranes are ruptured using a low salt buffer. The cytosolic fraction and intact nuclei are separated by centrifugation. (**c**) The intact nuclei are subsequently ruptured with a high salt buffer. The soluble nuclear proteins and nuclear debris are separated by centrifugation. (**d**) For the nextPBM experiment, the nuclear extract is incubated on the microarray and allowed to come to equilibrium. To detect your DNA-bound target protein, a primary antibody against your target (e.g., p65) is applied to the array, followed by a fluorescently conjugated secondary antibody. Using the 4 × 180K microarray format from Agilent, four separate experiments can be performed in parallel on a single microarray

NextPBMs are an extension of the basic PBM method [11–13] where cell nuclear extracts are used in the experiment instead of purified or in vitro-produced protein samples. Nuclear extracts are applied to a DNA microarray containing 60 base pair-long, double-stranded DNA probe sequences (Figs. 1 and 2a). DNA-bound proteins of interest are probed using a primary antibody followed by a fluorescently labeled secondary antibody (Fig. 1d). After antibody labeling, the microarray is imaged with a microarray scanner and the amount of target protein bound to each DNA probe sequence can be quantified by the fluorescence intensity of each microarray spot. To quantify the specificity of the signal above background, we transform the logarithm of our fluorescence intensity values into a *z*-score based on the binding intensity to a set of background probes.

The flexibility of the basic PBM platform has allowed a variety of DNA probe sequence designs to be used to study protein-DNA binding, from purely synthetic DNA sequences [11] to genome-derived sequences [31, 32]. Given the molecular complexity of the nuclear extracts used in the nextPBM approach, we have found a "seed plus single-nucleotide variant" (Seed-SV) probe design to be particularly useful for generating TF-DNA binding motifs and analyzing our nextPBM data (Fig. 2b). In this Seed-SV probe design, target sequences for a particular TF (i.e., the "seed") are included in the microarray along with all possible SV probes (Fig. 2b). This Seed-SV design is analogous to the seed-and-wobble approach used in the universal PBM design [11], however, we

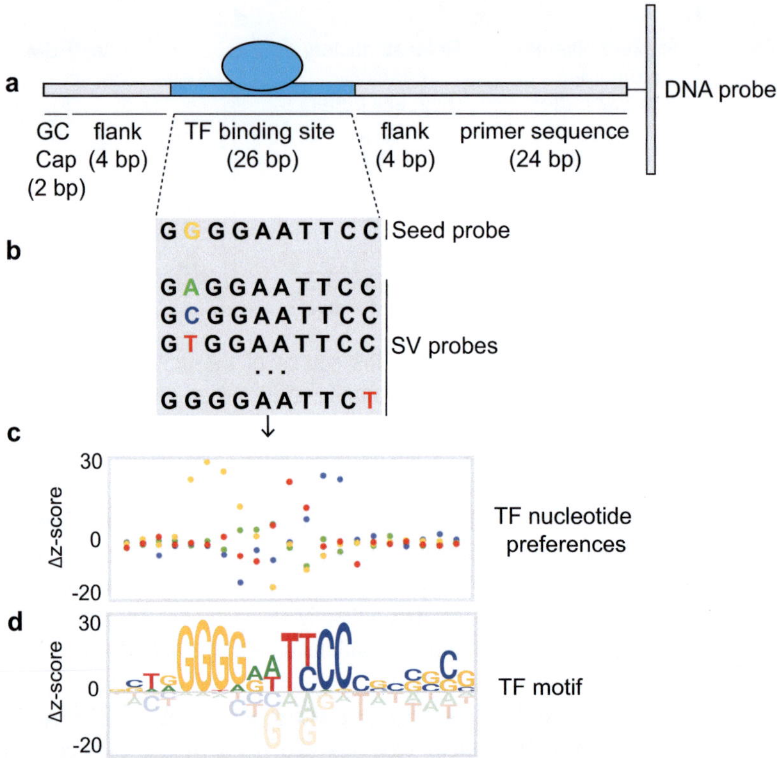

Fig. 2 TF motif generation using the Seed-SV probe design. (**a**) Schematic of the microarray probe design. Each 60 base pair (bp) probe has a conserved structure: a 24-bp primer sequence used for double stranding the array; constant 4-bp, 5-prime and 3-prime flanking sequences; a 26-bp customizable TF binding site; and a 2-bp GC cap for probe stability. (**b**) TF binding sites are represented on the microarray by a seed probe (i.e., synthetic or genomic-derived target site) and SV probes that include all single-nucleotide variations of each seed. (**c**, **d**) TF nucleotide preferences at each binding site position are quantified by determining a Δz-score for each nucleotide variant. Δz-scores are measured from the median z-score for each nucleotide at that position (i.e., median z-score for the seed and the three SV probes). The nucleotide preferences can be shown as "confetti plot" (**c**) or as a TF binding motif (**d**)

distinguish it to avoid confusion with this microarray design that is based on profiling binding to all k-mers as opposed to individual target sequences. By assaying the differential binding of a protein to Seed-SV probe sets, we can directly determine the nucleotide preferences at each position (Fig. 2c) and determine a TF binding motif (Fig. 2d). These binding motifs, which indicate both favorable and unfavorable nucleotide preferences, are analogous to the widely used position-weight matrix (PWM) models and can be directly compared. We note that when using the Seed-SV approach we are probing the nucleotide preferences for binding at *an individual DNA sequence*, which is distinct from motif generation in other contexts (e.g., ChIP-seq) based on averaging over many different DNA sites. This mode of motif generation is useful when examining cooperative binding [30] or indirect TF interactions (discussed below). NextPBMs can be used to examine NF-κB

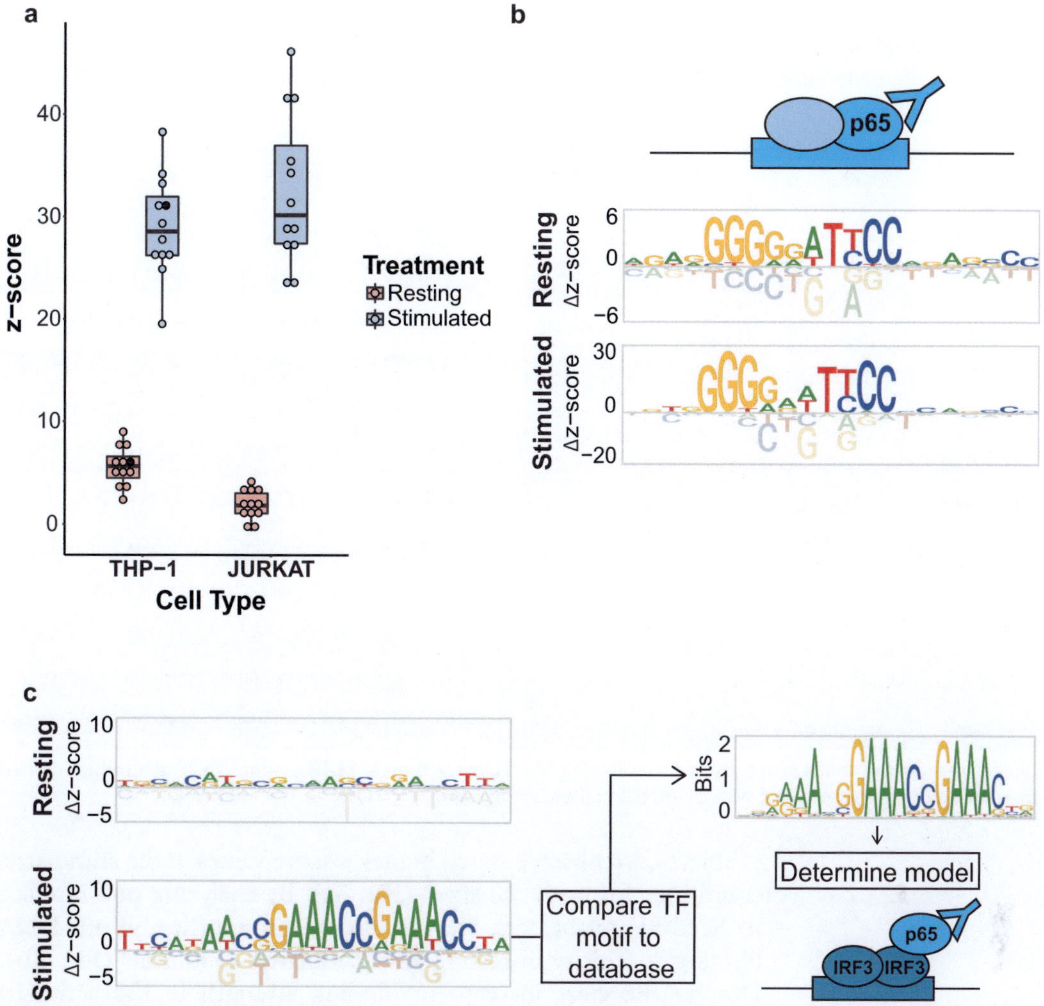

Fig. 3 Inducible NF-κB binding and recruitment in stimulated immune cell nuclear extracts. (**a**) Inducible p65 binding to twelve NF-κB seed probes (Table 1) across different immune cell types and stimulus conditions (THP-1 macrophages stimulated with lipopolysaccharide (LPS) and Jurkat T cells stimulated with ImmunoCult T Cell Activator). Data points shown as filled black circles correspond to the seed probe used to generate the TF motif in panel **b**. (**b**) p65 binding motifs derived from Seed-SV probe sets (highlighted as filled black circles in part **a**) (Table 1) in resting and LPS-stimulated conditions. Motifs are identical in both conditions and closely match established NF-κB binding motifs. (**c**) TF motifs showing the indirect recruitment of p65 to a consensus IRF binding site in stimulated THP-1 macrophages. The TF motif showing inducible, indirect p65 recruitment can be compared to a TF motif database to identify the underlying TF (i.e., matching the IRF3 motif suggests that p65 was recruited to DNA by IRF3). A model can subsequently be deduced by integrating the nextPBM data and database inference

binding activity and specificity in a cell type- and stimulus-dependent manner. Using nextPBMs to monitor binding of NF-κB-subunit p65 in LPS-stimulated THP-1 macrophages or T cell receptor-stimulated Jurkat T cells, we can quantify the stimulus-dependent increase in NF-κB binding activity (Fig. 3a). Profiling p65 binding to a set of twelve NF-κB "seed" probes

Table 1
DNA probe sequences containing NF-kB binding sites. Consensus binding sites are derived from universal PBM experiments (synthetic; [6]) or identified binding sites from the CXCL10 promoter (genomic; [36])

Source	Seed sequence
Synthetic	TGGAGGGGGATTTCCCCTTAGAAGTG
Synthetic	GCTAGGGGGATTTTCCCATCAAAGTG
Synthetic	GTGAGGGGGAATTTCCATGGTAAGTG
Synthetic	TAGTGGAATTTCCCTCGCCCGAAGTG
Synthetic	TCCGGGAAATTCCCTGGTGGAAAGTG
Synthetic	GGTGGGAAATTCCCCCAGACAAAGTG
Synthetic	CAGAGGGGATTTCCGTTAATTAAGTG*
Synthetic	GAAGGGGGAATTCCTCGACGTAAGTG
Synthetic	TGTGGGGGATTCCCGTAAATAAAGTG
Genomic	CCTCCAAGTTACGGAATTTCCCTCTG
Genomic	CCTGCTGGCTGTTCCTGGGGAAGTCC
Genomic	TGGCTGTTCCTGGGGAAGTCCCATGT

Binding sites are embedded in background regions of open chromatin from the hg19 genome. TF binding motifs derived from the seed sequence denoted with an asterisk is shown in Fig. 3b

(Table 1), we observe much higher *z*-score values in the stimulated conditions for both cell types (Fig. 3a). By analyzing p65 binding to Seed-SV probe sets, we can also determine the NF-κB-DNA binding specificity in resting and stimulated conditions (Fig. 3b). Despite the clear increase in binding strength in the activated conditions, we observe near identical binding specificities, demonstrating that similar p65-based dimers are present in the nucleus under both conditions. TF specificity differences can also be examined by comparing the relative binding across all probes in aggregate (i.e., as a scatter plot), and can sometimes reveal more subtle differences than by motif analysis alone (as in [10]).

An attractive feature of the nextPBM and Seed-SV probe design is that it provides a novel approach to study how interactions with other proteins can alter TF binding specificity, including indirect interactions in which a TF does not bind to DNA directly. p65 has previously been shown to bind DNA indirectly via DNA-bound IRF proteins in LPS-stimulated macrophages [29]. Using nextPBMs to profile the binding of p65 from stimulated macrophages to Seed-SV probes representing an IRF binding site resulted not in an NF-κB motif, but in a motif that matched the known IRF3 binding motif, allowing us to infer the recruitment of NF-κB by IRF3 to this DNA site (Fig. 3c). In a traditional binding experiment to a single DNA sequence, it would not be clear whether NF-κB was binding directly to the DNA or being recruited indirectly by an underlying TF. However, we can infer the indirect interaction by generating a site-specific binding motif which, when compared to TF database motifs, clearly matches IRF3 and not NF-κB. Given that hundreds of potential TF binding sites and corresponding SV sequences can be included on a single nextPBM microarray, this approach provides a platform to screen indirect recruitment of NF-κB by many TFs in a cell-dependent manner, opening up the possibility to discover other such interactions. Cooperative binding of TFs, such as PU.1 and IRF8, have also been analyzed using this basic approach [30], demonstrating that nextPBMs can also be used to analyze cooperative TF binding.

2 Materials

2.1 Preparation of Cell Nuclear Extracts

1. Cell cultures (*see* **Note 1**).
2. Tabletop centrifuge.
3. 50 mL Falcon tubes.
4. Tray filled with ice.
5. Ice bucket filled with ice.
6. 1.5 mL Eppendorf tubes.
7. 2.0 mL Eppendorf tubes.
8. Hemocytometer.
9. Hula Mixer (Thermo Fisher).
10. Vortex.
11. Gel-loading pipette tips.
12. 1× phosphate-buffered saline (PBS), pH 7.4: 1137 mM NaCl, 2.7 mM KCl, 10 mM Na_2HPO_4m, 1.8 mM KH_2PO_4, H_2O to 1 L.
13. 0.4% Trypan Blue.
14. Microscope.
15. Liquid nitrogen.
16. 5 M NaCl (0.2 μm filter sterilized).
17. 10% IGEPAL (0.2 μm filter sterilized).
18. Cell Scraping Buffer: 1× sterile PBS, 1× Protease Inhibitor.
19. Buffer A (low salt buffer): 10 mM HEPES, pH 7.9, 1.5 mM $MgCl_2$, 10 mM KCl, 0.5 mM DTT, 1× Protease inhibitor, 1× Phosphate Inhibitor.
20. Buffer C (high salt buffer): 20 mM HEPES, pH 7.9, 1.5 mM $MgCl_2$, 25% Glycerol, 0.5 mM DTT, 0.2 mM EDTA, 420 mM NaCl, 1× Protease Inhibitor, 1× Phosphatase Inhibitor.

2.2 Double Stranding Microarray Probes

1. Agilent CGH microarray (*see* **Note 2**).
2. GenePix 4400A scanner (with blue laser kit) (Molecular Devices).
3. GenePix Pro 7 acquisition and analysis software (Molecular Devices).
4. Microarray hybridization chamber kit (Agilent).
5. Gasket slide, 1 chamber per slide (Agilent) (*see* **Note 3**).
6. Two glass staining dishes (0.7 L), with cover (Wheaton).
7. Glass staining slide rack (Wheaton).
8. Microscope slide box.
9. Hybridization incubator (Thermo Fisher).
10. Vacuum desiccator.
11. Dust cover (*see* **Note 4**).
12. Water bath (set at 37 °C).
13. Stir plate and stir bars.
14. Dust Off XL canned air.
15. 70% ethanol solution in spray bottle.
16. Water in spray bottle.
17. Wash Buffer: 1 mL 10% Triton X-100, 1 L 1× PBS.
18. Thermo Sequenase Cycle Sequencing kit.
19. 10× Thermo Sequenase Buffer: 260 mM Tris, 65 mM $MgCl_2$, pH 9.5.
20. 100 mM dNTPs (25 mM per nucleotide).
21. 1 mM Cy3-conjugated dUTP (GE Healthcare).
22. 100 μM primer (24-nt, desalted) (EuroFins). Primer sequence 5′-CAGCAGCGTCAAGCGAATCAAGAC-3′ (*see* **Note 5**).
23. Primer Extension Mixture: 1× Thermo Sequenase Buffer, 1.17 mM primer, 163 nM dNTPs, 0.55 nM Cy3-conjugated dUTP, 0.036 U/mL Thermo Sequenase polymerase, sterile water to final volume of 900 μL.

2.3 Nuclear Extract Protein-Binding Microarray (nextPBM) Experiment

1. Equipment and reagents listed above (in Subheadings 2.1 and 2.2) are not listed again here.
2. Gasket slide, four chambers per slide (Agilent) or eight chambers per slide (Agilent) (*see* **Note 3**).
3. Coplin staining jars (VWR).
4. 1× HEPES-buffered saline (HBS): 20 mM HEPES, 137 mM NaCl, 1.5 mM $MgCl_2$, pH 7.4.
5. NextPBM Wash Buffer 1: 200 μL 10% Triton-X-100, 200 mL 1× HBS.

6. NextPBM Wash Buffer 2: 875 μL 20% Tween-20, 350 mL 1× HBS.
7. Blocking Buffer (0.45 μm filter sterilized): 2% NFDM (non-fat dry milk), 10 mL 1× HBS.
8. NextPBM Binding Buffer: 2% NFDM, 20 mM HEPES, 0.02% Triton X-100, 1 mM DTT, 0.2 mg/mL BSA, 1.11 mM $MgCl_2$, 55 mg/mL salmon testes DNA, sterile water to final volume.
9. Nuclear extract samples.
10. Primary antibodies (*see* **Note 6**).
11. Fluorophore-conjugated secondary antibodies (*see* **Note 6**).
12. Benchtop orbital shaker.

3 Methods

3.1 Preparation of Cell Nuclear Extracts

In this step, nuclear extracts are prepared from cultured cells of interest. Briefly, cells are harvested by scraping (adherent cells) or by centrifugation (suspension cells) (Fig. 1a). After harvesting, cells are treated with a low salt buffer to rupture the outer plasma membrane, and intact nuclei are collected (Fig. 1b). Next, nuclei are treated with a high salt buffer to extract the nuclear proteins (Fig. 1c). Finally, nuclear debris is pelleted, and the soluble nuclear proteins are collected for analysis on protein-binding microarrays (Fig. 1d). Preparation of nuclear extracts takes ~3–4 h.

3.1.1 Cell Harvesting for Adherent Cells

1. Before beginning the nuclear extraction, cool down a tabletop centrifuge to 4 °C.
2. If you are using multiple flasks of cells for the nuclear extraction (*see* **Note 1**), obtain a large tray and fill it with ice to place the cells on.
3. Prepare the Cell Scraping Buffer. 15 mL of buffer is needed per T175 flask. Place the buffer on ice. (*see* **Note 7**).
4. Aspirate off the cell culture media.
5. Wash the cells with 10 mL of 1× PBS. Gently rock the flask a few times, aspirate off the 1× PBS, replace with 10 mL of Cell Scraping Buffer, and place the flask on ice (*see* **Note 8**).
6. To harvest the cells, use a cell scraper to detach the cells from the wall of the flask.
7. Observe the cells under a microscope to ensure cells are detached. Once the cells have been fully dislodged, collect the detached cells and transfer them into a 50 mL Falcon tube. Place on ice.

8. Add another 5 mL of Cell Scraping Buffer to the flask to wash the flask and collect any remaining cells. Add this volume to the 50 mL Falcon tube the cell mixture was previously pipetted into.
9. Collect the cells by centrifugation at 4 °C at 500 × *g* for 5 min in the tabletop centrifuge. Aspirate the PBS, and keep the cell pellet on ice.
10. Continue to nuclear extraction, **step 1** (*see* **Note 9**).

3.1.2 Cell Harvesting for Suspension Cells

1. Collect the cell volume from the culture flask and transfer it to a 50 mL Falcon tube. Place on ice.
2. Collect the cells by centrifugation at 4 °C at 500 × *g* for 5 min. Aspirate the media.
3. Wash the cell pellet gently with cold 1× PBS.
4. Pellet the cells once more by centrifuging at 4 °C at 500 × *g* for 5 min. Aspirate the PBS, and keep the cell pellet on ice.
5. Continue to nuclear extraction, **step 1** (*see* **Note 9**).

3.1.3 Nuclear Extraction

1. Prepare Buffers A and C. 1 mL of Buffer A and 100 μL of Buffer C is needed for 40 to 60 million cells. This volume can be adjusted based on the number of cells used for the nuclear extraction. The phosphatase inhibitor, protease inhibitor, and DTT should be added to the buffers immediately before use. Keep the buffers on ice.
2. Obtain one 2 mL Eppendorf tube per sample and place on ice.
3. To the cell pellets, add 500 μL of Buffer A to the pellet at a time. Pipette up and down several times to resuspend the pellet. Transfer the volume to the Eppendorf tube on ice.
4. Repeat with the remaining 500 μL of Buffer A. Make sure to wash the sides of the Falcon tube to ensure that the entire pellet has been resuspended. Transfer the volume to the Eppendorf tube on ice. Incubate the Buffer A mixture on ice for 10 min.
5. After the 10 min incubation, add 20 μL of 10% IGEPAL detergent (*see* **Note 10**), by adding half of the detergent volume to the Buffer A mixture in the tube and the other half to the inside of the cap. Vortex for 10 s. Check for sufficient lysis of the plasma membrane using a hemocytometer (*see* **Note 11**).
6. Pellet the nuclei by centrifuging at 4 °C at 500 × *g* for 5 min.
7. Transfer the supernatant into a new 1.5 mL Eppendorf tube on ice. This is the cytosolic fraction and can be saved for other downstream analyses if desired. To avoid disturbing the nuclear pellet, use a gel-loading tip to remove the remaining, small amount of supernatant.

8. To wash the nuclear pellet and remove any residual cytosolic fraction, gently add 100 μL of cold Buffer A to the top of the nuclear pellet. Centrifuge at 4 °C at 500 × *g* for 1 min. Remove the Buffer A. Repeat this step for two additional washes.
9. Add 100 μL of Buffer C to the nuclear pellet. Resuspend the nuclear pellet, and vortex for 30 s. The nuclear pellet should be dislodged and floating in solution.
10. Nutate the floating nuclei at 4 °C for 1 h (*see* **Note 12**). To do this, use a Hula Mixer with the following settings: Orbital—25/off, Reciprocal—90°/30, Vibro—5°/5.
11. Pellet the nuclear debris by centrifuging at max speed for 20 min at 4 °C.
12. Transfer the supernatant containing the soluble nuclear proteins to a new tube. This is the nuclear extract sample that will be used for nextPBM experiments.
13. Quantify the protein concentration of nuclear extract samples by measuring $A_{280\ \mathrm{nm}}$ using a NanoDrop spectrophotometer. A typical nuclear extract sample of 40–60 million cells yields a nuclear protein concentration of 10–15 mg/mL.
14. Snap freeze the nuclear extract and cytosolic fractions using liquid nitrogen (*see* **Note 13**). Store samples at −80 °C.

3.2 Double Stranding Microarray Probes

In this step, the single-stranded DNA Agilent microarrays are made double-stranded by primer extension from a 24-nt primer sequence common to each probe sequence (Fig. 2a). Briefly, the microarray slide and primer extension mixture are heated to 85 °C in a hybridization oven, and the temperature is then slowly reduced in a stepwise fashion to the final primer extension temperature of 60 °C. When the primer extension has been completed, the microarray slide is washed and imaged using a microarray scanner. The double stranding process takes ~4 h; the microarray scan takes ~15 min.

1. Place a microcentrifuge tube rack into the hybridization oven set to a stable temperature of 85 °C. The hybridization oven exterior should be covered to avoid light exposure and photobleaching of the Cy3 fluorophore.
2. On ice, thaw the primer, dNTPs, and Cy3-conjugated dUTP.
3. Prepare 900 μL Primer Extension Mixture.
4. Warm Primer Extension Mixture, steel hybridization chamber, and single-chamber gasket slide in the 85 °C hybridization oven for 20 min.
5. Warm Agilent microarray slide in the 85 °C hybridization oven for 5 min with the DNA side (Agilent side) of the array facing up.

6. Assemble the microarray, gasket slide, and Primer Extension Mixture in the steel hybridization chamber (*see* **Note 14**), and return to the 85 °C oven. This step should be done quickly to maintain the heated temperature of the equipment and reagents. (*see* **Note 15**).
7. Manually decrease the oven temperature in a stepwise fashion to the final primer extension temperature, as follows: 10 min at 85 °C; 10 min at 75 °C; 10 min at 65 °C; 90 min at 60 °C.
8. Warm 1 L of Wash Buffer in a 37 °C water bath.
9. After the 90 min incubation, fill two glass staining dishes with 500 mL of pre-warmed Wash Buffer. Place dish #1 on the benchtop. Place dish #2 on a magnetic stir plate, and insert glass slide rack and small stir bar.
10. Remove the hybridization chamber from the oven. Disassemble the steel chamber, and separate the microarray slide from the gasket slide while immersed in Wash Buffer (dish #1) (*see* **Note 16**).
11. Place microarray into the slide rack in dish #2 with the DNA side (Agilent side) facing toward the stir bar in the center. Cover the staining dish with an empty ice bucket to limit light exposure. Wash with gentle stirring for 10 min.
12. Fill a new glass dish (dish #3) with 500 mL room temperature 1× PBS, a slide rack, and stir bar. Transfer microarray into dish #3, place the dish on a stir plate, and wash for 3 min.
13. Remove the microarray from the slide rack, and de-wet the array (*see* **Note 17**).
14. Scan the microarray in GenePix scanner at 2.5 μm resolution using the 535 nm laser and green filter (*see* **Note 18**). Save scan as .TIF image file. The scan should have relatively uniform fluorescence across probes and chambers. If irregularities are observed (e.g., due to air bubbles, scratches, or debris), manually flag those spots to be excluded from downstream analysis.
15. The microarray can be stored in a microscope slide box in the dark (e.g., in a drawer) until used for nextPBM experiments.

3.3 Nuclear Extract Protein-Binding Microarray (nextPBM) Experiment

In this step, the nuclear extracts are mixed with a PBM binding buffer and applied to the DNA microarray slide. The DNA-bound proteins are detected using fluorescent antibodies specific to a protein of interest, followed by imaging using a microarray scanner. All incubation and wash steps are performed at room temperature. The nextPBM experiment takes about 5 h, and the microarray scans take an additional 1.5 h (assuming scans are done at 3–4 different intensity levels).

1. Prepare a 10 mL of solution of Blocking Buffer. These solutions will be used as blocking reagents in protein-binding and antibody-binding steps, respectively. To allow milk to fully

dissolve, it is recommended to prepare these solutions the day before the experiment and allow them to dissolve overnight at 4 °C.

2. To clarify the Blocking Buffer before use, centrifuge the solutions at maximum speed for 10 min in a tabletop centrifuge. Filter the supernatant solution using a 0.42 μm syringe filter.
3. Prepare three separate nextPBM wash solutions for use in Coplin jars: 70 mL HBS, 200 mL, nextPBM Wash Buffer 1, 350 mL nextPBM Wash Buffer 2.
4. Set up three staining dishes for rinsing the arrays. Fill two glass staining dishes with 500 mL HBS (staining dishes #1 and #2) and a third staining dish with 500 mL 0.05% Tween-20 in HBS (staining dish #3). Keep these dishes covered throughout the nextPBM experiment to avoid accumulation of dust and debris.
5. On ice, thaw BSA, salmon testes DNA, and DTT to prepare the nextPBM Binding Buffer.
6. Pre-wet the double-stranded Agilent microarray in a Coplin jar filled with 0.01% Triton X-100/HBS on a rotating shaker at 125 rpm for 5 min.
7. While the array is pre-wetting, prepare two microarray-gasket slides. Rinse the gasket slides first with sterile water, followed by a rinse with 70% ethanol. Dry the gasket slide using Dust Off XL canned air (*see* **Note 19**).
8. While the array is pre-wetting, prepare the hybridization chamber and gasket slide with Blocking Buffer. The volume needed for each microarray chamber depends on the microarray design format. For an 8 × 60K microarray, 80 μL of solution is required; for a 4 × 180K microarray, 180 μL of solution is required. Keep the chamber and gasket slide covered to prevent the accumulation of dust and debris.
9. Once the pre-wetting is complete, remove the microarray from the Coplin jar. De-wet the array in staining dish #1 (*see* **Note 17**).
10. Carefully place the microarray with the DNA side (Agilent side) facing down, on top of the prepared gasket slide with the Blocking Buffer. Assemble the hybridization chamber (*see* **Note 14**). Incubate the assembled chamber in the dark (e.g., in a drawer) for 1 h.
11. During the blocking step, prepare nuclear extract binding mixtures in 1.5 mL Eppendorf tubes. Prepare the nextPBM binding buffer (*see* **Note 20**). The volume of nuclear extract that can be applied to each chamber of the PBM depends on the array format. The maximum volume of nuclear extract that can be applied to a single chamber for the 8 × 60K format is 21 μL and for the 4 × 180K format is 47 μL. If less than the

maximum volume is used per chamber, bring the volume up to the maximum volume with Buffer C (*see* **Note 21**). Combine nuclear extract, NaCl, and binding buffer for a final volume of 80 μL (8 × 60K format) or 180 μL (4 × 180K format). Allow the mixture to incubate at room temperature for 30 min before applying the mixture to the gasket slide.

12. When there are 5–10 min left on the blocking step, clean a gasket slide (*see* **Note 19**), and place it into the hybridization chamber base. Apply the prepared nuclear extract binding mixture into each chamber of the gasket slide. Cover the gasket slide to prevent dust accumulation before hybridization.
13. When the blocking step is complete, disassemble the steel hybridization chamber (*see* **Note 16**). Immerse the microarray-gasket slide "sandwich" in dish #3 and separate the microarray from the gasket slide underwater. Rinse the microarray briefly in dish #1, and de-wet the microarray (*see* **Note 17**). Place the microarray on top of the prepared gasket slide containing the nuclear extract mixture. Assemble the hybridization chamber (*see* **Note 14**). Incubate in the dark for 1 h.
14. During the nuclear extract binding step, prepare the primary antibody mixtures in 1.5 mL Eppendorf tubes (*see* **Note 6**). Dilute primary antibodies to a final concentration of 20 μg/mL in blocking buffer.
15. When there are 5–10 min left in the nuclear extract binding step, clean a gasket slide (*see* **Note 19**), and place the cleaned gasket slide into the hybridization chamber base. Apply 80 μL (8 × 60K format) or 180 μL (4 × 180K format) of primary antibody mixture into each chamber of the gasket slide. Cover the gasket slide to prevent dust accumulation.
16. When the nuclear extract binding step is complete, disassemble the steel hybridization chamber (*see* **Note 16**). Immerse the microarray-gasket slide "sandwich" in dish #3 and separate the microarray from the gasket slide underwater. Rinse the microarray briefly in dish #1, and de-wet the microarray (*see* **Note 17**). Place the microarray on top of the prepared gasket slide. Assemble the hybridization chamber (*see* **Note 14**). Incubate in the dark for 20 min.
17. During the primary antibody-binding step, prepare the secondary antibody mixtures. Dilute secondary antibodies to a final concentration of 20 μg/mL in 2% NFDM/HBS at a final volume of 80 μL (8 × 60K format) or 180 μL (4 × 180K format).
18. When there are 5–10 min left in the primary antibody incubation step, clean a gasket slide (*see* **Note 19**), and place the cleaned gasket slide into the steel chamber. Apply 80 μL

(8 × 60K format) or 180 μL (4 × 180K format) of secondary antibody mixture into each chamber of the gasket slide. Cover the gasket slide to prevent any dust accumulation.

19. When the primary antibody incubation step is complete, disassemble the steel hybridization chamber (*see* **Note 16**). Immerse the microarray-gasket slide "sandwich" in dish #3 and separate the microarray underwater. Rinse the microarray briefly in dish #1, and de-wet the microarray (*see* **Note 17**). Place the microarray on top of the gasket slide containing secondary antibody mixture. Assemble the hybridization chamber (*see* **Note 14**). Incubate in the dark for 20 min.
20. After the secondary antibody incubation step, disassemble the steel hybridization chamber (*see* **Note 16**).
21. Wash the microarray three times in Coplin jars on an orbital shaker at 125 rpm: two 3 min washes in 0.05% Tween-20/HBS, followed by one 2 min wash in HBS.
22. After the HBS wash, de-wet the array before scanning (*see* **Note 17**).
23. Scan the microarray using the GenePix scanner at a 2.5 μm resolution. Set the laser and filter settings according to whichever fluorophore is used. It is recommended that scans be taken at multiple sensitivities to ensure that all spots are properly visualized. When taking scans, start at a low sensitivity where the brightest spots are below the signal saturation window and increase the gain in each scan until the dimmest spots are visible. When using the GenePix 4400A scanner, set the laser power to 50 for all scans, and adjust the photomultiplier gain (usually starting at 500 and preceding upward to 700–800) for each scan. Save each scanned image as a .TIF file for subsequent analysis.
24. The microarray chambers can be reprobed with a second set of primary and secondary antibodies (*see* **Note 6**). This can be done if the species of the primary antibodies are different (e.g., mouse primary antibody followed by rabbit primary antibody) and the secondary antibodies are highly-cross adsorbed.
25. Before reprobing the array with a second set of antibodies, pre-wet the array (**step 6**) (*see* **Note 19**).
26. Repeat **steps 15–23**.
27. Store used microarrays in a microscope slide box in the dark.

3.4 Data Analysis

This step involves analyzing the microarray images and quantifying the protein binding to each DNA probe sequence on the array. Several normalization procedures are used to process the microarray-derived fluorescence values. In this chapter, we detail the use of masliner adjustment, spatial normalization, probe sequence alignment, replicate probe averaging, and data matrix generation. Additionally, we discuss several approaches for representing the nextPBM data, and describe how to generate energy matrices from seed and SV probes.

1. After scanning the microarrays, use GenePix 7.0 software to grid microarray images by opening a .TIF image file and a .GAL (GenePix Array List) file. .GAL files can be downloaded through the Agilent website for each microarray design format. Align the .GAL grid with the microarray spots on each .TIF image. Irregular spots (due to scratches, dust, or bubbles) can be manually flagged and excluded from analysis.
2. Use the "Analyze" function on GenePix to save and export the data for each microarray chamber as an individual .GPR file containing the background-subtracted median fluorescence intensity values for each spot on the microarray.
3. Repeat gridding and data export (**steps 1** and **2**) for each scan taken at different intensities.
4. Use masliner software (*see* **Note 22**) to combine microarray scans taken at different intensities. This software performs a linear regression to quantify the fluorescence intensities of saturated spots in higher intensity scans using values from lower intensity scans. This produces masliner-adjusted, background-subtracted, median fluorescence intensities for each microarray probe spot.
5. Spatial normalization across all probes normalizes the fluorescent intensities across different regions of the scan to a common global median. This is based on the concept that large regions of the scan should, on average, have the same median values. This normalization is performed by scaling the median intensity of each 15 × 15 grid section of the microarray chamber to the global median of the entire chamber. Software for spatial normalization is available at http://thebrain.bwh.harvard.edu/PBMAnalysisSuite/indexSep2017.html [33].
6. Next, each probe spot on the array is assigned to its corresponding DNA probe sequence using an analysis file, which relates the spot coordinates of the microarray with each DNA probe sequence (*see* **Note 23**).
7. Once probe sequences are assigned, probe intensities are averaged across probe orientations and probe replicates. Possible averaging methods include averaging across both replicates and orientations, averaging over replicates while keeping orientations separate, or averaging over replicates while keeping orientations separate and then taking the top orientation for each probe (best orientation). Typically, best-orientation averaging is preferred for downstream analysis as some TFs exhibit an orientation-specific binding preference.

8. The last step in processing nextPBM data is to export the data in matrix format to facilitate downstream processing. Normalized and averaged fluorescence intensities for each unique probe sequence and for each experiment/chamber are collated into a matrix. Different matrices are produced for each type of averaging, if multiple forms of averaging are performed in **step 8**.
9. A useful transformation of the data is from probe fluorescence values (v) into z-scores (z). To do this, all probes fluorescence values (v) are first log transformed ($f = \ln(v)$), and then further transformed into z-scores using the mean (u_{BG}) and standard deviation (s_{BG}) of the log fluorescence values for some user-defined set of background (BG) probe sequences (i.e., $z = (f - u_{BG})/s_{BG}$). z-scores are helpful in interpreting the data as they provide an intuitive measure of "strength above background" for each binding event and are related to DNA--binding energies because the probe fluorescence values are proportional to the DNA-binding constants (K_d) [31].
10. A variety of analyses can be performed using the processed data matrices. To examine TF binding specificity, binding motifs can be determined for each Seed-SV probe set [30, 34, 35]. Briefly, z-scores of each Seed-SV probe set are assembled into a $4 \times n$ matrix, where n is the number of base pairs of a target transcription factor binding site covered by SV probes. The raw z-score matrix is transformed using the column-wise medians to generate delta(z-scores), which can be visualized as an energy logo (*see* **Note 24**). We refer to these as "energy" logos as the delta z-scores are related to binding energy differences associated with each nucleotide. These energy logos show the nucleotide binding preference along each position of a binding site (Fig. 2c, d).
11. Binding affinity of TFs for different sequences, or to the same sequences under different conditions, can be analyzed using box and whisker plots of the z-scores of different probe sequences, or the same probes under different conditions (Fig. 3a).
12. To analyze binding behavior of dimers, pairwise comparison of different binding partners can be analyzed using scatter plots. Here, we illustrate the DNA binding of a single NF-κB family TF p65; however, to study the binding behavior of p65 with partner proteins (ex p65:p50 dimers), a pairwise comparison can be performed by probing for two different NF-κB family proteins in parallel experiments and constructing pairwise scatter plots [6].

4 Notes

1. This cell nuclear extraction protocol can be performed using either suspension or adherent cell cultures. The protocol described here has been successfully used for samples of 40–60 million cells from THP-1 and Jurkat cell lines. This protocol produces concentrated nuclear extracts (10–15 mg/ml) that are desirable for use on nextPBM experiments. The nuclear yield from different cell types can vary slightly, so it is recommended to use these numbers for preliminary experiments and adjust the cell numbers as needed. For lower or higher cell numbers, the volumes of buffers indicated in this protocol can be scaled accordingly.
2. NextPBM experiments are performed with Agilent CGH microarrays. These microarrays are available in different formats, with different number of unique probe features (spots on the array) and different numbers of chambers. The multi-chamber format on a single microarray slide allows for different binding experiments to be performed in parallel. Typical formats for nextPBM experiments include the 4 × 180K format, which contains 4 identical chambers each with 177,400 possible probe features per chamber, and the 8 × 60K format, which contains 8 identical chambers each with 61,979 possible probe features per chamber.
3. Agilent CGH microarrays are manufactured with 60 nucleotide (nt) single-stranded DNA probes, which are made double-stranded in the lab following a primer extension protocol. Each 60 nt probe feature contains a constant 24 nt primer sequence used in the primer extension reaction, a 26 nt variable sequence of interest containing different transcription factor binding sites embedded within constant flanking regions on each end (Fig. 2a).

   Protein-binding microarrays can be performed using a universal array design containing all possible 8 bp sequences, thus giving an unbiased analysis of transcription factor binding motifs. However, transcription factors often bind longer sequences of DNA as dimers with other partner proteins; for example, canonical NF-κB dimer binding sites can be 9–12 bp in length. To study transcription factors like NF-κB, we choose target sequences containing a single NF-κB binding site per probe (Fig. 2b). These binding sites can be synthetically derived consensus sites as previously characterized [6] or genomic NF-κB binding sites. Here, we show the successful characterization of p65 binding to a set of probes containing synthetic consensus sites as well as genomic NF-κB sites from the CXCL10 promoter [36] After choosing target binding sites (seed probes), additional probe sequences containing single-nucleotide variants (SV probes) at each position along the binding site are generated and included in the array design. This allows for the generation of TF binding site motifs to study binding specificity. Additionally, each probe sequence is included in both forward and reverse complement directions with respect to the microarray slide, to account for possible orientation bias. Finally, five replicates of each probe orientation are included, for a total of ten replicates per probe sequence.

4. Agilent gasket slides are available in several different chamber formats. A single-chamber gasket slide is used to double-strand the microarrays, to allow for uniform double stranding across different chambers. Four- or eight-chamber gasket slides are used for microarray experiments, depending on the format of the array (4×180K or 8×60K, respectively).

5. Care should be taken to prevent dust and debris from accumulating on the microarray. As such, the microarray, gasket slides, and hybridization chambers should be covered with a dust cover (i.e., a pipette box lid) at all times during the experiment. Additionally, all solutions should be filtered before using in the microarray experiment.

6. The primer is used to double-strand the microarray using a primer extension reaction. The primer sequence used in these experiments to study NF-κB is 5′- CAGCAGCGTCAAGC GAATCAAGAC-3′.

7. Proteins of interest are detected by labeling with antibodies conjugated with fluorophores (Alexa Fluor 488 or Alexa Fluor 647). Labeling can be done using either a fluorophore-conjugated primary antibody (one-step labeling) or via a primary antibody specific to the protein of interest followed by a fluorophore-conjugated secondary antibody (two-step labeling). Using the GenePix 4400A scanner with both red and blue laser kits also allows for two different proteins to be probed for within a single chamber of the microarray, if antibodies raised in different species are used for detection. For example, one protein can be detected using a primary antibody raised in mouse followed by an Alexa Fluor 488-conjugated anti-mouse antibody, and a second protein can be detected using a primary antibody raised in rabbit followed by an Alexa Fluor 647-conjugated anti-rabbit antibody. It is recommended that the fluorophore-conjugated secondary antibodies are highly cross-adsorbed to reduce potential cross-reactivity.

8. As indicated, this protocol is written for adherent cells cultured in T175 culture flasks. For this size flask, 10 mL of cell scraping buffer is needed for scraping cells and an additional 5 mL of

scraping buffer is needed for washing the flask after scraping. This volume can be adjusted based on the size of the culture flasks being used.

9. After removing growth media and washing the cells, flasks should be placed on ice immediately. This is especially important if cells are subjected to different treatments prior to harvesting (i.e., LPS stimulation), in order to preserve sensitive cell state and cell dynamics at the time of harvesting.
10. All the following steps can be done outside of the cell culture hood. All samples and reagents should be kept on ice to prevent protein degradation.
11. This volume of 10% Igepal detergent is used for 1 mL of Buffer A. Adjust this volume accordingly.
12. To ensure that the outer plasma is sufficiently lysed, take 0.5 μL of the vortexed cell and Buffer A mixture, and add 9 μL of Buffer A and 1 μL of 0.4% Trypan Blue. Observe the cells under the microscope using a hemocytometer. You should see round, blue nuclei. The 10 min incubation time is used to lyse most cell samples of 40–60 million cells. However, depending on cell type, the incubation time may have to be adjusted if there is insufficient lysis of the plasma membrane. If the time needs to be increased, allow the lysis to proceed in 1–2 min intervals while checking the cells under the hemocytometer to prevent prematurely lysing the nuclei.
13. Use a 2 mL Eppendorf tube for this step to ensure that the nuclei in solution move freely and do not get stuck on the underside of the cap while rotating. This helps ensure efficient lysis of the nuclei.
14. Care should be taken to preserve protein structure and activity of nuclear extracts when storing samples before use on protein-binding microarray experiments. Aliquot the nuclear extract sample into "single-use" volumes, to minimize the amount of freeze-thaw cycles. Snap freezing the nuclear extract samples helps reduce formation of water crystals and preserves the structure of proteins. When thawing nuclear extracts for nextPBM experiments, quickly thaw them by placing the Eppendorf tubes with the nuclear extract into room temperature water and then place on ice.
15. Refer to the Agilent G2434A Microarray Hybridization Chamber User Guide and (PBM methods Siggers et al. [14]) for detailed instructions and pictures on microarray handling. Briefly, to assemble the chamber, place the gasket slide facing up onto the steel hybridization chamber base. Load samples onto the gasket slide chambers, and slowly lower the microarray, DNA side facing down (Agilent side), carefully onto the gasket slide avoiding bubbles, creating a "sandwich" of the two

slides. Place the chamber cover carefully on top of the slides, slide the clamp assembly onto the assembled chamber, and tighten the clamp using the thumbscrew. Note that instead of vertically rotating the hybridization after assembly as per the User Guide, we recommend avoiding bubbles while loading and assembling the chamber, and instead gently tapping the assembled chamber on the benchtop to remove bubbles if they persist.

16. It is important to load the primer extension mixture and assemble the hybridization chamber quickly to keep the temperature of the materials as close to 85 °C as possible, as a drop in temperature may result in mis-priming. Additionally, keep the oven door closed when assembling the hybridization chamber on the benchtop to maintain the internal temperature of the oven.

17. Refer to the Agilent User Guide for detailed instructions on hybridization chamber disassembly. Briefly, loosen the thumbscrew and slide the clamp assembly off the chamber. Carefully remove the chamber cover, making sure to avoid breaking the seal of the "sandwiched" gasket slide and microarray slide. Carefully remove the "sandwich" from the chamber base by holding the short edges of the slides, and immerse into the wash buffer dish. Gently, use the plastic blunt-edged tweezers to pry open the bottom corner of the "sandwich," and hold the microarray slide, while allowing the bottom gasket slide to drop the bottom of the dish.

18. To de-wet the microarray, hold the short edges of the microarray with the DNA side (Agilent side) facing down, and immerse into a dish of wash buffer (dish #1, dish #2) without detergent. Gently agitate the microarray slide to rinse off residual debris or detergent. Slowly lift the microarray at an angle so that one edge lifts from the buffer before the other. Continue slowly lifting the microarray from the buffer so that the slide "de-wets" and is free of buffer. If liquid droplets persist on the slide, repeat the process until the array is dry.

19. Cy3 scans are used to manually inspect the efficiency of the primer extension reaction. The fluorescence intensity of scans should appear relatively strong and uniform across the chamber. Cy3 scans are often used to quantify the double-stranding efficiency and incorporated into binding analyses of universal PBMs, due to the sequence variability across probes. However, in our custom microarrays based on the Seed-SV approach, sequence variability is much lower and the impact of Cy3 scans in analysis is negligible.

20. Two microarray-gasket slides are required throughout a PBM experiment. Gasket slides can be used up to ~10 times with gentle cleaning and proper storage. Be sure to clean and dry

gasket slides thoroughly before use, and keep cleaned gasket slides under dust covers to avoid accumulation of debris. Gasket slides should be stored in their original packaging between uses to prevent scratching.

21. These buffer conditions (buffering reagent, salt, pH) can be adjusted and optimized for your own experiments, as is done for most biochemical binding experiments. We have used this buffer in many nextPBM experiments to successfully study NF-κB family proteins, as well as other TFs, across different cell types.
22. One important buffer component to consider in binding experiments is salt concentration. Our experiments routinely use a final salt concentration of 110 mM NaCl for the protein-binding step. Since the nuclear extraction protocol uses a high salt (420 mM NaCl) solution to extract nuclear proteins, the volume of nuclear extract that can be used per chamber on a microarray is limited by salt concentration, unless adjusted. 21 μL max volume of nuclear extract can be used for an 8 × 60K array format and 47 μL max volume of nuclear extract can be used for a 4 × 180K format. Typically, in nextPBM experiments we use sufficient nuclear extract to reach a final total protein concentration of >2.5 mg/mL.
23. Masliner software can be downloaded for free from the Church Lab website (http://arep.med.harvard.edu/masliner/supplement.htm).
24. An analysis file relating the probe spots to each DNA sequence can be generated using files downloaded from Agilent for each custom array design.

Acknowledgments

Research in the authors' laboratory on the nextPBM and NF-kB experiments was supported by NIH grant R01AI151051 (T.S.). We thank Melissa Inge and Rebekah Miller for comments and suggestions.

References

1. Pahl HL (1999) Activators and target genes of Rel/NF-κB transcription factors. Oncogene 18(49):6853–6866
2. Ghosh S, May MJ, Kopp EB (1998) NF-κB and Rel proteins: evolutionarily conserved mediators of immune responses. Annu Rev Immunol 16(1):225–260
3. Hoffmann A, Natoli G, Ghosh G (2006) Transcriptional regulation via the NF-κB signaling module. Oncogene 25(51):6706–6716
4. Williams LM, Gilmore TD (2020) Looking down on NF-κB. Mol Cell Biol. https://doi.org/10.1128/MCB.00104-20
5. Natoli G (2006) Tuning up inflammation: how DNA sequence and chromatin organization control the induction of inflammatory genes by NF-κB. FEBS Lett 580(12):2843–2849
6. Siggers T, Chang AB, Teixeria A, Wong D et al (2012) Principles of dimer-specific gene regulation revealed by a comprehensive

characterization of NF-κB family DNA binding. Nat Immunol 13(1):95–102

7. Wong D, Teixeira A, Oikonomopoulos S, Humburg P et al (2011) Extensive characterization of NF-κB binding uncovers non-canonical motifs and advances the interpretation of genetic functional traits. Genome Biol 12(7):R70
8. Carey MF, Peterson CL, Smale ST (2012) Experimental strategies for the identification of DNA-binding proteins. Cold Spring Harb Protoc 7(1):18–33
9. Carey MF, Peterson CL, Smale ST (2013) Electrophoretic mobility-shift assays. Cold Spring Harb Protoc 8(7):636–639
10. Andrilenas KK, Penvose A, Siggers T (2015) Using protein-binding microarrays to study transcription factor specificity: homologs, isoforms and complexes. Brief Funct Genomics 14(1):17–29
11. Berger MF, Philippakis AA, Qureshi AM, He FS, Estep PW, Bulyk ML (2006) Compact, universal DNA microarrays to comprehensively determine transcription-factor binding site specificities. Nat Biotechnol 24(11):1429–1435
12. Bulyk ML, Gentalen E, Lockhart DJ, Church GM (1999) Quantifying DNA-protein interactions by double-stranded DNA arrays. Nat Biotechnol 17(6):573–577
13. Linnell J, Mott R, Field S, Kwiatkowski DP, Ragoussis J, Udalova IA (2004) Quantitative high-throughput analysis of transcription factor binding specificities. Nucleic Acids Res 32 (4):1–7
14. Siggers T, Gilmore TD, Barron B, Penvose A (2015) Characterizing the DNA binding site specificity of NF-κB with protein-binding microarrays (PBMs). Methods Mol Biol 1280:609–630
15. Mansfield KM, Carter NM, Nguyen L, Cleves PA, Alshanbayeva A, Williams LM, Crowder C, Penvose AR, Finnerty JR, Weis VM, Siggers T, GIlmore TD (2017) Transcription factor NF-κB is modulated by symbiotic status in a sea anemone model of cnidarian bleaching. Sci Rep 17(1)
16. Ryzhakov G, Teixeira A, Salibra D, Blazek K, Muta T, Ragoussis K, Udalova I (2013) Cross-species analysis reveals evolving and conserved features of the nuclear factor κb (NF-κB) proteins. J Biol Chem 288(16):11546–11554
17. Copley RR, Totrov M, Linnell J, Field S, Ragoussis J, Udalova IA (2007) Functional conservation of Rel binding sites in drosophilid genomes. Genome Res 17(9):1327–1335
18. Siggers T, Gordân R (2011) Protein-DNA binding: complexities and multi-protein codes. Nucleic Acids Res 42(4):1551–1556
19. Slattery M, Zhou T, Yang L, Dantas Machado AC, Gordân R, Rohs R (2014) Absence of a simple code: how transcription factors read the genome. Trends Biochem Sci 39(9):381–399
20. Slattery M, Riley T, Liu P, Abe N, Gomez-Alcala P, Dorr I, Zhou T, Rohs R, Honig B, Bussemaker HJ, Mann RS (2011) Cofactor binding evokes latent differences in DNA binding specificity between hox proteins. Cell 147 (6):1270–1282
21. Badis G, Berger MF, Philippakis AA, Talukder S et al (2009) Diversity and complexity in DNA recognition by transcription factors. Science 324(5935):1720–1723
22. Jolma A et al (2013) DNA-binding specificities of human transcription factors. Cell 152 (1–2):327–339
23. Fang B, Mane-Padros D, Bolotin E, Jiang T, Sladek FM (2012) Identification of a binding motif specific to HNF4 by comparative analysis of multiple nuclear receptors. Nucleic Acids Res 40(12):5343–5356
24. Huang B, Yang XD, Lamb A, Chen LF (2010) Posttranslational modifications of NF-κB: another layer of regulation for NF-κB signaling pathway. Cell Signal 22(9):1292–1290
25. Perkins ND, Edwards NL, Duckett CS, Agranoff AB, Schmid RM, Nabel GJ (1993) A cooperative interaction between NF-kappa B and Sp1 is required for HIV-1 enhancer activation. EMBO J 12(9):3551–3558
26. Betts JC, Cheshire JK, Akirav S, Kishimotov T, Woo P (1993) The role of NF-KB and NF-IL6 transactivating factors in the synergistic activation of human serum amyloid A gene expression by interleukin-1 and interleukin-6. J Biol Chem 268(34):25624–25631
27. Stein B, Cogswell PC, Baldwin AS (1993) Functional and physical associations between NF-kappa B and C/EBP family members: a Rel domain-bZIP interaction. Mol Cell Biol 13(7):3964–3974
28. Stein B, Baldwin AS, Ballard DW, Greene WC, Angel P, Herrlich P (1993) Cross-coupling of the NF-kappa B p65 and Fos/Jun transcription factors produces potentiated biological function. EMBO J 12(10):3879–3891
29. Ogawa S et al (2005) Molecular determinants of crosstalk between nuclear receptors and toll-like receptors. Cell 122(5):707–721
30. Mohaghegh N, Bray D, Keenan JL, Penvose A, Andrilenas KK, Ramlall V, Siggers T (2019) NextPBM: a platform to study cell-specific transcription factor binding and cooperativity. Nucleic Acids Res 47(6):e31
31. Siggers T, Duyzend MH, Reddy J, Khan S, Bulyk ML (2011) Non-DNA-binding cofactors enhance DNA-binding specificity of a

transcriptional regulatory complex. Mol Syst Biol 7:555

32. Gordân R et al (2013) Genomic regions flanking E-box binding sites influence DNA binding specificity of bHLH transcription factors through DNA shape. Cell Rep 3 (4):1093–1104

33. Berger MF, Bulyk ML (2009) Universal protein-binding microarrays for the comprehensive characterization of the dna-binding specificities of transcription factors. Nat Protoc 4(3):393–411

34. Andrilenas KK, Ramlall V, Kurland J, Leung B, Harbaugh AG, Siggers T (2018) DNA-binding landscape of IRF3, IRF5 and IRF7 dimers: Implications for dimer-specific gene regulation. Nucleic Acids Res 46(5):2509–2520

35. Penvose A, Keenan JL, Bray D, Ramlall V, Siggers T (2019) Comprehensive study of nuclear receptor DNA binding provides a revised framework for understanding receptor specificity. Nat Commun 10(1):2514

36. Majumder S, Zhou LZ, Chaturvedi P, Babcock G, Aras S, Ransohoff RM (1998) p48/STAT-1alpha-containing complexes play a predominant role in induction of IFN-gamma-inducible protein, 10 kDa (IP-10) by IFN-gamma alone or in synergy with TNF-alpha. J Immunol 161(9):4736–4744

Chapter 5

Molecular and Biochemical Approaches to Study the Evolution of NF-κB Signaling in Basal Metazoans

Pablo J. Aguirre Carrión, Leah M. Williams, and Thomas D. Gilmore

Abstract

Extensive genomic and transcriptomic sequencing over the past decade has revealed NF-κB signaling pathway homologs in organisms basal to insects, for example, in members of the phyla Cnidaria (e.g., sea anemones, corals, hydra, jellyfish) and Porifera (sponges), and in several single-celled protists (e.g., *Capsaspora owczarzaki*, some choanoflagellates). Therefore, methods are required to study the function of NF-κB and its pathway members in early branching organisms, many of which do not have histories as model organisms. Here, we describe a combination of cellular, molecular, and biochemical techniques that have been used for studying NF-κB, and related pathway proteins, in some of these basal organisms. These methods are useful for studying the evolution of NF-κB signaling, and may be adaptable to the study of NF-κB in other non-model organisms.

Key words NF-kappaB, Evolution, Metazoan, Phylogeny, Protein function, Western blotting, EMSA, Immunofluorescence

1 Introduction

Transcription factor NF-κB is a critical regulator of immunity and development in advanced metazoans [1], and NF-κB activity is regulated by many upstream signaling pathways in organisms from flies to vertebrates. In the past decade, it has become apparent that homologs to NF-κB pathway proteins are conserved across diverse phyla, including many organisms at the base of the metazoan lineage, including single-celled protists, such as choanoflagellates and *Capsaspora owczarzaki*, sponges, and cnidarians (hydra, corals, sea anemones, and jellyfish) [2] (Fig. 1). Understanding the properties of NF-κB in these organisms will provide insights into the evolutionary origins of NF-κB, the regulation of its activity, and its biological functions, in particular in terms of immunity.

Pablo J. Aguirre Carrión and Leah M. Williams contributed equally to this work.

Guido Franzoso and Francesca Zazzeroni (eds.), *NF-κB Transcription Factors: Methods and Protocols*, Methods in Molecular Biology, vol. 2366, https://doi.org/10.1007/978-1-0716-1669-7_5,

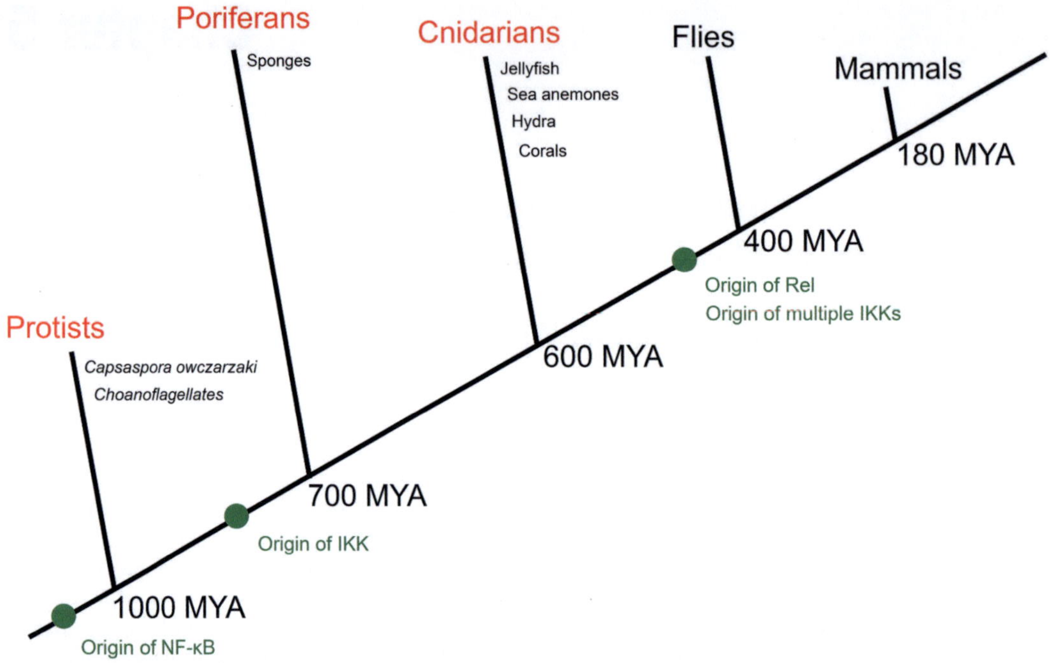

Fig. 1 A simplified phylogeny of select species to exemplify the organisms that are discussed in this manuscript. *MYA* million years ago

All proteins in the NF-κB superfamily are related by the presence of the Rel Homology Domain (RHD), which contains sequences important for dimerization, DNA binding, and nuclear localization [1]. The two subfamilies of NF-κB proteins comprise the traditional NF-κBs (p52/p100, p50/p105, *Drosophila* Relish) and the Rel proteins (RelA, RelB, c-Rel, *Drosophila* Dif and Dorsal). These subfamilies differ in their DNA-binding site preferences and their C-terminal domains [1]. That is, NF-κB proteins contain C-terminal inhibitory sequences consisting of Ankyrin (ANK) repeats, whereas Rel proteins possess C-terminal transactivation domains. The NF-κB subfamily appears to have arisen first evolutionarily, in that organisms basal to insects have NF-κB-like proteins but not Rel proteins [2].

NF-κB pathway activation is often initiated by the binding of various upstream ligands to their specific receptors (e.g., Toll-like receptors (TLRs), Interleukin-1 receptors (IL-1Rs), tumor necrosis factor receptors (TNFRs)), which then promotes downstream signaling cascades [1]. Activation of NF-κB invariably results in translocation of NF-κB from the cytoplasm to the nucleus, as a consequence of regulated degradation of an NF-κB inhibitor called IκB. In vertebrates, there are two primary NF-κB pathways, which are termed the canonical and non-canonical pathways. In the canonical pathway, an NF-κB dimer (such as p50-RelA) is activated when IκB kinase-beta (IKKβ) phosphorylates a bound IκB protein, causing its ubiquitination and degradation. In the non-canonical pathway, p100 (in a p100-RelB dimer) is phosphorylated by IKKα on a cluster of serine residues located C-terminal to the ANK repeats. p100 is then ubiquitinated, and the C-terminal ANK repeat sequences are degraded by the proteasome to release an active p52-RelB dimer [3].

While organisms from flies to humans have multiple NF-κB/Rel, IκB and IKK proteins, most basal organisms have single members of each family. In most cases, the basal NF-κB proteins resemble mammalian NF-κB p100, by virtue of containing an N-terminal RHD and a C-terminal ANK repeat domain; however, some cnidarians and choanoflagellates contain NF-κB proteins that consist primarily of an RHD [2]. Several studies have shown that cnidarians contain single homologs of IKK [2]. Moreover, several basal NF-κBs have been shown to undergo C-terminal proteasomal processing when co-expressed with any of several IKKs in human cells in cell culture. Furthermore, these processed basal NF-κB proteins enter the nucleus, are capable of binding conserved κB sites, and can activate transcription of reporter genes containing consensus NF-κB binding sites [2, 4–9]. Nevertheless, in tissue extracts from their native organisms (e.g., a sea anemone or a sponge), most of the NF-κB protein is constitutively in its active, processed nuclear form [6, 8]. Thus, it is not yet certain how processing of these basal NF-κB proteins is regulated in vivo. For example, constitutively active NF-κB proteins might be generated due to the pressures of living in a "soup" of microbiota, requiring a chronically active immune system [2].

Progress has been made in describing the biochemical properties, regulation, and biological functions of basal NF-κB proteins [2]. At least in sponges and cnidarians, current evidence suggests that NF-κB proteins have roles in early development and in adult immunity [2], and in the case of one sea anemone, NF-κB is expressed in and required for the development of body cnidocytes (specialized cells of cnidarians used for prey capture, among other things; *see* **Note 1**). We have previously described methods for comparing the sequence phylogeny and DNA-binding activities of evolutionarily diverse NF-κB proteins [9, 10]. Herein, we describe a combination of cellular, molecular, and biochemical techniques that have been used to investigate NF-κB function in basal metazoans, including demosponges [8], sea anemones (*Nematostella vectensis* [4, 5], *Aiptasia* sp. [6]), and a stony coral (*Orbicella faveolata* [7]). The methods described herein enable one to explore basal NF-κB properties—including DNA binding, subcellular localization, and cell type-specific expression—in a stepwise fashion using in vitro and in vivo experimental approaches (Fig. 2).

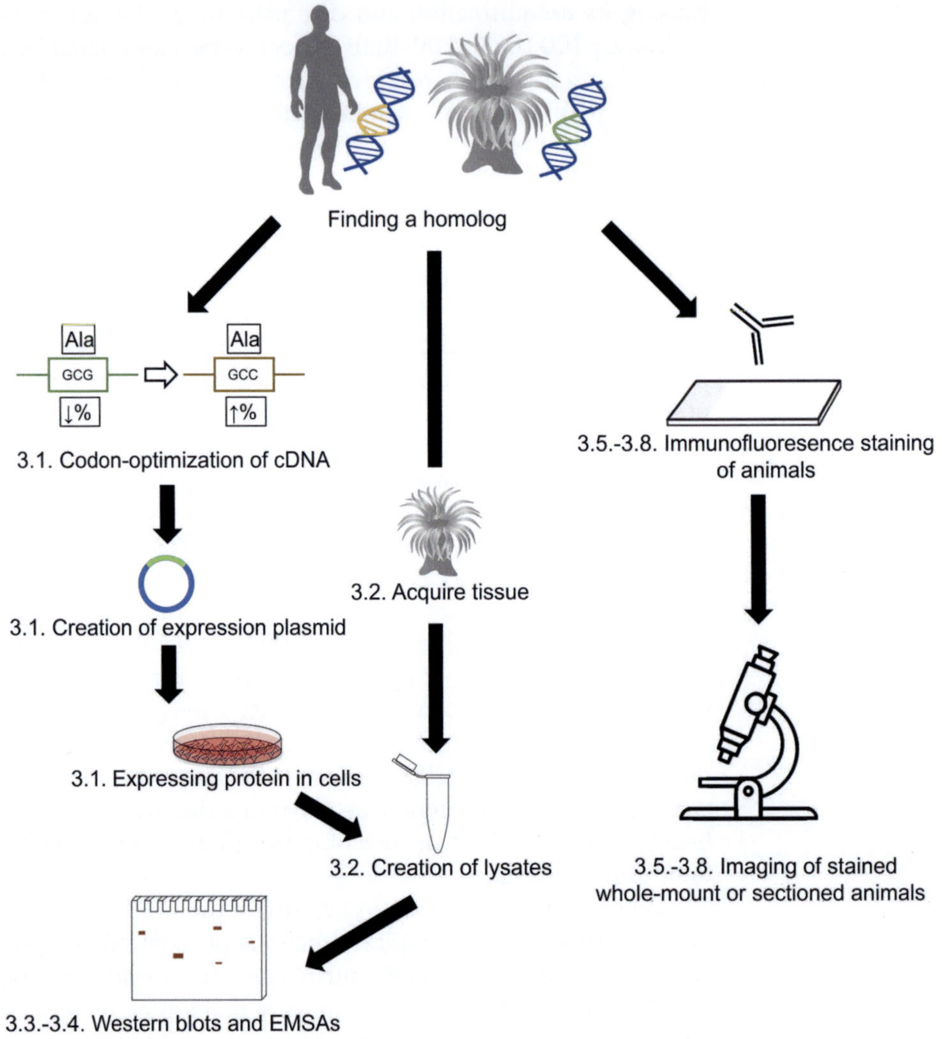

Fig. 2 A flowchart demonstrating the project design and capabilities of the methods described in this manuscript. From finding an NF-κB homolog on NCBI (top) to exploring its properties in cell culture or in vivo, our methods of investigating the molecular and biochemical properties of these homologs are streamlined. Although not described here, the abilities of basal NF-κB proteins to activate transcription of a reporter gene, to be processed by IKK-dependent proteasomal processing, to be phosphorylated by IKKs, and to localize to the nucleus have also been described [4–8]

2 Materials

2.1 Codon Optimization of DNA Sequences for Expression in Tissue or Bacterial Cells

1. A computer with internet access that can run a modern browser.
2. Sequence database: National Center for Biotechnology Information.
3. DNA synthesis company, GenScript (www.genscript.com).

2.2 Creation of Tissue Lysates

1. Animals: anemones, corals, and sponges (*see* **Note 2**).
2. Artificial seawater (ASW) (Instant Ocean).
3. Glass Dounce tissue grinder (Wheaton) (for *Aiptasia*).
4. 1.5-mL microcentrifuge tubes.
5. Aluminum foil (for stony corals).
6. Hammer (for stony corals).
7. Plastic pestle (Thermo Fisher Scientific) (for *N. vectensis*).
8. 10× AT lysis buffer stock: Mix 500 μL of 1 M DTT, 1 mL of 0.5 M EDTA, 2 mL of 250 mM EGTA, 0.42 g NaF, 2.66 g $Na_4P_2O_7$ in 80 mL dH_2O until dissolved. Adjust final volume to 100 mL with water. Store at −20 °C indefinitely.
9. 1× AT lysis buffer: Mix 10 mL of 10× AT lysis buffer stock, 20 mL glycerol, 10 mL 10% Triton X-100, 2 mL 1 M HEPES (pH 7.9), and 58 mL dH_2O. This mixture can be stored at −20 °C indefinitely. When ready to use, add PMSF, leupeptin, and aprotinin to a final concentration of 1 μg/mL (*see* **Note 3**).
10. Nutator.
11. Vortex.
12. Sonicator (for *N. vectensis*).
13. 1 mL syringe (1 × 100) (Henkesasswolf).
14. 18, 27.5, and 30 G needles.
15. Microcentrifuge capable of 15,000 × *g*.
16. 5 M NaCl.

2.3 Western Blotting

1. SDS-polyacrylamide gel short plates and 1.5-mm spacer glass plates.
2. Mini-PROTEAN Tetra Cell Casting Stand and Clamps.
3. 30% acrylamide, 29:1 acrylamide:bis-acrylamide.
4. 4× separating (lower) gel buffer: 0.5 M Tris, 0.4% (w/v) SDS, pH 8.8.
5. 4× stacking (upper) gel buffer:1.5 M Tris, 0.4% (w/v) SDS, pH 6.8.
6. Ammonium persulfate (APS): 10% (w/v) solution in water (*see* **Note 4**).
7. *N,N,N′,N′*-Tetramethylethylenediamine (TEMED).
8. 1.5 mm comb for creating wells in gel.
9. Isopropanol.
10. Electrophoresis tank.
11. 10× SDS gel electrophoresis running buffer: 0.25 M Tris–HCl, 1.92 M glycine, 1% (w/v) SDS.

12. 4× SDS sample buffer: 0.625 M Tris–HCl, pH 6.8, 2.3% (w/v) SDS, 10% (w/v) glycerol, 5% (v/v) β-mercaptoethanol, 0.1% (w/v) Bromophenol blue.
13. Heating block.
14. Protein molecular weight standards.
15. Power supply.
16. 10× Transfer Buffer: Combine 24.22 g Tris–HCl, 112.6 g glycine, and ~800 mL dH_2O. Mix and when in solution, bring volume to 1 L with water.
17. Methanol.
18. Transfer gel holder cassette.
19. Foam pads.
20. Nitrocellulose paper.
21. Whatman 3MM filter paper.
22. Forceps.
23. Gel transfer cell.
24. Cooling unit (such as an ice pack).
25. Magnetic stir bar.
26. 28. Magnetic stir plate.
27. 10× Tris-buffered saline (TBS): combine 6.05 g Tris and 8.76 g NaCl in 800 mL of H_2O. Adjust pH to 7.5 with HCl. Mix, and then bring volume to 1 L with water.
28. TBS-T: TBS, containing 0.05% (v/v) Tween-20.
29. Blocking solution: 5% (w/v) fat-free powdered milk in TBS-T.
30. Plastic staining trays.
31. Primary and secondary antisera (Table 1).
32. SuperSignal™ West Dura Extended Duration Substrate (Thermo Fisher Scientific) (equal parts Stable Peroxide solution and Luminol/Enhancer Solution).
33. Sapphire Biomolecular Imager (Azure Biosystems).

2.4 Electrophoretic Mobility Shift Assay

1. NF-κB double-stranded DNA oligonucleotides (underlined nucleotides are the palindromic consensus κB site binding motif): 5′-TCGAGAGGTCGGGGAATTCCCCCCCG-3′; 5′-TCGACGGGGGGGAATTCCCCGACCTC-3′.
2. 4 M NaCl.
3. 1 M Tris–HCl, pH 8.0.
4. 2.5 mM EDTA.
5. PCR tubes.
6. Thermocycler.

Table 1
Antisera used in these experiments

Name	Procedure	Host of origin	Dilution	Buffer
Primary antisera				
Nv-NF-κB	WB	Rabbit	1:10,000	5% (w/v) milk in TBS-T
Nv-TLR	WB	Guinea pig	1:25,000	8% (w/v) milk, 5% (v/v) NGS, 1% (v/v) BSA in TBS-T
Ap-NF-κB (for *Aiptasia)*	WB	Rabbit	1:10,000	5% (w/v) milk in TBS-T
Ap-NF-κB (for sponge*)*	WB	Rabbit	1:1000	5% (w/v) milk in TBS-T
Ap-NF-κB (for coral*)*	WB	Rabbit	1:10,000	5% (w/v) milk in TBS-T
Nv-NF-κB	IF	Rabbit	1:100	5% (v/v) NGS, 1% (w/v) BSA in PTx
Ap-NF-κB (for sponge)	IF	Rabbit	1:5000	0.3% (v/v) Triton X-100, 5% (v/v) NGS in PBS
Secondary antisera				
HRP-α-rabbit IgG	WB	Goat	1:4000	5% milk (w/v) in TBS-T
HRP-α-guinea pig IgG	WB	Goat	1:50,000	8% (w/v) milk, 5% (v/v) NGS, 1% (w/v) BSA in TBS-T
Alexa 488-α-rabbit IgG (for NF-κB detection in sponge)	IF	Goat	1:500	0.3% (v/v) Triton X-100, 5% (v/v) NGS in PBS
Alexa 488-α-rabbit IgG (for *Nv*-NF-κB detection)	IF	Goat	1:160	5% (v/v) NGS, 1% (w/v) BSA in PTx

BSA bovine serum albumin, *NGS* normal goat serum, *WB* Western blotting, *IF* immunofluorescence

7. T4 polynucleotide kinase and 10× buffer (New England BioLabs).
8. [γ-^{32}P]-ATP (Perkin Elmer).
9. TE buffer: Combine 0.2 mL 0.5 M EDTA, 1 mL 1 M Tris–Cl (pH 8.0), and 98.8 mL dH_2O. Mix to combine.
10. Bio-Spin 6 Chromatography Column (Bio-Rad).
11. 14 mL Falcon polypropylene tubes.
12. Centrifuge capable of spinning 14 mL Falcon tubes at 1000 × *g*.
13. 1.5-mL microcentrifuge tubes.
14. Radioactivity-safe plexiglass container.
15. 10× TGE buffer: Mix 35.66 g of glycine, 7.56 g of Tris–HCl, and 0.93 g of EDTA (*see* **Note 5**). Add water to approximately 225 mL. Bring pH to 8.3 with 12 N HCl. Bring total volume to 250 mL. Buffer can be stored at 4 °C and is stable for up to 2 weeks. Discard if precipitate forms.

16. Glycerol.
17. Thirty percent acrylamide/bis-acrylamide solution (29:1) (Bio-Rad).
18. 10% (w/v) APS solution in dH_2O.
19. TEMED.
20. Glass plates for electrophoresis.
21. 2-mm-width plastic glass plate spacers (to pour a 2-mm thick gel).
22. EMSA electrophoresis apparatus.
23. Power supply.
24. EMSA running buffer: Dilute 10× TGE with dH_2O to make 1 L 1× TGE.
25. 5× HEPES DNA-binding buffer: 50 mM HEPES pH 7.8, 250 mM KCl, 5 mM DTT, 5 mM EDTA, 20% (w/v) glycerol.
26. 1 mg/mL poly dI/dC.
27. 10 mg/ml BSA, 100× (New England BioLabs).
28. Primary antisera (optional, only needed for supershifting).
29. Bromophenol blue solution (*see* **Note 6**).
30. Whatman 3MM filter paper.
31. Plastic wrap (e.g., Saran Wrap).
32. Gel drier capable of drying with vacuum at 50 °C.
33. Phosphorus screen and phosphorus screen cassette or X-ray film, two intensifying screens, and X-ray film developer.
34. Sapphire Biomolecular Imager (Azure Biosystems).

2.5 Immunohistochemistry of Whole Mount and Sectioned *Nematostella vectensis* Tissue

1. 12-well plate tissue culture plate.
2. Filtered 1/3 ASW: Combine 333 mL artificial seawater (ASW) (Instant Ocean) and 667 mL dH_2O. Filter through a 22 μm filter.
3. $MgCl_2$ in ASW: slowly add 0.7 g $MgCl_2$ per to 8 mL filtered 1/3 ASW in fume hood, allow solution to cool down and bring final volume to 10 mL. Allow any black precipitate to settle and remove from solution.
4. 37% (v/v) Formaldehyde.
5. Fixative solution: Combine 892 μL filtered 1/3 ASW and 108 μL 37% (v/v) formaldehyde. Store solution at 4 °C for up to 6 months.
6. Nutator.
7. 30% (w/v) sucrose in full-strength ASW.
8. Shandon Peel-A-Way Disposable Embedding Molds (Thermo Fisher Scientific).

9. Parafilm.
10. Optimal cutting temperature compound (OCT) (Sakura Tissue-Tek).
11. Microtome.
12. Urea.
13. Microwave-safe glass dish that can hold ~10 mL.
14. Standard commercial microwave.
15. Phosphate-buffered saline (PBS), may be prepared from a 10× stock with distilled water.
16. PTx: Combine 0.5 mL of 20% (v/v) Triton X-100, 5 mL 10× PBS, and 44.5 mL dH_2O. Store indefinitely at 4 °C.
17. Blocking buffer: combine 0.5 mL of 20% (v/v) Triton X-100, 5 mL 10× PBS, 2.5 mL normal goat serum, 0.5 g BSA, and 40 mL dH_2O. Mix well and bring total volume to 50 mL with water. Store indefinitely at −20 °C.
18. Primary and secondary antisera (Table 1).
19. Humidified chamber (can use a large petri dish with damp paper towels).
20. Vectashield mounting medium (Vector Laboratories, Inc.).
21. Superfrost plus microscope slides.
22. Coverslips (no. 1.5; VWR).

2.6 Cnidocyte Staining

1. 12-well tissue culture plate.
2. 1/3 ASW: Combine 333 mL artificial seawater (ASW) (Instant Ocean) and 667 mL dH_2O. Filter through a 22 μm filter.
3. Cnidocyte fixative solution: Combine 872 μL 1/3 ASW, 108 μL 37% (v/v) formaldehyde, and 20 μL 0.5 M EDTA. Store solution at 4 °C for up to 6 months.
4. Nutator.
5. Cnidocyte wash buffer: Mix 0.6 g Tris–HCl, 0.29 g NaCl, and 1.86 g EDTA in 400 mL dH_2O. Adjust pH to 7.6 with 12 N HCl. Bring volume to 500 mL with dH_2O. Store indefinitely at room temperature.
6. DAPI (4′,6-diamidino-2-phenylindole) staining solution: Combine 49 μL 1 mg/mL DAPI with 951 μL cnidocyte wash buffer. Alternatively, use Acridine Orange (AO) staining solution: Combine 10 μL 110 mM AO with 990 μL cnidocyte wash buffer (to a final AO concentration of 1.1 mM).
7. Vectashield mounting medium (Vector Laboratories, Inc.).
8. Superfrost plus microscope slides (Thermo Fisher Scientific).
9. Coverslips (no. 1.5; VWR).

2.7 Immunohistochemistry of Sectioned Sponge Tissue

1. Filtered ASW (Instant Ocean).
2. Ethanol.
3. 4% paraformaldehyde (for example, 40 μL paraformaldehyde in 960 μL filtered full-strength ASW).
4. 30% (w/v) sucrose in full-strength ASW.
5. Shandon Peel-A-Way Disposable Embedding Molds (Thermo Fisher Scientific).
6. Optimal cutting temperature compound (OCT).
7. Microtome.
8. Superfrost plus microscope slides.
9. PBS (pH 7.4) (may be prepared from a 10× stock with sterile distilled water).
10. Forceps.
11. 48-well plate.
12. Nutator.
13. Blocking solution: 0.3% Triton X-100, 5% normal goat serum (v/v) in PBS.
14. Primary and secondary antisera and buffer diluent (*see* Table 1).
15. Humidified chamber (can use a large petri dish with damp paper towels).
16. PBS-T: 0.3% Triton X-100 in PBS.
17. Hoechst 33342 for nuclear staining.
18. Prolong Gold (Molecular Probes by Life Technologies).
19. Confocal microscope.

3 Methods

3.1 Codon Optimization of cDNA Sequences for Expression in Tissue Culture or Bacterial Cells

1. Search NCBI for your gene of interest (e.g., NF-κB) in your species of interest or use private genomic or transcriptomic sequence information.
2. Obtain the full cDNA coding sequence.
3. Using GenScript, start by imputing your cDNA sequence into the website.
4. Select "Add flanking sequences" to add restriction sites for subcloning into another plasmid, if necessary.
5. Choose your preferred production speed for synthesizing your gene.
6. Select codon optimization and choose the species in which you will be expressing your plasmid (e.g., *Homo sapiens* for HEK 293 cells) (*see* **Note 7**).

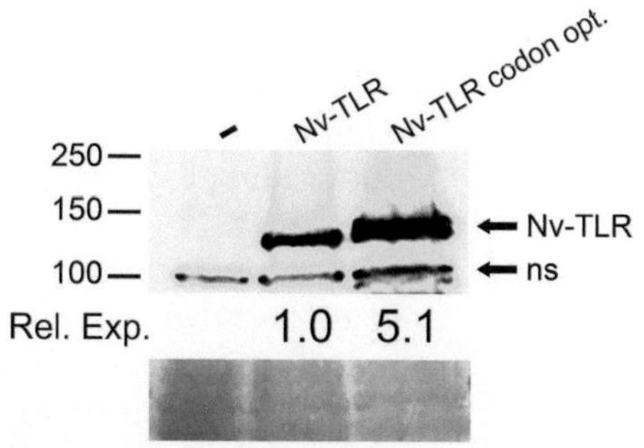

Fig. 3 Western blot demonstrating the benefit of codon optimization. HEK 293 T cells were transfected with cDNAs subcloned into the expression vector pcDNA FLAG with either the native sequence of Nv-TLR or the human cell codon-optimized sequence of Nv-TLR. Lysates of these cells were electrophoresed on a 7.5% SDS-polyacrylamide gel and subjected to anti-Nv-TLR Western blotting (top) (as in [12]) or the filter was stained with Ponceau S as a protein loading control (bottom). Molecular weight markers in kDa are shown to the left of the top panel. *Rel. Exp.* relative expression, *ns* non-specific

7. Choose your target vector (e.g., pUC57), and the preferred cloning sites.
8. Choose your quantity (usually 4 μg).
9. Proceed to order sequence through GenScript. You will receive plasmid DNA as a dry DNA pellet.
10. Rehydrate plasmid, transform bacteria, grow up plasmid, and use as needed for subcloning.
11. Once subcloned into an expression, the plasmid can be used to express in your cell line of choice (e.g., A293T cells). Codon optimization will provide better protein expression (e.g., for Western blotting (Subheading 3.3), *see* Fig. 3).

3.2 Creation of Tissue Lysates for Western Blotting and EMSAs

3.2.1 Sponge Lysates

1. Wash sponge tissue (approximately 2 cm by 2 cm) with full-strength ASW three times. Tissue can be immediately flash frozen in a 1.5-mL microcentrifuge tube and stored at −80 °C until use.
2. To prepare lysate, take tissue and place in a glass Dounce tissue grinder.
3. Add 1 mL of 1× AT Lysis Buffer containing protease inhibitors.
4. Grind approximately ten times by hand.
5. Transfer sample into a 1.5-mL microcentrifuge tube and gently rotate for 1 h at 4 °C with occasional vortexing.
6. Pass lysates five times through a 27.5-G needle, and then centrifuge sample at 15k × *g* for 15 min to pellet debris.

7. Transfer supernatant (protein lysate) into a clean 1.5-mL microcentrifuge tube and store lysate at −80 °C until use.
8. Samples are now ready to use for either Western blotting (*see* Subheading 3.3) or EMSA (*see* Subheading 3.4).

3.2.2 N. vectensis *Lysates*

1. Wash adult anemones (~2 cm long) three times with 1/3 ASW in a glass bowl. Transfer a single anemone to a 1.5-mL microcentrifuge tube containing ice-cold 150 μL 1× AT buffer (containing protease inhibitors) and homogenize with a small glass or plastic pestle. Incubate sample on ice for 20 min with vortexing every 5 min.
2. Sonicate samples five times for 10 s each on setting 3 on a sonicator, with 1 min on ice between each sonication.
3. Pass lysates five times through a 30-G needle (a 27.5-G needle may be used at first if lysates are too thick).
4. Add NaCl to a final concentration of 150 mM (i.e., 4.5 μL of 5 M NaCl for 150 μL total volume), mix by inversion several times, and spin at 15k × *g* for 30 min at 4 °C. Collect and store supernatant at −80 °C (e.g., most proteins, including *N. vectensis* (Nv) NF-κB are in the supernatant).
5. Other proteins of interest might be in pellet and can be recovered by resuspending the pellet in 75 μL 1× AT lysis buffer with a 30-G needle. Then add 2.25 μL 5 M NaCl, mix, and store at −80 °C.
6. Lysates are ready to use for either Western blotting (*see* Subheading 3.3) or EMSA (*see* Subheading 3.4).

3.2.3 Aiptasia *Lysates*

1. Wash adult anemones (~2 cm long) with 1/3 ASW three times. Transfer a single anemone to a 1.5-mL microcentrifuge tube containing ice-cold 100 μL 1× AT buffer (with protease inhibitors) and homogenize with a small glass or plastic pestle. Incubate on ice for 20 min with vortexing every 5 min.
2. Pass lysates five times through an 18-G needle or until tissue breaks apart.
3. Pass lysates five times again through a 27.5-G needle.
4. Add NaCl to a final concentration of 150 mM (i.e., 3 μL 5 M NaCl for 100 μL total volume), mix by inversion, and spin at 15k × *g* for 30 min at 4 °C. Collect and store supernatant at −80 °C (e.g., most proteins, including *Aiptasia* (Ap) NF-κB are in the supernatant).
5. Lysates are ready to use for either Western blotting (*see* Subheading 3.3) or EMSA (*see* Subheading 3.4).

3.2.4 Coral Lysates

1. Wash coral branch (approximately 1 cm by 1 cm) with full-strength ASW three times. Tissue can be immediately flash frozen in a 1.5-mL microcentrifuge tube and stored at −80 °C until use, but it is recommended to prepare lysates for same-day use.
2. To prepare lysate, take coral branch and place on a small square of clean aluminum foil (approximately 9 cm by 9 cm).
3. Place another small square of clean aluminum foil on top of the coral branch.
4. Carefully hammer the coral piece between the aluminum foil pieces ~4–5 times, or until the branch has been broken apart into smaller pieces.
5. Place ground coral pieces into a 1.5-mL microcentrifuge tube and add 1 mL of 1× AT buffer (with protease inhibitors).
6. Rotate for 1 h at 4 °C, with occasional vortexing.
7. Remove large tissue bits with forceps and discard them.
8. Pass lysate five times through a 27.5-G needle.
9. Centrifuge sample at 15k × *g* for 15 min.
10. Transfer supernatant to a fresh 1.5-mL microcentrifuge tube and store lysate at −80 °C until use.
11. Lysates are ready to use for either Western blotting (*see* Subheading 3.3) or EMSA (*see* Subheading 3.4).

3.3 Western Blotting of Protein From Animal Lysates (See Note 8)

3.3.1 Preparing an SDS-Polyacrylamide Gel

1. Assemble gel pouring apparatus.
2. For a 1.5 mm, 7.5% acrylamide gel (*see* **Note 9**): Combine 1.9 mL 30% acrylamide, 1.9 mL 4× separating buffer, and 3.65 mL dH_2O. Mix to combine.
3. Add 112.5 μL 10% APS and 22.5 μL TEMED. Mix to combine and quickly pour or pipette between glass panes, leaving ~3 cm of empty space between top of the liquid gel and the inner smaller plate.
4. Gently add 500 μL isopropanol to top of gel. Allow gel to solidify (~15 min).
5. Pour isopropanol out and allow any remaining isopropanol to air dry (~5 min).
6. Combine 0.75 mL 30% acrylamide, 1.25 mL 4× stacking buffer, and 3 ml water, Mix.
7. Add 75 μL 10% APS and 15 μL TEMED. Mix to combine and quickly pour or pipette between glass panes to the top of the separating gel.

8. Carefully insert well comb, pushing out extra liquid, and avoid forming any bubbles. Allow gel to solidify (~20 min).
9. Rinse gel with dH_2O and remove well comb. Gel is ready for use.

3.3.2 Preparing and Running Samples for Western Blotting

1. Insert gel in gel running apparatus.
2. Fill both chambers with 1× SDS gel electrophoresis running buffer.
3. Prepare sample for electrophoresis by combining 100 μg animal lysate (for example, if animal lysate is 10 μg/μL, combine 10 μL lysate, 10 μL dH_2O, and 20 μL 4× SDS sample buffer) (*see* **Note 10**). Mix to combine.
4. Heat samples at 95 °C for 10 min.
5. Load the entire sample (40 μL) into an individual well. Also load one well with 10 μL protein standards (Bio-Rad). Load any empty wells with 10 μL 4× SDS sample buffer (*see* **Note 11**).
6. Run gel at 80 V until bromophenol blue has migrated to just past the stacking gel and into the separating gel.
7. Raise voltage to 125 V. Allow gel to run until the bromophenol blue of the loading buffer reaches the bottom of the gel (*see* **Note 12**).

3.3.3 Western Blotting

1. About 15 min before SDS-PAGE has finished, prepare 1× Transfer buffer as follows: 700 mL dH_2O, 100 mL 10× Transfer buffer, and 200 mL methanol (for proteins of higher weight, methanol can be reduced to 100 mL and replaced with dH_2O). Pre-cool 1× Transfer buffer at 4 °C.
2. Once SDS-PAGE has finished, turn off the power supply, and carefully remove gel from the apparatus.
3. Gently break apart the front glass plate from the back plate. Separate and discard the stacking gel and the bottom of the gel containing the bromophenol blue.
4. Place gel into the filter paper sandwich as follows, which should be built in a large container (such as a glass tray) containing enough 1× Transfer buffer to completely cover the sandwich: Black side of gel holder cassette.

   Foam pad.

   Whatman 3MM filter paper.

   SDS-polyacrylamide gel.

   Nitrocellulose membrane.

   (!!!) Only handle with forceps.

   Whatman 3MM filter paper.

Foam pad.

Clear side of gel holder cassette.

5. Carefully roll the sandwich (can use a 14 mL tube or small glass pipet) to remove bubbles between the membrane and the gel, and when finished, close the sandwich firmly and slide lock into place.
6. Slide filter paper sandwich into a gel transfer cell, and place cell into a transfer tank.
7. Place cooling unit (such as an ice pack) into the tank, and add a magnetic stir bar.
8. Place tank on a magnetic stir plate, connect tank lid, and plug in lid cables to power supply and transfer proteins to the nitrocellulose membrane toward the anode at 250 mA for 4 h, and then 170 mA overnight, at 4 °C.
9. The next day, disassemble the gel-nitrocellulose membrane sandwich, and remove membrane with forceps. Place the nitrocellulose membrane in a small dish protein-side up (this is the side that was in contact with the gel). Give the membrane a quick wash with dH_2O to remove transfer buffer and methanol.
10. Add 10 mL blocking solution to the membrane and allow to rotate for 1 h at room temperature.
11. Remove blocking solution and add appropriate primary antibody diluted in blocking solution (*see* **Note 13**), and incubate overnight at 4 °C.
12. The next day, remove primary antibody (can save at −20 °C for several uses), and wash 5 × 5 min with TBS-T.
13. Add secondary antibody diluted in blocking solution and incubate for 1 h at room temperature.
14. Remove secondary antibody and wash 4 × 10 min with TBS-T.
15. After the last TBS-T wash, wash 2 × 5 min with TBS.
16. After the last TBS wash, mix equal parts Stable Peroxide solution and Luminol/Enhancer Solution to make a Working solution. Use 0.1 mL of Working solution per 1 cm^2 of membrane.
17. Lay membrane on a piece of plastic wrap and, using a pipette, gently add mixed Working solution onto membrane so that it is evenly coated, and incubate for 5 min. Cover the top of the membrane with another piece of plastic wrap to help spread Working solution onto membrane.
18. Expose membrane onto X-ray film or image using a Sapphire Biomolecular Imager.

3.4 EMSA Using Animal Lysates

3.4.1 Radiolabeling κB Site Probe

1. Combine 1 μL of each 1 μg/μL oligonucleotide stock with 0.5 μL 4 M NaCl, 0.2 μL 1 M Tris–HCl (pH 7.5), 4 μL 2.5 mM EDTA, and 13.3 μL dH_2O in a PCR tube.
2. Place sample in a thermocycler with the following steps for annealing: 88 °C for 2 min, 65 °C for 10 min, 37 °C for 10 min, 25 °C for 10 min.
3. Dilute annealed oligonucleotides to a final concentration of 1 μM by bringing total volume to 60 μL with dH_2O. Mixture can be stored at −20 °C.
4. Label probe by combining 2 μL annealed probe, 1.5 μL 10× T4 polynucleotide kinase buffer, 1 μL T4 polynucleotide kinase, and 20 μCi [γ-^{32}P]-ATP (Perkin Elmer) in a final volume of 15 μL and incubating at 37 °C for 2 h.
5. Bring reaction volume to 100 μL with TE buffer.
6. Snap bottom and remove cap from a Bio-Spin 6 Chromatography Column (Bio-Rad) to allow extra buffer to drain by gravity.
7. Spin column into a collection tube in a 14-mL Falcon polypropylene tube at 1000 × *g* for 2 min.
8. Transfer column to new collection tube and apply entire oligonucleotide labeling reaction to it.
9. Spin again at 1000 × *g* for 3.5 min.
10. Dispose of column in a radioactivity-safe container.
11. Transfer clear liquid in the collection tube to a clean microcentrifuge tube.
12. Measure cpm/μL. Store labeled probe at −80 °C in appropriate radioactivity-safe container (*see* **Note 14**).

3.4.2 DNA-Binding Reaction and Running EMSA Gel

1. Mix 8.33 mL 30% acrylamide, 1.25 mL glycerol, 5 mL 10× TGE, and 35.5 mL dH_2O. Add 400 μL 10% APS and 85 μL TEMED, mix well, and pour into assembled glass plates, avoid forming bubbles. Allow gel to polymerize for at least 45 min (*see* **Note 15**).
2. Assemble electrophoresis apparatus with gel. Pour EMSA running buffer into top and bottom chambers. Use a syringe to remove any bubbles that block contact between buffer and gel. Pre-run gel for 30 min at 50 mAmp.
3. Assemble binding reactions by combining 10 μL 5× HEPES binding buffer, 2 μL 1 mg/ml poly dI/dC, and 1 μL 100× BSA, 100 μg of animal tissue lysate, 200,000 cpm of labeled κB site probe, and water to a final volume of 50 μL. Incubate for 30 min at 30 °C (*see* **Note 16**). We have optimized our assays to the reagents described above and note that these reagents maximize NF-κB binding with animal lysates and the clarity of the bound protein-DNA complex on the EMSA (Fig. 4).

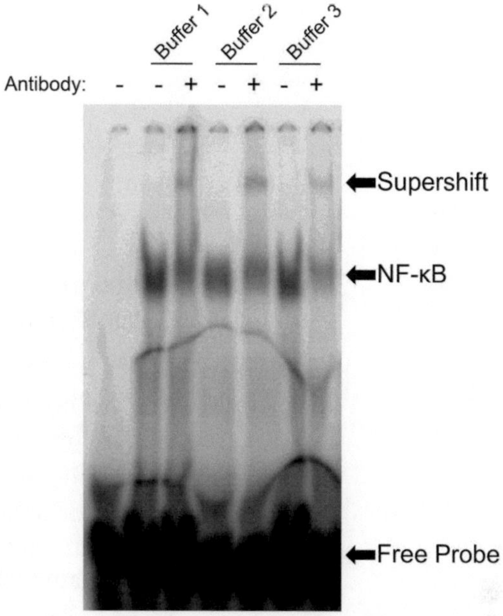

Fig. 4 An EMSA using the same *Nematostella vectensis* (*Nv*) lysate, but with three different DNA-binding reaction buffers (*see* **Note 17**). All lanes contain a κB probe and the indicated binding buffer. Lane 1 contains the probe alone. Lanes 2–7 contain the probe with the same *Nv* lysate. Lanes 2 and 3 contain Buffer 1, lanes 4 and 5 contain Buffer 2, and lanes 6 and 7 contain Buffer 3. (Buffer 1 is the preferred 5× HEPES binding buffer, described in the text.) Lanes 3, 5, and 7 also contain *Nv*-NF-κB antiserum. As indicated by the top arrow, all lysates from *Nv* bind the κB site probe to varying degrees, and bands supershift with the addition of the antiserum. Free probe is indicated at the bottom of the gel with an arrow

4. (Optional) If a supershift is needed, add 2 μL of preferred antibody (e.g., anti-Nv-NF-κB antiserum for *N. vectensis* lysates) after incubating binding reactions at 30 °C and then incubate on ice for 1 h.
5. Unplug gel apparatus. Load samples in individual lanes and load one lane with 1× bromophenol blue. Electrophorese gel at 30 mAmp until bromophenol blue travels approximately ¾ of the way through the gel.
6. Disassemble gel, carefully lift top plate using a metal spatula. Cover gel with two sheets of 3MM Whatman filter paper. Invert gel, paper, and lower plate so that the paper is on the bottom and plate is on the top. Carefully remove glass plate, so that the gel remains on the filter paper, and cover gel with plastic wrap.
7. Place gel on a gel dryer with plastic wrap-side up. Dry at 50 °C for 2 h under vacuum.

8. Expose gel to X-ray film at −80 °C overnight with an intensifying screen on each side forming a sandwich, and then develop using a standard X-ray film developer. Alternatively, expose gel to a phosphor screen at room temperature overnight, and image on a Sapphire Biomolecular Imager.

3.5 Immunohistochemistry of Whole Mount N. vectensis (Nv) Tissue

3.5.1 Fixation of Anemones

1. Allow *Nv* to relax in 500 μL filtered 1/3 ASW in one well of a 12-well plate.
2. Carefully, add 500 μL 7% (w/v) $MgCl_2$ in ASW drop by drop using a p200 Pipetman into the well.
3. Allow *Nv* to relax undisturbed for 10 min.
4. Pipette most of the liquid out (~900 μL), leaving just enough liquid to cover *Nv*. Be careful not to touch *Nv* as some stimuli can cause the anemone to contract.
5. Slowly add 1 mL of ice-cold fixative solution and incubate overnight at 4 °C in the dark on a nutator (~80 rpm). *Nv* may be stored in fixative solution for up to 1 month at 4 °C.

3.5.2 Indirect Immunofluorescence

1. To perform antigen retrieval (Optional, *see* **Note 18**) prepare 10 mL of 5% (w/v) urea in dH_2O and add to microwave-safe container.
2. Heat the urea solution until it reaches 80 °C, do not boil.
3. Remove fixative solution and wash anemones three times with PTx for 5 min each.
4. Replace PTx with warmed urea solution, making sure the anemone is completely covered.
5. Heat samples in a microwave at the lowest setting for 5 min. Make sure solution does not boil and plate does not heat.
6. Remove container from microwave and allow to cool for 20 min at room temperature.
7. Remove urea and wash three times with PTx for 5 min each.
8. Wash anemones three times with 1 mL PTx for 5 min (unless antigen retrieval was performed). If you intend to co-stain for cnidocytes with DAPI or AO (*see* **Note 1**), supplement each buffer solution and antiserum with EDTA to a final concentration of 10 mM.
9. Remove PTx and replace with 500 μL immunofluorescence blocking buffer.
10. Place on a nutator (~80 rpm) at 4 °C overnight in the dark.
11. Remove immunofluorescence blocking buffer.
12. Add 500 μL diluted primary antiserum (e.g., anti-Nv-NF-κB antiserum (1:100)) and incubate at 4 °C overnight in the dark.
13. Remove primary antiserum and wash anemones four times with 1 mL PTx for 10 min each.

14. Remove PTx and add 500 μL of appropriate secondary antiserum (Table 1). After addition of secondary antiserum, all of the following steps should be done in low-light conditions to avoid decay of the signal.
15. Remove secondary antiserum and wash anemones four times with 1 mL PTx for 10 min each. If co-staining for cnidocytes using DAPI or AO, proceed to Subheading 3.7 before mounting on slides.
16. Remove anemones from PTx, carefully blot excess liquid onto Whatman 3MM filter paper and transfer to slides. Add an appropriate amount of Vectashield mounting medium depending on size of anemone (can be as little as 20 μL).
17. Mount a coverslip and seal edges with nail polish. Allow nail polish to completely dry in the dark.
18. Image samples on a confocal microscope (Fig. 5).

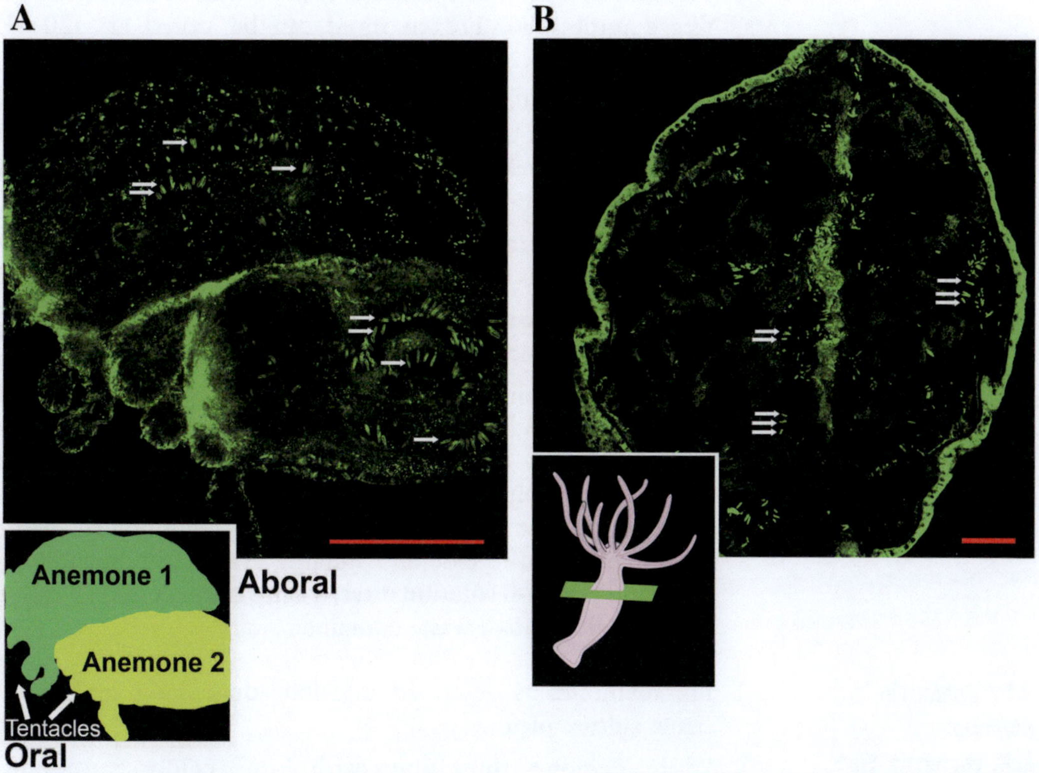

Fig. 5 Immunohistochemistry detection of *Nv*-NF-κB in *Nematostella vectensis* tissue. (**a**) Whole mount imaging of two, young (approximately six months old) *N. vectensis* anemones, stained with anti-*Nv*-NF-κB primary antiserum, and Alexa 488-conjugated secondary antibody, and were imaged under a confocal microscope (Nikon C2 Si). The inset outlines the general structure and axes of the two anemones. White arrows indicate NF-κB-positive cnidocytes. Red bar = 500μm. (**b**) Sectioned tissue slice of an adult *N. vectensis* anemone stained with anti-Nv-NF-κB primary antiserum, and Alexa 488-conjugated secondary antibody, and were imaged under a confocal microscope (Nikon C2 Si). The inset shows approximately where the anemone was sliced to obtain the image. White arrows indicate NF-κB-positive cnidocytes. Red bar = 500 μm

3.6 Immunohistochemistry of Sectioned N. vectensis *Tissue*

1. Fix anemones with paraformaldehyde as described in Subheading 3.5.1 in 12-well tissue culture plates.
2. Wash anemones three times with 1 mL PTx for 5 min each wash.
3. Remove PTx and replace with 1 mL 5% (w/v) sucrose in PBS.
4. Perform five washes in sucrose/PBS for 20 min each, increasing the concentration of sucrose by 5% for each new wash. (i.e., 10%, 15%, 20%, 25%, and 30%).
5. Remove anemones from sucrose and carefully blot excess liquid. Transfer to a sheet of parafilm.
6. Cover anemones with Optimal Cutting Temperature Compound (OCT).
7. Transfer anemones to a plastic sectioning mold filled with OCT, orienting aboral toward the bottom or long ways, depending on what view of the anemone you are interested in.
8. Place mold on top of a mixture of dry ice and ethanol to flash freeze until solid. Frozen mold can be stored at −20 °C indefinitely.
9. Remove individual frozen block from the mold. Cryosection tissue on a microtome into 12-μm-thick slices, then place onto Superfrost Plus Microscope Slides (Fisher Scientific). Slides may be stored at −20 °C indefinitely.
10. Preceding immunohistochemical staining, heat slides in a dry incubator for 30 min at 50 °C.
11. Wash slides once with PTx by pipetting ~1 mL of PTx onto the slide and allowing excess to drain.
12. Follow immunohistochemical staining procedure as described in Subheading 3.5.2. Blocking and antiserum incubation steps can be done by either submerging slides in the appropriate solution or by pipetting the solution onto the slide and incubating it in a humidified chamber. If using a humidified chamber, do not put the slide on a nutator. Washes can be done by pipetting the wash solution directly onto the slide and allowing excess to flow into a waste container.

3.7 Cnidocyte Staining in N. vectensis *(See Note 1)*

1. Fix anemones as described in Subheading 3.4.1 in 12-well tissue culture plates.
2. Wash anemones three times with 1 mL cnidocyte washing buffer on a nutator at about 80 rpm for 5 min each.
3. Remove cnidocyte washing buffer and replace with 500 μL cnidocyte stain. After addition of stain, all of the following steps should be done in low-light conditions to avoid degradation of signal.
4. Put anemones back on nutator for 12 min (if using DAPI) or 15 min (if using AO).

5. Remove stain and wash anemones three times with cnidocyte washing buffer for 5 min each on the nutator.
6. Transfer anemones to slides. Add an appropriate amount of Vectashield mounting medium depending on size of anemone (can be as little as 20 μL).
7. Place a coverslip on top of the specimen and seal edges with nail polish.
8. Image on a fluorescent microscope. DAPI-stained cnidocytes fluoresce green (~515 nm wavelength), AO-stained cnidocytes fluoresce red (~615 nm wavelength).

3.8 Immunohistochemical Staining of Sectioned Sponge Tissue

1. After collecting, tissue can be flash frozen and stored at −80 °C.
2. Place tissue into ice-cold 4% paraformaldehyde in full-strength filtered ASW to fix overnight at 4 °C.
3. Wash fixed tissue three times with filtered ASW.
4. Dehydrate tissue in 30% sucrose in ASW at 4 °C overnight.
5. Embed tissue samples in OCT using plastic molds.
6. Cryosection tissue on a microtome into 45 μM slices, then place onto Superfrost Plus Microscope Slides.
7. Sectioned tissue can be store at −20 °C on slides until ready for immunofluorescence.
8. Rehydrate tissue slices with approximately 300 μL of room temperature PBS.
9. Place tissue slices in individual wells of a 48-well plate using forceps.
10. Wash tissue slices with agitation three times with PBS.
11. Block and permeabilize tissue slices with 0.3% Triton X-100 + 5% goat serum in PBS for 1 h at room temperature.
12. Incubate slices with primary antiserum (e.g., anti-Ap-NF-κB antiserum (1:5000) [5]) diluted in the buffer diluent (*see* Table 1) overnight at 4 °C in a humidified chamber, with rotation.
13. Wash slides with agitation three times in PBS-T.
14. Incubate with secondary antiserum (e.g., goat-anti-rabbit Alexa Fluor 488 (1:500)) diluted in the buffer diluent (*see* Table 1).
15. Wash tissue slices three times with PBS-T, and on the last wash add Hoechst (1:5000 of 20 μM) for 10 min.
16. Wash slices three more times with PBS-T.
17. Mount tissue onto Superfrost Plus Microscope Slides with Prolong Gold with coverslip.
18. Image samples on a confocal microscope.

4 Notes

1. Cnidocytes are harpoon-like ectodermal cells characteristic of the phylum Cnidaria. Cnidocytes play various roles in defense, prey capture, environment sensing [11], and possibly immunity [5, 12]. In *N. vectensis*, NF-κB is mainly expressed in cnidocytes [5], and this can be shown by co-staining of NF-κB (with antiserum) and cnidocytes. When cnidocytes are stained with DAPI, poly-γ-glutamate in the cnidocytes will fluoresce in the green channel, or if stained with AO, cnidocytes will fluoresce red [13].
2. The sea anemones *N. vectensis* and *Aiptasia* are fairly common model organisms within the cnidarian community and most labs have populations big enough to share. Therefore, anemones can usually be requested. *Aiptasia pallida* is considered an aquarium pest and can be found in many commercial aquariums, and wild *Aiptasia* can be found in tropical seas where corals are common. Alternatively, *Aiptasia* can be purchased from Carolina Biological Supply Company (https://www.carolina.com/marine-and-saltwater-animals/sea-anemone-aiptasia-living/162865.pr). Adult anemones can easily be transported or shipped in a sterile 50 mL Falcon tubes filled with seawater. *See* [14] for information on the collection of wild specimens and maintenance of *N. vectensis.* Coral and sponge tissue collections usually require special government permits, and therefore, necessitate collaboration with a certified marine biologist to collect fragments of these animals in their natural habitats. Nevertheless, many commercial aquarium supply companies sell a range of corals and sponges, although caution should be taken to ensure that the seller is partaking in safe and sustainable collection practices. Plenty of materials exist for information on raising, propagating, and experimenting with cnidarians in aquaria (e.g., https://cdhc.noaa.gov/education/protocols.aspx).
3. Protease inhibitors should be added fresh each time AT Buffer is made.
4. 10% APS can be used for up to one week when kept at 4 °C.
5. You can replace powdered EDTA with 5 mL 0.5 M EDTA to speed up the process.
6. Bromophenol blue, traditionally used in loading buffer for DNA in gel electrophoresis, is used here to track the movement of the EMSA samples through the gel, as the binding buffer for the samples is clear.
7. Be sure to check that the codon-optimized version of your gene does not contain the flanking restriction enzyme sites that you have chosen for subcloning. Further optimization of protein

expression can be performed for the cell types that are used. For example, for bacterial expression, if after codon optimizing your gene of interest, there are still a few rare codons present in the sequence, *E. coli* strains such as *B834 (DE3) pRARE* (B834 is the parental strain for BL21) can be used to increase expression due to their capacity to allow expression of genes encoding tRNAs for rare codons.

8. For a detailed description of Western blotting procedures *see* also [15].
9. The described recipe is for a 7.5% acrylamide gel. Other percentages can be made by adjusting the ratios of 30% acrylamide to water. This might be necessary for better visualization of other proteins of interest. We find that 7.5% polyacrylamide gels are suitable for NF-κB proteins in cnidarians and poriferans in the range of 50–120 kDa.
10. When performing Western blotting or EMSAs, it is recommended to use 100–150 μg of tissue lysate for the detection or DNA-binding activity, respectively, of NF-κB proteins from cnidarians and poriferans. The amount of protein can be determined by a Bradford Assay (Bio-Rad).
11. Adding 4× SDS sample buffer into empty wells encourages your samples to run straight across the gel, i.e., to avoid gel "smiling."
12. For other proteins of interest, it might be necessary to run the gel longer. It is okay to allow bromophenol blue to run out of the gel. Use protein standards to measure where your protein of interest might be.
13. The membrane can be incubated for 1 h at room temperature or overnight at 4 °C. If reusing the primary antibody, it is best to use overnight at 4 °C. The blocking solution may be modified if excessive background occurs; for example, adding goat serum or BSA can decrease nonspecific binding.
14. The half-life of ^{32}P is 14.29 days. Therefore, cpm/μL should be measured on the day of usage or can be calculate using the formula:

$$\Upsilon(t) = c \times (0.5)t/14.29\text{days}$$

where $\Upsilon(t)$ equals the current cpm/μL of the probe, c equals the measured radioactivity in cpm/μL on the day of initial measurement and t equals the days that have passed since the initial reading. It is preferable to use freshly labeled probe or up to one half-life, as long as the probe is >100,000 cpm/μL.

15. It is recommended to make 10× TGE buffer and to pour gel the day before you run the EMSA. The gel should be stored by wrapping in paper towels that have been wetted with dH_2O and then wrapped in plastic wrap at 4 °C.

16. It is recommended to create a master mix containing the HEPES buffer, poly dI/dC, and BSA to make aliquoting easier and more evenly dispersed between samples.
17. Several recipes for buffers used in cnidarian NF-κB DNA--binding assays with slight variations exist in the literature [4, 6, 7, 16]. We have optimized these using *Nv* protein extracts and have found that Binding Buffer 1 (provided in the Materials) will yield the strongest signal and give the clearest supershift when antibody is added. Other binding buffer recipes are as follows, and can be tried if Binding Buffer 1 does not work for your organism of choice: Binding Buffer 2: 25 mM HEPES (pH 7.5), 100 mM NaCl, 1 mM EDTA, 1 mM DTT, 0.025% (v/v) Triton; Binding Buffer 3: 50 mM NaCl, 10 mM Tris–HCl (pH 7.4), 1 mM DTT, 1 mM EDTA, 4% (w/v) glycerol.
18. Antigen retrieval is recommended to enhance signal detection and to lower background signaling; however, antigen retrieval is not necessary for all proteins of interest (in case the antigen retrieval is not necessary, proceed to **step 8**).

Acknowledgments

The authors' research on the evolution and basal functions of NF-κB was supported by the following National Science Foundation grants (to T.D.G.): MCB-0924749, IOS-1557804, and IOS-1937650. L.M.W. was supported by an NSF Graduate Research Fellowship.

References

1. Gilmore TD (2006) Introduction to NF-κB: players, pathways, perspectives. Oncogene 25:6680–6684
2. Williams LM, Gilmore TD (2020) Looking down on NF-κB. Mol Cell Biol 40: e00104–e00120
3. Sun S-C (2011) Non-canonical NF-κB signaling pathway. Cell Res 21:71–85
4. Wolenski FS, Garbati MR, Lubinski TJ, Traylor-Knowles N, Dresselhaus E, Stefanik DJ et al (2011) Characterization of the core elements of the NF-κB signaling pathway of the sea anemone *Nematostella vectensis*. Mol Cell Biol 31:1076–1087
5. Wolenski FS, Bradham CA, Finnerty JR, Gilmore TD (2013) NF-κB is required for cnidocyte development in the sea anemone *Nematostella vectensis*. Dev Biol 373:205–215
6. Mansfield KM, Carter NM, Nguyen L, Cleves PA, Alshanbayeva A, Williams LM et al (2017) Transcription factor NF-κB is modulated by symbiotic status in a sea anemone model of cnidarian bleaching. Sci Rep 7:16025
7. Williams LM, Fuess LE, Brennan JJ, Mansfield KM, Salas-Rodriguez E, Welsh J et al (2018) A conserved Toll-like receptor-to-NF-κB signaling pathway in the endangered coral *Orbicella faveolata*. Dev Comp Immunol 79:128–136
8. Williams LM, Inge MM, Mansfield KM, Rasmussen A, Afghani J, Agrba M et al (2020) Transcription factor NF-κB in a basal metazoan, the sponge, has conserved and unique sequences, activities, and regulation. Dev Comp Immunol 104:103559
9. Siggers T, Gilmore TD, Barron B, Penvose A (2015) Characterizing the DNA binding site

specificity of NF-κB with protein-binding microarrays (PBMs). Methods Mol Biol 1280:609–630

10. Finnerty JR, Gilmore TD (2015) Methods for analyzing the evolutionary relationship of NF-κB proteins using free, web-driven bioinformatics and phylogenetic tools. Methods Mol Biol 1280:631–646
11. Kass-Simon G, Scappaticci AA Jr (2002) The behavioral and developmental physiology of nematocysts. Can J Zool 80:1772–1794
12. Brennan JJ, Messerschmidt JL, Williams LM, Matthews BJ, Reynoso M, Gilmore TD (2017) Sea anemone model has a single Toll-like receptor that can function in pathogen detection, NF-κB signal transduction, and development. Proc Natl Acad Sci U S A 114: E10122–E10131
13. Szczepanek S, Cikala M, David C (2002) Poly-γ-glutamate synthesis during formation of nematocyst capsules in *Hydra*. J Cell Sci 115:745–751
14. Stefanik DJ, Friedman LE, Finnerty JR (2013) Collection, rearing, spawning and inducing regeneration of the starlet sea anemone, *Nematostella vectensis*. Nat Protoc 8:916–923
15. Kurien BT, Scofield RH (2015) Multiple immunoblots by passive diffusion of proteins from a single SDS-PAGE gel. Methods Mol Biol 1312:77–86
16. Ramaswamy S, Hayden MS (2015) Electrophoretic mobility shift assay analysis of NF-κB DNA binding. Methods Mol Biol 1280:3–13

Part II

Methods for Studying the Activation of NF-κB Downstream of Distinct Signaling Pathways

Chapter 6

Methods for Modulating the Pathway of NF-κB Using Short Hairpin RNA (ShRNA)

Marta Moretti, Barbara Di Francesco, Mauro Di Vito Nolfi, Annapaola Angrisani, and Enrico De Smaele

Abstract

Target gene silencing is a strategy that can be used to turn off pathways or genes which are difficult to turn off pharmacologically, both because of lack of targeting drugs, or because of the risk of wider off-target effects. Here we describe the design and use of short hairpin RNA (ShRNA) and lentiviral vectors as an efficient technique for silencing NF-kappaB (NF-κB) pathway in cultured cells. This method can be used also in hard to transfect primary cell cultures.

Keywords NF-kB, RelA, ShRNA, Lentiviral vectors, Transduction, Gene silencing

1 Introduction

In order to study the biological role played by genes belonging to specific signaling pathway, is often useful to abolish expression of the genes and monitor the consequences on the cells and organisms of interest. The most common approaches to achieve this result are currently based on the principle of RNA interference, delivering into the target cells either short double-stranded (ds) interfering RNA oligonucleotides (siRNA) or short hairpin RNA (shRNA) which, after pairing with specific sequences of the target mRNA interfere with translation and generally induce mRNA degradation by the cell machinery. While cytoplasmic delivery of siRNA results only in transient gene silencing effect; nuclear delivery of gene expression cassettes that shRNA, which are processed like endogenous interfering RNA leads to stable gene downregulation [1]. The techniques for an efficient delivery of the interfering RNA sequences are constantly evolving, from standard transfection methods to the use of nanoparticles [2]. While the standard transfection is generally not very efficient, especially in sensitive cells or non-replicating cells, the use of nanoparticles is still under

Guido Franzoso and Francesca Zazzeroni (eds.), *NF-κB Transcription Factors: Methods and Protocols*, Methods in Molecular Biology, vol. 2366, https://doi.org/10.1007/978-1-0716-1669-7_6,

development, presenting several critical aspects to be solved, from the size and material of the nanoparticle, to the specificity, immunomodulation and toxicity of the delivery [3].

In the most recent years, the development of viral vectors for the delivery of shRNA is becoming a "gold standard" for efficient and stable interference of the genes of interest [4–6]. Lentiviral vectors in particular are the most efficient system for hard to transfect and non-replicating cells [7].

Here we describe the protocol that can be used to silence the NF-kappaB (NF-κB) pathway, and in particular the most relevant member of NF-κB transcription factors (RelA), which form dimers with most of the other family members and whose depletion reduces in most cases dramatically NF-κB transcriptional activity [8, 9]. Of course, identical approach could be used to silence other NF-κB members to evaluate the particular contribution of the other family members in specific contexts.

2 Materials

Prepare all solutions using deionized, distilled water.

2.1 Generation of the Lentiviral-ShRNA Vector

1. pLentilox3.7-GFP lentiviral vector (*see* **Note 1**).
2. Sense and antisense oligos designed to generate the desired shRNA (60 pmol/μl).
3. Oligos Annealing Buffer: 100 mM K-Acetate, 30 mM HEPES-KO pH 7.4 and 2 mM Mg-acetate.
4. XhoI, HpaI, and NotI restriction enzymes.
5. Alkaline phosphatase, Calf intestinal (CIP).
6. T4 DNA Ligase.
7. Phenol.
8. Chloroform.
9. Ethanol.
10. Sodium Acetate.
11. Agar plates.
12. Ampicillin.
13. LB Broth.
14. 1.5 ml polypropylene centrifuge tubes.
15. 0.2 ml PCR tubes.
16. High efficiency chemically competent bacterial cells, specifically designed for cloning unstable inserts cells (e.g., STBL-2).
17. 2% Agarose gel in Tris-Acetate-EDTA buffer (TAE).

18. Primers for the screening by PCR and identification of cloned vectors containing the desired shRNA sequence (10 μM).
19. PCR Thermal cycler.
20. Kit for extraction and purification of plasmidic DNA from bacteria (miniprep and maxiprep) (*see* **Note 2**).

2.2 Calcium-Phosphate Transfection and Lentivirus Production

1. HEK 293T cell lines (*see* **Note 3**).
2. DMEM—Dulbecco's Modified Eagle Medium completed with 10% Fetal Bovine Serum (FBS), 1% Ampicillin/Streptomycin, 1% L-Glutamine.
3. 10 cm tissue culture dishes.
4. Sterile 50 ml conical tubes.
5. 0.45 and 0.22 μm pore size filters units 250 ml PES sterile.
6. 50 ml Luer Lock Tip Syringes without needle.
7. Plasmids used:
   - Lentiviral vectors: pLL3.7-GFP-shns, pLL3.7-GFP-shRelA (*see* **Note 1**).
   - Packaging vectors for pLentiLox3.7-GFP: pCMV-VSV-G, pMDLg/pRRE, pRSV-Rev (*see* **Notes 1** and **4**).
8. $CaCl_2$ (1 M) solution: 9.4 g of $CaCl_2{\cdot}2H_2O$ (MW 147.02) in 200 ml of water. Filter sterilize through 0.22 μm nitrocellulose filter. Store at 4 °C (*see* **Note 5**).
9. BBS (2×) pH 7.00: Dissolve all the reagents in 300 ml of water add 5.3 g *N*,*N*-Bis-(2-hydroxyethyl)-2-ethanesulphonic acid (BES) (MW213.25) (0.05 M), 8.18 g NaCl (MW58.44) (0.28 M) and 0.106 g $NaHPO_4$ (MW141.96) (0.0015 M); adjust pH to 7.00 with 1 M NaOH (*see* **Note 6**). Filter sterilize through 0.22 μm nitrocellulose filter. It can be stored up to 1 year at −80 °C in 50 ml aliquots (*see* **Note 7**).
10. Arrange a vortex under the biological safety cabinet.
11. Open-Top Thin wall Polypropylene tubes, 38.5 mL volume, 25 × 89 mm.
12. Ultracentrifuge with SW 28 rotor or similar able to run up to 50,000 × *g*.
13. Sterile 1.8 ml polypropylene tubes (Cryovials).
14. Rotating wheel.

2.3 Lentiviral Transduction

1. Cells to be transduced.
2. Appropriate complete Medium for the cells.
3. 96-well tissue culture plates for virus titration.
4. 6-well tissue culture plates.

5. LentiViruses preparations subdivided in aliquots, stored at −80 °C and tested for the efficiency of transduction (*see* **Note 8**).
6. Polybrene (Hexadimethrine Bromide) (*see* **Note 9**). Dissolve polybrene with an appropriate amount of water to a final concentration of 5 mg/ml. Make 1 ml aliquots in 1.5 ml polypropylene tubes and store at −20 °C.
7. Benchtop centrifuge equipped with multiwell plate adaptors.
8. Puromycin (*see* **Note 10**).
9. Sterile Polystyrene Round Bottom Tube (FACS tube).

3 Methods

Conduct all the experiment involving lentiviral vectors in a BSL-2 facility, equipped with a Class II Biosafety Cabinet (BSC) and following institutional safety guidelines. Always use the prescribed Personal Protective equipment (PPE).

Diligently follow all waste disposal regulations when disposing waste materials.

3.1 Generation of ShRelA-Carrying Lentivirus

Generate a Lentiviral vector codifying for a shRelA by cloning a stem-loop sequence in the vector of interest (*see* **Notes 11** and **12**).

3.1.1 Design a Stem-Loop Sequence for pLenti Lox 3.7-GFP

- Select several potential 18 bases long target sequences within the gene to be silenced.
- The stem-loop sequences should be designed following the consensus sequence: AAGN_{18}TT. A 5′ guanine is required due to the constraints of the U6 promoter (as for the commercial ones *see* **Note 13**).
- Test several G N_{18} along the relA gene (*see* **Note 14**).

3.1.2 Cloning pLenti Lox3.7-GFP Vectors for shRelA

Clone the Short hairpin sequence in pLL3.7-GFP as follow:

- Resuspend the oligos containing the complementary forward (sense) and reverse (antisense) sequences of the desired shRNA in water at a final concentration of 60 pmol/μl.
- Mix 1 μl of sense oligo with 1 μl of antisense oligo in 48 μl of Annealing buffer (total volume of 50 μl), in a 200 μl thin-wall PCR tube.
- Incubate the annealing mix at 95 °C for 4 min followed by 10 min at 70 °C (*see* **Note 15**).
- Allow the tube cool down slowly (ideally, they should reach room temperature in about 45 min.

- Put the mix in ice for 10 min.
- kept at 4 °C for temporary storage.
- Sequentially digested the pLL3.7-GFP with XhoI followed by HpaI.
- Treat the digestion with CIP to remove the 5′ phosphate groups and avoid self-annealing of partially digested vectors.
- Purify all by standard phenol chloroform and Ethanol/Sodium Acetate precipitation.
- Following the concentration estimation, ligates with annealed oligos at equimolar concentration, following the ligase enzyme protocol.
- Transform the reaction in STBL-2 bacteria. Plate the bacteria on agar plates containing ampicillin.
- The next day, identify individual bacterial colonies and grow it in small volumes of LB broth.
- Extract the plasmid DNA and purify it by miniprep.
- Test the minipreps for the insertion of the stem-loop by restriction digestion (the presence of the insert causes a band shift of 60 bp in an XhoI/NotI fragment compared to the parental vector, which can be identified following electrophoresis o 2% agarose gel).
- Sequence the positive clone using the sequencing primer which anneals to the U6 promoter sequence: 5-′-cagtgcaggggaaagaatagtagac-3′.
- Once identified the correct clone, use it to prepare a DNA Maxiprep (*see* **Note 16**).

3.2 Lentivirus Production

- Day 1: Seed 2 × 10^6 293T cells in 10 cm plates with 10 ml complete DMEM.
- Day 2: Transfect cells by using Calcium-phosphate method.

1. Preparation of the transfection mix, for a 10 cm tissue culture plate mix together: 3.6 μg pCMV-VSV-G, 6.5 μg pMDLg/pRRE, 2.5 μg pRSV-Rev, 10 μg pLL3.7-GFP-shns or pLL3.7-GFP-shRelA (*see* **Note 17**).
2. Add 375 μl sterilized water, followed by 125 μl $CaCl_2$ (1 M).
3. Place the 50 ml tube on a vortex and vortexing continuously (*see* **Note 18**) add drop by drop 500 μl of BBS2X pH 6.95 (*see* **Note 19**).
4. Let the solution at RT for 20 min, leaving the cap unscrewed to allow the passage of air.

5. In the meantime, replace the culture media of the cells with 9 ml of complete DMEM.
6. Add 1 ml of the transfection mix into each plate dropwise.
7. Place the cells overnight (O/N) in the CO_2 incubator at 37 °C (*see* **Note 20**).

- Day 3: Following 16 h, Change the media to the cells by adding 10 ml of complete DMEM (*see* **Note 21**).
- Day 4: 24 h after the medium change.

1. Harvest all the medium from the dishes and replaced it with 10 ml of fresh completed DMEM and place again the dishes in the incubator for another O/N at 37 °C with 5% CO_2.
2. Filter the medium with 50 ml syringe using a PVDF 0.45 μm filter and place it in a polypropylene tube.
3. Centrifuge at 4 °C, 50,000 × *g* for 1 h 30 min.
4. Discard the Supernatant (SN) and resuspend the pellet with 8 ml of complete DMEM (*see* **Note 22**).
5. Store it at 4 °C.

- Day 5: 24 h from the first collection.

1. Harvest the medium as in day 4.
2. Pool together, in polypropylene centrifuge tubes, the media collected in day 4 and day 5.
3. Ultracentrifuge for 1 h 30 min at 4 °C, 50,000 × *g* in order to precipitate the viral particles.
4. Discard the SN and resuspend the viral pellet in complete DMEM (*see* **Note 23**) and transfer it to a cryovials (*see* **Note 24**).
5. Resuspend the precipitated virus by rotating wheel for 1 h at 4 °C (*see* **Note 25**).
6. Aliquot and store it at −80 °C (*see* **Note 26**).
7. In order to check for transfection efficiency, detach and collect the transfected cells from the plate and analyze for GFP expression by flow cytometry (FACS) (*see* **Note 27**).

3.3 Virus Titration

- Day 1: Seed the appropriate number of cells in a 96-well tissue culture plate (*see* **Note 28**).
- Day 2: 24 h later, transduce the cells with different concentrations of viral stock.

1. Gently remove the culture medium from the wells containing the cells and replace it with 70 μl of serum free media in each well.

2. Number the wells from 1 to 5.
3. For the first viral concentration add 30 μl of the original viral stock to the first well. Then, after pipetting carefully the medium in the well a couple of time, transfer 30 μl from the first well to the second well, generating a further diluted viral solution. Continue in the same way to generate further dilutions (generally up to five dilutions).
4. Centrifuge the plate in a benchtop centrifuge with biocontainment caps at 1100 × *g*, RT for 1 h.
5. Add 130 μl of growth medium to each well (100 μl in the last well) (final volume 200 μl/well).
6. Incubate the plate in the CO_2 incubator at 37 °C.

- Day 3: Following 16 h of incubation, replace the medium with 200 μl of fresh medium and continue to incubate the cells at 37 °C.
- Day 5: 72 h later.

1. Count the GFP expressing cells in the wells by using an inverted florescence microscopy.
2. Calculate the Transducing Units per ml (TU/ml) for each cell line using the following formula: # of GFP positive cells counted × dilution factor × 33.3 (where 33.3 = 1000/30) = #TU/ml (*see* **Note 29**).

3.4 Lentiviral Transduction of Mammalian Cells

Cells infection can be started once the virus has been titrated.

- Day 1: Seed the cells the day before in six wells plate. The number of cells to be plated needs to be assessed for any different cell line and should be sufficient to have cells still growing exponentially the next day (cells in proliferation have better transduction efficiency) (*see* **Note 30**).
- Day 2: prepare a solution of Polybrene 3.2 μl/ml in sterile culture medium (RPMI) (prepare 4 ml for each well) (*see* **Note 31**).

1. Thaw the lentivirus on ice.
2. Mix the chosen amount of virus, according to the TU/ml and the desired multiplicity of infection (MOI) (*see* **Note 32**), to the polybrene/medium solution.
3. Add the mix to each well and centrifuge the six wells plate for 1 h in a benchtop centrifuge with containment cap at 1100 × *g*, 32 °C (*see* **Note 33**).
4. Incubate the cells at standard cell culture conditions.

– Day 3: After 20–24 h change the medium with fresh medium. The fluorescence will be visible after the following 2–3 days by fluorescence microscopy or by FACS (*see* **Note 34**).

3.5 Selection of Stably Infected Clones

In case of GFP expressing vectors, infected cells can be detached from the culture plate, and selected by use of FACS sorting (*see* **Note 35**). The GFP positive subpopulation can then be replated, thus achieving a nearly 100% infected cell population (*see* **Note 36**).

3.6 Identification of the Cells Which Present Efficient Gene Silencing

The effective silencing of the target gene is verifiable by Western Blot analysis as shown in Fig. 1 (*see* **Note 37**). The silenced cells can be used immediately for further experiments or expanded in culture for few days in larger wells and then aliquoted and frozen in sterile 10% DMSO FBS solution in liquid nitrogen tank (*see* **Note 38**).

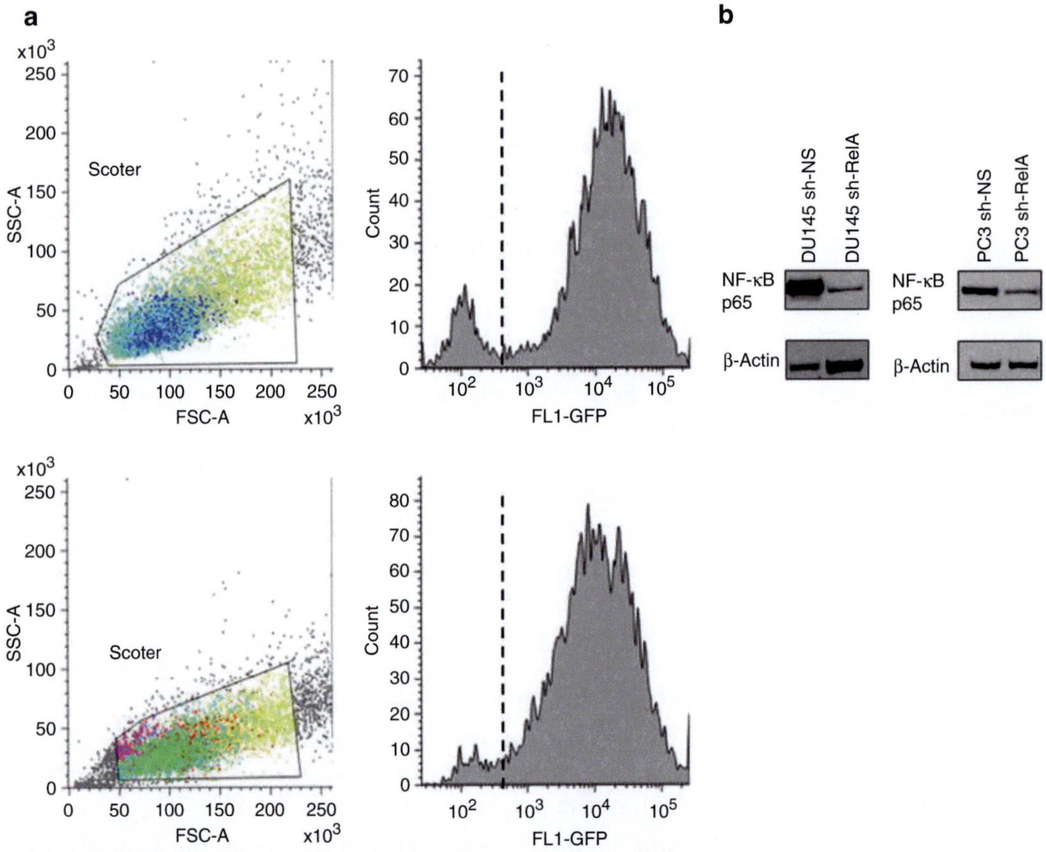

Fig. 1 Evaluation of efficiency of transduction and p65 silencing. (**a**) FACS analysis and cell sorting was performed 72 h post transduction. Here we show dot plot (left) and histogram (right) of DU145 (upper) and PC3 (down) prostate cancer cells. (**b**) Silencing efficiency in DU145 and PC3 cells transduced with pLL3.7-GFP-shns and pLL3.7-GFP-shRelA was evaluated by assessing NF-κB p65 expression level by western blot analysis. β-actin is shown as loading control

4 Notes

1. pLL3.7 was a gift from Luk Parijs (Addgene plasmid # 11795; http://n2t.net/addgene:11795; RRID: Addgene_11795).

   pCMV-VSV-G: VSV-G- envelope expressing plasmid. pCMV-VSV-G was a gift from Bob Weinberg (Addgene plasmid # 8454; http://n2t.net/addgene:8454; RRID: Addgene_8454).

   pMDLg/pRRE: Packaging vector contains Gag and Pol. pMDLg/pRRE was a gift from Didier Trono (Addgene plasmid # 12251; http://n2t.net/addgene:12251; RRID: Addgene_12251).

   pRSV-Rev: Packaging vector contains Rev. pRSV-Rev was a gift from Didier Trono (Addgene plasmid # 12253; http://n2t.net/addgene:12253; RRID:Addgene_12253).
2. It is recommended to use a kit containing RNAse A, to get rid of the RNA contamination.
3. Keep in mind that the 293T should be in exponential growth and at 70% confluence on the day of transfection.
4. Alternatively, the pLL3.7 vectors can be substituted by analogous lentiviral vectors, such as pLKO.1-puro-shCtr and pLKO.1-puro-shRelA, and the corresponding envelope and packaging vectors: pMD2.G, pCMV-dR8.2 (or psPAX2).
5. Make $CaCl_2$ fresh each time. It is possible to make aliquots and store it at −20 °C for a longer period.
6. It is critical that the pH be adjusted accurately; below 6.95, the precipitate will not form; above 7.05, the precipitate will be coarse, and the transfection efficiency will be low. For this reason, it is recommended to prepare BBS (2×) at several pH, between 6.95 up to 7.04 and then test it for the best transfection.
7. It is possible to freeze and thaw a tube multiple time.
8. Freeze and thawing the virus more than halves the viability of the virus and its ability to infect cells. Avoid freezing and thawing by making small aliquots of the virus and thaw them only once at the time of the transduction experiment.
9. Polybrene is also known as Hexadimethrine Bromide. It is a cationic polymer that can greatly enhance the efficiency of the lentiviral infection to the mammalian cells. Since this powder is really hydrophilic, it is recommended to resuspend the entire content of the bottle, otherwise the weight could be altered. The amount of polybrene to be used for each cell lines should be carefully assessed, since it can be toxic for sensitive cells.

10. Suggested initial Puromycin concentration 10 mg/ml. The optimal concentration should be assessed for any different cell line. Storage condition −5 to −20 °C up to 12 months. Puromycin can be used if the Lentivirus have an antibiotic resistance.
11. Lentivirus vectors are based on the Human Immunodeficiency Virus (HIV) which is the virus responsible for the development of Acquired Immunodeficiency Syndrome (AIDS). Lentiviruses are a subclass of retroviruses which are able to infect both proliferating and non-proliferating cells. These Lentivirus vector have been modified to provide a safer version of the HIV virus in which the viral replication genes have been removed. The major risks to be considered for research with HIV-1 based lentivirus vectors are: potential for generation of replication-component lentivirus (RCL) and potential for oncogenesis. Lentivirus may be transmitted by: Penetration of the skin via puncture or absorption (though scratches, cuts, abrasions, dermatitis or other lesions). Mucous membrane exposure of the eyes, nose, and mouth (through direct contact or aerosol).
12. In our case we choose the vector pLL3.7-GFP to clone in the short hairpins of our interest and generate a pLL3.7-GFP shRelA. There are also several companies which sell commercial Lentivirus codifying for shRelA (e.g., pLKO.1-puro). We suggest testing at least 3–5 different target sequences, in different location along the sequence of the gene of interest.
13. It is possible to obtain the SHC001 pLKO.1-puro (empty vector) and follow the pLKO.1—TRC protocol reported in www.addgene.org/protocols/plko/.
14. From all the GN_{18} sequences tested we found that the best working for our cell lines was a sequence corresponding to relA sequence 538–556 bp which correspond to:

    5′-GCATCCAGACCAACAACAA-3′.

    Read also [10–12].
15. Heating allows the breaking of the hydrogen bonds and elimination of secondary structures. Slow cooling allows the formation of new bonds between complementary sequences.
16. Read also [7].
17. For pLKO.1 we suggest to use: 10 μg pMD2.G; 12 μg pCMV-dR8.2 (or psPAX2) and 20 μg pLKO.1-puro-shCtr or pLKO.1-shRelA.
18. It is important to do not stop the vortex because the air bubbles produced by the vortex is fundamental for a proper reaction formation. If you do not have a vortex in BL3, it is possible to produce bubbles using a mechanical pipettor.

19. The BBS2X pH depends on the best previously tested via transfection (as reported in material and methods and in **Note 4**).
20. Do not leave the precipitate longer than 14/16 h.
21. Do not wash the cells with PBS. Do not let the cells detach from the plate.
22. Make a first wash with 4 ml medium and then wash again with other 4 ml and poll the resuspension together.
23. We suggest 850 μl for 4× 10 cm tissue culture dishes.
24. If cryovials are not available, transfer into sterile 1.5 ml tubes.
25. DO NOT PIPET to resuspend.
26. Make aliquots not more than 200 μl.
27. For FACS Analysis to pellet the cells by centrifuge at 600 × *g* 5 min at RT. Discard the SN and wash the pellet with 1 ml of PBS. Centrifuge again at 600 × *g* 5 min at RT. Discard the SN and resuspend the pellet with 1 ml of PBS. Remind to use non-transduced 293T cell as negative control. Then proceed to FACS acquisition.
28. For human prostate cancer cell lines DU-145 and PC-3 cells we seeded 16×10^4 cells/well in 100 μl of growth medium.
29. Alternatively, it is also possible to purchase a commercial virus titration kit from several companies.
30. Usually for DU-145 and PC-3 we seed 4.5×10^5 cells per well.
31. Polybrene acts to neutralize the charge repulsion between virions and the cell surface, thereby increasing infection efficiency. Not all the cells need Polybrene. It is important to test the suitable amount able to enhance the infection without kill the cells.
32. Multiplicity of infection (MOI) refers to the ratio of the number of virus particles to the number of target cells present in a defined space. It is related to #TU by the following formula: MOI = #TU of virus used for infection/number of cells. Theoretically a MOI of 1 (1 infectious viral particle per cell) is required to achieve a 63.2% of infection, that means that to achieve a 99% of infection is required a MOI of 4.6.
33. This step is essential for suspension cells, however, for some adherent cells (as DU-145 and PC-3) it can improve the success of infection.
34. Florescence intensities of stable transduced cells have to be monitored daily under the microscope. Usually, the maximal intensity of fluorescence signals is reached 3–6 days after transduction.

35. For FACS sorter analysis, tripsinize the cell 72 h after transduction, centrifuge at 800 × *g* for 5 min RT, discard the SN and wash the pellet with PBS. Centrifuge at 800 × *g* 5 min at 4 °C, discard the SN and resuspend the cell pellet in 1 ml of PBS and place it in FACS tube for the sorting. Remind to use non-transduced cells for instrument setting and as negative control. Performed the sorting for PLL3.7-GFP selecting the FITC Channel (488 nm); Follow culturing the positive cells in growth medium.
36. Alternatively, if the lentiviral vector encodes for an antibiotic resistance gene, it is possible to select stable cells by use of antibiotics. In case of PLKO.1 Puromycin (concentration) can be added to the culture medium. Typically, the final concentration ranging from 0.1 to 10 μg/ml (it depends on cell lines). Puromycin quickly kills the eukaryotic cells that do not contain the pac gene. Dying cells detach from plates allowing easy identification of transformant clones. Puromycin is normally used 24 h after transduction. Replace culture medium 48–72 h after the transduction process with fresh puromycin-containing medium every 3–4 days until resistant colonies can be identified. After about 7 days growing cells surviving selection of will start to form visible foci. It may require additional 7 or more days to fully develop foci (it depends on the host cell line). After that you can consider having a stable shRelA cell line. For best results, puromycin needs to be titrated on non-transduced cells.
37. It is also possible to verify the silencing by RT-Q-PCR to double check the absence of the total mRNA.
38. We suggest verifying the silencing by WB analysis every vials of cells when are thaw and to double check every week. It is not suggested to maintain in culture the cells for more than 3 weeks, since with time cells may change their characteristics. In any case effective silencing need to be verified periodically.

Acknowledgments

This work was supported by Sapienza University research grants to E.D.S and A. A.

References

1. Sandy P, Ventura A, Jacks T (2005) Mammalian RNAi: a practical guide. BioTechniques 39 (2):215–224
2. Yonezawa S, Koide H, Asai T (2020) Recent advances in siRNA delivery mediated by lipid-based nanoparticles. Adv Drug Deliv Rev 2020:S0169-409X(20)30106-X
3. Filion MC, Phillips NC (1997) Toxicity and immunomodulatory activity of liposomal vectors formulated with cationic lipids toward immune effector cells. Biochim Biophys Acta Biomembr. https://doi.org/10.1016/S0005-2736(97)00126-0

4. Sliva K, Schnierle BS (2010) Selective gene silencing by viral delivery of short hairpin RNA. Virol J 7:248
5. Paddison P, Caudy AA, Bernstein E, Hannon GJ, Conklin DS (2002) Short hairpin RNAs (shRNAs) induce sequence specific silencing in mammalian cells. Genes Dev 16:948–958
6. Paddison PJ, Caudy AA, Hannon GJ (2002) Stable suppression of gene expression by RNAi in mammalian cells. Proc Natl Acad Sci 99:1443–1448
7. Rubinson DA, Dillon CP, Kwiatkowski AV et al (2003) A lentivirus-based system to functionally silence genes in primary mammalian cells, stem cells and transgenic mice by RNA interference. Nat Genet 33(3):401–406
8. Mauro C, Leow SC, Anso E et al (2011) NF-κB controls energy homeostasis and metabolic adaptation by upregulating mitochondrial respiration. Nat Cell Biol 13 (10):1272–1279
9. Yang J, Feng S, Yi G et al (2016) Inhibition of RelA expression via RNA interference induces immune tolerance in a rat keratoplasty model. Mol Immunol 73:88–97
10. Moore CB, Guthrie EH, Huang MT, Taxman DJ (2010) Short hairpin RNA (shRNA): design, delivery, and assessment of gene knockdown. Methods Mol Biol 629:141–158
11. Paddison P, Cleary M, Silva J et al (2004) Cloning of short hairpin RNAs for gene knockdown in mammalian cells. Nat Methods 1:163–167
12. Knott SRV, Maceli A, Erard N et al (2014) A computational algorithm to predict shRNA potency. Mol Cell 56(6):796–807

Chapter 7

Immunoblot Analysis of the Regulation of TNF Receptor Family-Induced NF-κB Signaling by c-IAP Proteins

Eugene Varfolomeev, Tatiana Goncharov, and Domagoj Vucic

Abstract

Proper maintenance of organismal homeostasis, development, and immune defense requires precise regulation of survival and signaling pathways. Inhibitor of apoptosis (IAP) proteins are evolutionarily conserved regulators of cell death and immune signaling that impact numerous cellular processes. Although initially characterized as inhibitors of apoptosis, the ubiquitin ligase activity of IAP proteins is critical for modulating various signaling pathways (e.g., NF-κB, MAPK) and cell survival. Cellular IAP1 and 2 regulate the pro-survival canonical NF-κB pathway by ubiquitinating RIP1 and themselves thus enabling recruitment of kinase (IKK) and E3 ligase (LUBAC) complexes. On the other hand, c-IAP1 and c-IAP2 are negative regulators of noncanonical NF-κB signaling by promoting ubiquitination and consequent proteasomal degradation of the NF-κB-inducing kinase NIK. Here we describe the involvement of c-IAP1 and c-IAP2 in NF-κB signaling and provide detailed methodology for examining functional roles of c-IAPs in these pathways.

Key words IAP, Inhibitor of apoptosis, NF-kB, IAP antagonist, TNF, c-IAP1/2, RING domain, Ubiquitin, RIP1, NIK, TRAF2, Proteasomal degradation

1 Introduction

1.1 Regulation of NF-κB Signaling by c-IAP Proteins

The inhibitor of apoptosis (IAP) family of evolutionarily conserved proteins contains structurally related regulators of diverse cellular processes [1]. Among the human IAP proteins, cellular IAP1 and 2 (c-IAP1 and c-IAP2) and X chromosome-linked IAP (XIAP) are probably the most-studied, although other IAP proteins (NAIP, ML-IAP, survivin, ILP2, and Apollon) also play important roles in cell survival, cell cycle, inflammation and overall homeostasis. IAP proteins contain one to three copies of a signature baculovirus IAP repeat (BIR) domain that mediates protein-protein interactions that are necessary for IAP function [2]. XIAP, c-IAP1, c-IAP2, ML-IAP and ILP2 also contain a carboxy-terminal really interesting new gene (RING) domain, which gives them ubiquitin ligase activity [3]. Ubiquitination is a posttranslational modification that

Guido Franzoso and Francesca Zazzeroni (eds.), *NF-κB Transcription Factors: Methods and Protocols*, Methods in Molecular Biology, vol. 2366, https://doi.org/10.1007/978-1-0716-1669-7_7,

involves covalent attachment of 76-amino acid protein ubiquitin to the substrate protein, and requires the enzymatic activity of a ubiquitin-activating enzyme (E1), a ubiquitin-conjugating enzyme (E2), and a ubiquitin ligase (E3) [4]. Attachment of a single ubiquitin molecule to a lysine residue of the substrate protein generates monoubiquitination [4]. However, since ubiquitin contains seven lysine residues and a free amino-terminus, polyubiquitin chains can be synthesized through eight different isopeptide linkages [5]. Lysine 48-linked polyubiquitin chains predominantly target proteins for proteasomal degradation, whereas lysine 63-, amino-terminal methionine- and, in some cases, lysine 11-linked chains provide a scaffolding platform for the assembly of signaling complexes [4, 6].

Even though IAP proteins were initially characterized as anti-apoptotic factors, the regulation of cell survival and homeostasis by IAP proteins is not limited to cell death pathways. A number of studies conducted in recent years have established IAP proteins as important regulators of MAPK, and in particular, NF-κB signaling pathways [7–11]. The NF-κB family of transcription factors regulates expression of numerous genes involved in cell survival, inflammation and immunity [12]. NF-κB family transcription factors NF-κB1 (p105/p50), NF-κB2 (p100/p52), RelA (p65), RelB and c-Rel act as homodimers or heterodimers whose activation is regulated by phosphorylation and ubiquitination [13]. In canonical NF-κB signaling, the inhibitor of κB (IκB) binds NF-κB proteins RelA and p50 in the cytoplasm to block them from entering the nucleus in unstimulated cells [13]. Binding of TNF to TNFR1 triggers the recruitment of the proximal receptor-associated complex that includes TRADD, RIP1, TRAF2, and TRAF2-associated c-IAP1 and c-IAP2 proteins [10, 14] (Fig.1). The assembly of the receptor-associated complex results in c-IAP-mediated ubiquitination of RIP1, TRAF2, and the c-IAPs themselves. This promotes the recruitment of the linear ubiquitin chain assembly complex, LUBAC, transforming growth factor β-activating kinase 1 (TAK1), the TAK1-TAB2/3 (TAK1-binding protein 2/3) complex and IκB kinase (IKK) complex [15].

IKKγ or NEMO (NF-κB essential modifier) in the IKK complex and TAB2/3 in the TAK1-TAB2/3 complex bind linear and K63-linked polyubiquitin chains on RIP1, which brings the kinase TAK1 into proximity with IKKβ, thus allowing phosphorylation of IKKβ [16]. Subsequent phosphorylation of IκB by IKKβ is recognized by the E3 ligase complex $SCF^{\beta\text{-}TRCP}$, which promotes IκB ubiquitination and degradation [16]. Autoubiquitination of c-IAP1/2 proteins with K63-linked chains enables the recruitment of LUBAC, which assembles linear polyubiquitin chains on NEMO and RIP1 [15, 17]. LUBAC is comprised of two regulatory components: heme-oxidized IRP2 ubiquitin ligase 1 homolog (HOIL-1L) and SHANK-associated RH domain interactor (SHARPIN),

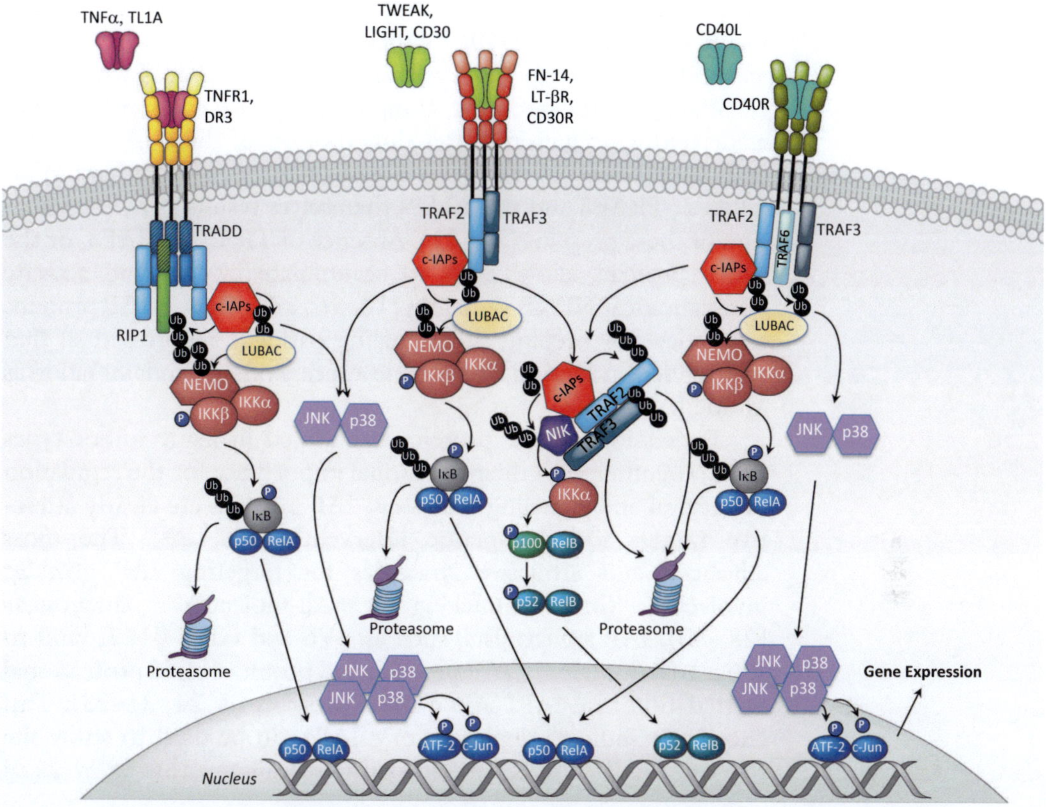

Fig. 1 Activation of NF-κB and MAP kinase pathways by TNF family ligands

and the E3 ligase HOIL-1-interacting protein (HOIP) [18–20]. Linear polyubiquitin chains stabilize TNFR-associated signaling complexes, whereas a reduction in the amount of LUBAC diminishes NF-κB signaling [21]. Besides the TNFR1-associated complex, the ubiquitin ligase activity of c-IAP1 and 2 is critical for linking a number of TNF family receptors to the distal kinase and E3 ligase complexes IKK, TAK1/TAB2/3 and LUBAC [10]. DR3, FN14, LT-βR, CD40 and CD30 all rely on the adaptor protein TRAF2 and the E3 ligases c-IAP1/2 to stimulate canonical NF-κB signaling [7, 8, 10, 11, 22].

Although they act as positive regulators of the canonical NF-κB pathway, c-IAP proteins are crucial negative regulators of noncanonical NF-κB signaling [11]. The NF-κB-inducing kinase, NIK, initiates noncanonical signaling by phosphorylating IKKα, which leads to phosphorylation and proteasomal processing of p100 to p52 [23]. NIK is present at low levels in unstimulated cells because it is ubiquitinated constitutively by c-IAP1 and 2 proteins. This posttranslational modification targets NIK for proteasomal degradation [24]. A cytoplasmic complex comprising the adaptor proteins TRAF2 and TRAF3 links the E3 ligases c-IAP1/2 to NIK

[10, 25]. Activation of a number of TNF family receptors (including FN14, LT-βR, CD40) by their respective ligands or agonistic antibodies leads to the recruitment of TRAF2, TRAF3 and c-IAP1/2 to the receptor complex [10, 25]. This membrane-associated aggregation causes dimerization of the c-IAP proteins and stimulation of their E3 ligase activity. Ubiquitination of TRAF2, TRAF3 and the c-IAPs themselves results in the degradation of these proteins [10, 22]. Absence of TRAF2, TRAF3, or the c-IAP proteins allows NIK to accumulate in cells and activate non-canonical NF-κB signaling [10, 26, 27]. Thus, c-IAP proteins are critical for keeping this signaling pathway suppressed so that unintended induction of cytokine expression and inflammation is avoided.

Expression of IAP proteins is elevated in many tumor types [28]. Together with their functional importance for the regulation of survival and signaling pathways, IAP proteins are clearly attractive targets for therapeutic intervention [3, 29]. The most advanced and attractive strategies for targeting IAP proteins involves SMAC-mimicking small-molecule antagonists [28, 30]. IAP antagonists, such as BV6 and GDC-0152, bind to select BIR domains of IAP proteins and promote rapid proteasomal degradation of c-IAP1 and c-IAP2 proteins [2, 24, 31–33]. This chemically induced depletion of c-IAPs can be used to study the role of c-IAP1/2 in NF-κB and other signaling pathways.

2 Materials

2.1 Equipment

1. PowerPac™ HC Power Supply (Bio-Rad).
2. XCell4 SureLock Midi-Cell Electrophoresis system (Thermo Fisher Scientific).
3. Criterion™ Blotter with Plate Electrodes (Bio-Rad).
4. Centrifuges.
5. Cell Lifter (Thermo Fisher Scientific).
6. Q55 Sonicator (QSonica LLC).
7. DynaMag2 magnet, Thermo Fisher Scientific.
8. X-ray cassettes.
9. Orbital shaker.

2.2 Reagents

1. Cells: HT1080 human fibrosarcoma cells, HT-29 and SW620 human colorectal adenocarcinoma cells, Ku812F human chronic myelogenous leukemia cells, Daudi and Ramos human Burkitt's lymphoma (ATCC).
2. Dulbecco's modified Eagle's and RPMI supplemented with 10% FBS, 150 U/mL penicillin, 200 U/mL streptomycin

Table 1
Sensitivity of different cell lines to the TNF family ligands

	TNF	CD40L	LIGHT	TWEAK	TL1A
Daudi	+	+			
HT1080	+		+	+	
HT-29	+		+	+	
Ku812F	+				+
Ramos	+	+			
SW620	+		+	+	

[a]All of these ligands activate p38, JNK, and ERK MAP kinases and canonical NF-κB signaling. In addition, CD40L, LIGHT and TWEAK induce noncanonical NF-κB signaling

and 2 mM L-glutamine. The sensitivity of these cell lines to the TNF family ligands is indicated in Table 1.

3. Recombinant proteins and antibodies for cells stimulation and immunoprecipitations: TNF (Enzo Life Sci., R&D Systems), CD40 (R&D Systems), LIGHT (R&D Systems), TWEAK (R&D Systems), TL1A (R&D Systems).
4. Antibodies to crosslink recombinant ligands and CD40 agonistic antibody: anti-His (Thermo Fisher Scientific), anti-Flag (Sigma Aldrich), and anti-mouse IgG 2b specific (Jackson Immunoresearch).
5. Lipofectamine RNAiMAX reagent (Thermo Fisher Scientific).
6. OPTI-MEM (Gibco).
7. 2 mM IAP antagonist BV6 [24] in DMSO.
8. 20 mM Proteasome inhibitor MG-132 (1000×) (Calbiochem) in DMSO.
9. 20 mM Lysosome inhibitor CA-074Me (1000×) (Calbiochem) in DMSO
10. 1.5 and 50 ml test tubes.
11. 10 or 15 cm dishes.
12. 225 cm^2 flasks.
13. 6 and 24 well plates.
14. CellSTACK Culture Chambers (Corning).
15. Triton Lysis Buffer (TLB): 20 mM Tris–HCl, pH 7.5, 150 mM NaCl, 4 mM EDTA, 1% Triton X-100 supplemented with protease and phosphatase inhibitor cocktail (Thermo Fisher Scientific).
16. Triton Immunoprecipitation Buffer (TIP): TLB supplemented with 10% glycerol.

17. SDS Lysis Buffer (SLB): TLB supplemented with 1% SDS.
18. Ubiquitin Binding Buffer (UBB): 20 mM Tris–HCl pH 7.5, 135 mM NaCl, 1.5 mM $MgCl_2$, 1 mM EGTA, 1% Triton X-100, 20 μM MG132, 4 mM NEM, 20 mM iodoacetomide and protease inhibitor cocktail (Roche).
19. Urea-UBB buffer: 6 M urea in UBB Buffer.
20. Cell Dissociation Buffer, Enzyme-Free (Gibco).
21. 20 μM stock concentrations sense siRNA oligos for *c-IAP1* and/or *c-IAP2* transfection experiments at: *c-IAP1*—<hcIAP112S> *UCGCAAUGAUGAUGUCAAA tt*; <hcIAP113S> *GAAUGAAAGGCCAAGAGUU tt*; *c-IAP2*—<hc25S> *UCTAACACAAGAUCAUUGA tt*; <hc29S> *AUUCGGUACAGUUCACAUGtt* [10].
22. 20 μM stock concentrations sense siRNA oligos for *TRAF2* transfection experiments: <HT2SIS7>—*CGACGTGACTTCATCCTCT tt;* <HT2SIAS7>—*AGAGGATGAAGTCACGTCGtt* [10].
23. Halt protease and phosphatase inhibitor cocktail (Thermo Fisher Scientific).
24. Pierce™ Protein A/G magnetic beads (Thermo Fisher Scientific).
25. Pierce™ Protein A/G agarose beads (Thermo Fisher Scientific).
26. BCA Protein Assay Kit (Thermo Fisher Scientific).

2.2.1 Materials for Western Blotting

1. 4× XT Sample Buffer(Bio-Rad).
2. 20× XT Reducing Agent (Bio-Rad).
3. 20× XT MOPS Running Buffer (Bio-Rad).
4. NuPAGE, 4–12% Bis-Tris polyacrylamide gel (Thermo Fisher Scientific).
5. Nitrocellulose/Filter Paper Sandwiches (Thermo Fisher Scientific).
6. 20× NuPAGE transfer buffer (Thermo Fisher Scientific).
7. Methanol.
8. Protein Marker.
9. Western Lightning Plus-ECL kit (Perkin-Elmer).
10. PBS-T buffer: PBS supplemented with 0.05% Tween 20.
11. BLOTTO (membrane blocking solution): PBS-T with 5% non-fat dry milk.
12. PBS-T BSA (membrane blocking solution for detection of phosphorylated proteins): PBS-T with 5% Bovine Serum Albumin (BSA).

2.2.2 Antibodies

1. Primary antibodies for Western Blotting (*see* Table 2).
2. HRP-conjugated secondary antibodies (Jackson Immunoresearch): anti-mouse IgG1 (1:10,000); anti-mouse IgG2a (1:5000); anti-rabbit (1:5000); anti-goat (1:10,000); anti-rat (1:10,000), streptavidin (1:1000). Reconstitute the antibodies with 0.5 ml of PBS, mixed with 0.5 ml of 100% glycerol.

3 Methods

3.1 Western Blot-Based Assessment of Activation of MAPKs and NF-κB Signaling Pathways by TNF Ligands

Downregulation of c-IAP1/2 and/or TRAF2 expression by siRNA knock downs is a powerful genetic tool to access the function of these molecules in TNF ligand signaling pathways.

3.1.1 siRNA Transfection

1. Rinse the cells once with PBS and trypsinize them for 3–5 min. Collect the cells and neutralize the trypsin with 20 ml of growth media without antibiotics. Count the cells and plate an appropriate number for transfection (*see* **Note 1**).
2. The next morning, check the confluence of the seeded cells (*see* **Note 1**). If the cells are too sparse, then one can wait for 8–24 h before proceeding with the transfection. However, if the cells are more than 80% confluent or grown in clumps, repeat the seeding with fewer cells.
3. Place 1 ml of OPTI-MEM in each marked 15 ml tube. Add 30 μl of stock siRNA duplexes. Mix RNAiMAX reagent with OPTI-MEM in a separate tube; use 30 μl of reagent and 1 ml of medium for each transfection. Add 1 ml of diluted RNAiMAX to each siRNA and gently mix. Incubate tubes for 10 min at room temperature and then add to the cells in 10 cm dishes. Swirl the dishes gently.
4. Grow transfected cells for 48 h before treating them with TNF ligands.

3.1.2 Cell Growth and BV6 Treatment

1. Grow an appropriate number of cells to semi-confluence (*see* **Note 2**).
2. Treat the cells with 1 μM BV6 for 4–12 h or DMSO as a control. Chemical elimination of c-IAP1/2 proteins is used as alternative method to siRNA knockdown technique. It is particularly useful in the case of immunoprecipitation of protein complexes from high number of cells.

3.1.3 Treatment of Cells with TNF Ligands or Agonistic Anti-TNF Receptor Antibodies

1. In the case of adherent cells, rinse the cells once with warm PBS and then apply warm Cell Dissociation buffer (*see* **Note 3**). When cells are grown in suspension, go directly to **step 3**. Incubate the cells at 37 °C for 5–10 min, periodically monitoring cell detachment under the microscope.

Table 2
List of primary antibodies used for Western Blotting

	Dilution/concentration	Vendor/source
p-IκBα	1:1000	Cell Signaling
IκBα	1:1000	Cell Signaling
p-p38	1:1000	Cell Signaling
p38	1:1000	Cell Signaling
p-JNK	1:2000	BD Biosciences
JNK	1:1000	Cell Signaling
Pan-cadherin	1:1000	Cell Signaling
HSP90	1:1000	Cell Signaling
SP1	1:1000	Cell Signaling
FN14	1:1000	Cell Signaling
tubulin	1:1000	Cell Signaling
RIP1	1:1000	BD Biosciences
TRADD	1:1000	BD Biosciences
TRAF2	1:1000	BD Biosciences
NEMO	1:1000	BD Biosciences
TRAF3	1:1000	Thermo Fisher Sci.
TRAF6	1:1000	Sigma Millipore
c-IAP1	1:1000	R&D Systems
IKK2	1:1000	R&D Systems
TAK1	1:1000	R&D Systems
Lymphotoxin βR	1:500	R&D Systems
CD40	1:500	R&D Systems
DR3	1:500	R&D Systems
c-IAP2	1:1000	Abcam
HOIP	1:1000	Novus
actin	1:5000	Sigma
K11-ub chain	2.5μg/ml	Genentech
K48-ub chain	2.5μg/ml	Genentech
K63-ub chain	2.5μg/ml	Genentech
Lin-ub chain	2.5μg/ml	Genentech

2. Transfer the cells into 50 ml conical tubes. Wash the dishes once with 10 ml of growth media and add it to the collected cells.
3. Centrifuge the cells for 5 min at 200 × *g*. Resuspend the cells in the appropriate growth media and seed them into 24 or 6 well dishes (*see* **Note 4**).
4. Prepare treatment reagents in serum-free media (*see* **Note 5**). The final concentration of TNFα should be 20 ng/ml; 100 ng/ml of other cytokines and 200 ng/ml of anti-CD40 agonistic antibodies. Before addition to the media, CD40L, LIGHT and TWEAK should be cross-linked with an anti-His antibody. Anti-CD40 antibody should be cross-linked with IgG2b. Mix the ligands with 2× amount of anti-His or IgG2b antibodies, and perform crosslinking at room temperature for 5–10 min in a minimal volume (20–100 μl).
5. Treat the cells for 5, 15, 30, and 60 min, or 12–24 h in the case of non-canonical NF-κB activation (*see* **Note 6**).
6. Detach the cells by pipetting (in the case of a short treatment) or with a cell scraper (for example, if adherent cells were treated for several hours). Transfer the cells into 1.5 ml test tubes, add 0.5 ml of cold PBS to the wells, and then transfer the remaining cells. Spin cells at 1000 × *g* for 3 min at 4 °C and discard all liquid.

3.1.4 Cell Lysis and Protein Sample Preparation

1. Lyse the cells in TLB (*see* **Note 7**).
2. Incubate the lysates for 20–30 min on ice.
3. Spin the lysates at 14,000 × *g* for 10 min at 4 °C.
4. Transfer supernatants to new 1.5 ml tubes.
5. Determine the protein concentration using BCA Protein Assay Reagent following the instruction manual.
6. Prepare protein samples for SDS-PAGE: in a new 1.5 ml tube, combine 1/20 of the final volume XT reducing agent, 1/4 XT sample buffer, appropriate amounts of protein lysates and TLB (*see* **Note 8**).
7. Heat the samples for 5 min at 90–95 °C and then spin for a short time (20–30 s) to collect the samples at the bottom of the tube.

3.1.5 SDS-PAGE and Membrane Transfer

1. Assemble gel units.
2. Prepare 1× MOPS XT running buffer (about 800 ml per one full unit). Use 26-well gels if the samples contain 1.5 μg/μl protein or less. Otherwise, use 20-well gels because they typically give better sample resolution.
3. Load 6 μl of protein standard and 12–15 μl of the protein samples on the gel.

4. Run the samples at 200 V for 1 h.
5. Prepare 1× transfer buffer supplemented with 20% methanol (about 1 l/U, *see* **Note 9**).
6. Assemble transfer chamber.
7. Perform the protein transfer at 4 °C 120 V for 90 min.

3.1.6 Immunoblotting

1. When the transfer is complete, place the membranes in plastic boxes (*see* **Note 10**). One may combine several membranes for blocking.
2. Incubate membranes with 50–100 ml of BLOTTO or PBS-T BSA at room temperature for 60 min or at 4 °C overnight on a rotator (60 rpm).
3. Once the membranes are blocked, add primary antibodies in BLOTTO or PBS-T BSA reagents diluted fivefold with PBS-T to achieve 1% blocking reagent (*see* **Note 11**). Take into consideration that one may utilize the same membrane to detect 2–3 different proteins (*see* **Note 12**).
4. Incubate membranes with primary antibodies for 8 h or longer at 4 °C on a rotator (60 rpm).
5. Wash membranes for 30 min with several (2–4) changes of PBS-T, then incubate them for 1 h at room temperature on a shaker (60 rpm) with secondary antibodies prepared in 1% BLOTTO.
6. Wash membranes as described above for 30 min.
7. Place the membranes on Saran Wrap, remove excess liquid with paper towels, and then incubate the membranes with ECL reagent for 1 min at room temperature.
8. Remove excess liquid with paper towels, cover membranes with Saran Wrap, and place membrane into X-ray cassettes.
9. Proceed with X-ray filmography performing several exposures starting from the longest (5–7 min) to obtain optimal band intensity.

3.2 Subcellular Fractionation of Proteins

Subcellular fractionation of the proteins is used to examine the changes in localization of the molecules upon receptor triggering. It was particularly helpful for studying proteins significantly modified post translationally.

1. Grow and treat cells as described above with the following modifications: (a) use four million adherent and eight million suspension cells per sample; (b) when detecting and analyzing proteins that are degraded by the proteasome or/and lysosome, pretreat the cells with MG-132 and/or CA-074Me for 30 min ahead of TNF ligand treatment.

2. Lyse the cells in TLB.
3. Cell pellets obtained by centrifugation at 20000 × *g* are dissolved in SLB (*see* **Note 13**).
4. Sonicate the samples using a microtip with amplitude 5, 1–2 short, 4–6 s bursts (*see* **Note 14**). Prepare the samples for SDS-PAGE, taking care to proceed to sample boiling as soon as possible after sonication.
5. Proceed with western blot detection of proteins as described above.

3.3 Immunoprecipitation of Endogenous Receptor-Associated Protein Signaling Complexes

1. For immunoprecipitation of endogenous receptor-associated protein complexes, grow cells (1–3 × 10^8 cells per time point) in 10-layer stackers (*see* **Note 15**). Treat half of the cells with 1 μM BV6 (as described in Subheading 3.1.2) and the other half with DMSO. Leave the cells in a 37 °C incubator overnight.
2. Wash cells once with room temperature PBS and then detach using Cell Dissociation buffer (*see* **Note 16**).
3. Collect the cells in a 500 ml bottle and centrifuge for 5 min at 4 °C at 200 × *g*.
4. Resuspend the cells in warm media and distribute between 50 ml conical tubes for treatment.
5. Leave cells untreated or treat them with 1 μg/ml TL1A, anti-CD40 antibody, or anti-Flag or anti-His antibody-cross-linked ligands (Flag-TNFα, His-TWEAK, His-LIGHT) for 5 or 30 min.
6. Wash cells with cold PBS and then lyse them in TIB (10 pellet volumes of lysis buffer). After incubation on ice for 30 min, centrifuge the lysates at 20000 × *g* for 10 min. Collect the supernatants and discard the pellets unless additional lysis procedures are planned.
7. Reserve 100 μl of soluble lysates to determine protein concentration (as described in Subheading 3.1.4) and run on SDS-PAGE to analyze protein levels by Western blotting (as described in Subheadings 3.1.5 and 3.1.6).
8. In the case of TL1A treatment, incubate supernatants with anti-DR3 antibody for 2 h and add 14 μl of protein A/G magnetic beads for another 3 h. For other tagged ligands and antibody combinations, add protein A/G magnetic beads for 3 h.
9. Place tubes on the magnet for 2–4 min.
10. Wash the beads with 1.4 ml of lysis buffer three times on magnet for 2 min.

11. After the last wash, remove supernatants, add 4× XT Sample buffer. Supplemented with reducing agent and heat for 5–7 min at 90–95 °C along with cleared lysates prepared at point #7 of this protocol.
12. Proceed to SDS-PAGE, Membrane Transfer and Western Blotting as described in Subheadings 3.1.5 and 3.1.6.

3.3.1 Secondary Immunoprecipitation of Ubiquitinated Proteins in Receptor Signaling Complexes

1. After **step 10** of the primary immunoprecipitation, disrupt immunoprecipitated endogenous receptor complexes by incubation in Urea-UBB buffer for 20 min at room temperature with constant soft rocking. After incubation, pellet disrupted immunoprecipitates by centrifugation at 1200 × *g* for 5 min and collect supernatants.
2. If using K11-, K48-, or K63-linkage-specific anti-ubiquitin antibodies, dilute supernatants twofold in UBB buffer. If using Linear ubiquitin (Lin-ub)-specific antibodies do not dilute the supernatants [34–41].
3. Incubate diluted supernatants with K11-, K48-, or K63-linkage-specific anti-ubiquitin antibodies overnight at 4 °C with rotation or undiluted supernatants with Lin-ub-specific antibodies at room temperature. The next day, add 55–75 μl of protein A/G agarose beads and rotate at 4 °C or room temperature for another 3 h.
4. Wash immunoprecipitated proteins five times as described in Subheading 3.3. Following washes, add 4× loading dye and resolve immunoprecipitates and lysates on SDS-PAGE and immunoblot as described in Subheadings 3.1.5 and 3.1.6.

4 Notes

1. Use one million cells per 10 cm dish. Cells should be 60–80% confluent by the next morning. One 10 cm dish allows the inclusion of 3–5 experimental points. However, the appropriate number of plates to transfect may vary between different cell lines and depend on protein contents of specific cell lines (it would be useful to get at least 2 μg/μl of protein lysates).
2. One may use 1.5–3 million adherent cells and 3–6 million suspension cells per one experimental point. The number of experimental points defines the total number of cells needed for the experiment.
3. Use 5 ml of detachment buffer for 10 cm dishes or 75 cm^2 flasks, 15 ml for 15 cm dishes or 175-cm^2 flasks.
4. Use 24 well plates for short (up to 2 h) treatment periods and 6 well dishes for longer times. Seed cells in 1 or 4 ml of growth media per well into 24 or 6 well dishes, respectively.

5. Use 0.1 ml of serum-free media supplemented with appropriate TNF ligand or anti-TNF agonist receptor antibody for the treatment of one well.
6. Use a 5 min time point to test for the phosphorylation of IκB; 15 or 30 min time points correspond to peak JNK and p38 activation for most cell lines; for IκB, 30 min corresponds to the lowest amount of total IκBα for most cell lines, while newly synthetized protein has maximally accumulated at 60 min.
7. Use 30–50 μl of TLB for small pellets (10 μl); 70 μl of TLB for medium pellets (>10, <25 μl); 100–140 μl for larger pellets (25–50 μl) -approximately 5× volume of the lysis buffer relative to the pellet volume. Before accruing experience, it is convenient to visually compare the size of cell pellets to defined volumes of liquid in 1.5 ml test tubes.
8. The final volume of protein samples to run can be determined by the number of loadings, loading volume (typically 12–15 μl), protein concentration and total amount of obtained protein. Protein concentrations below 2 μg/μl are low, while 3–8 μg/μl is normal. Try to avoid running samples with extremely low (less than 0.5 μg/μl) or high (5 μg/μl) protein concentrations since this can lead to increased errors in protein concentration determination, protein instability and incomplete cell lysis (high concentration). In general, the final protein concentration in the loaded sample should be between 3 and 4 μg/μl.
9. Prepare transfer buffer in advance—transfer buffer should not be warm, since warm liquids can damage nitrocellulose membranes.
10. Use a box whose size does not significantly exceed the membrane size, but one that is deep enough to accommodate at least 50 ml of liquid.
11. One may incubate several membranes together (up to 5) with an appropriate amount of antibody solution that can be defined based on the number and sizes of membranes used (10–20 ml per up to three membranes; 15–25 ml per five membranes). It is also important to ensure that all of the membranes are shaking and immersed in liquid during incubation.
12. The expression of two proteins that run at different molecular weights on SDS-PAGE (for example, 25, 40, and 60 kDa) can be detected simultaneously using the same membrane. In this case the membrane can be cut in lower and upper parts. It is also possible to cut membranes prior to blocking if one part of the membrane will be used for the detection of phosphorylated proteins.
13. Use at least 80 μl of SLB per sample (200 μl is the optimal amount). Lower volumes will lead to significant sample loss

during sonication, and volumes higher than 200 μl can cause significant sample dilution.

14. Try to avoid touching tube walls since it may result in bubbling of the samples. Avoid heating the sample, and if additional sonication is needed, cool the sample on ice.
15. For most adherent cell lines, one needs 8–10 confluent 225 cm^2 flasks to seed two 5-layer stacker with cells in 750 ml per stacker. We recommended seeding a single layer or 1-stack of cells per condition as the cells can be observed under the microscope.
16. Cells within 5-layer stackers need to be washed with warm PBS and incubated with 300 ml of Cell Dissociation buffer for 10–20 min in the TC incubator. To make sure all the cells are removed from the stacker, the stackers should be tapped by hand or against the hard surface. After the cells are removed, stackers can be rinsed with an additional 250 ml of PBS to collect remaining cells.

Acknowledgments

We thank researchers at Genentech who helped with suggestions, reagents, and comments. The authors are employees of Genentech.

References

1. Varfolomeev E, Vucic D (2011) Inhibitor of apoptosis proteins: fascinating biology leads to attractive tumor therapeutic targets. Future Oncol 7:633–648
2. Ndubaku C, Cohen F, Varfolomeev E et al (2009) Targeting inhibitor of apoptosis (IAP) proteins for therapeutic intervention. Future Med Chem 1:1509–1525
3. Vaux DL, Silke J (2005) IAPs, RINGs and ubiquitylation. Nat Rev Mol Cell Biol 6:287–297
4. Hershko A, Ciechanover A (1998) The ubiquitin system. Annu Rev Biochem 67:425–479
5. Pickart CM, Fushman D (2004) Polyubiquitin chains: polymeric protein signals. Curr Opin Chem Biol 8:610–616
6. Vucic D, Dixit VM, Wertz IE (2011) Ubiquitylation in apoptosis: a post-translational modification at the edge of life and death. Nat Rev Mol Cell Biol 12:439–452
7. Bertrand MJ, Milutinovic S, Dickson KM et al (2008) cIAP1 and cIAP2 facilitate cancer cell survival by functioning as E3 ligases that promote RIP1 ubiquitination. Mol Cell 30:689–700
8. Silke J, Brink R (2010) Regulation of TNFRSF and innate immune signalling complexes by TRAFs and cIAPs. Cell Death Differ 17:35–45
9. Varfolomeev E, Goncharov T, Fedorova AV et al (2008) c-IAP1 and c-IAP2 are critical mediators of tumor necrosis factor alpha (TNFα)-induced NF-κB activation. J Biol Chem 283:24295–24299
10. Varfolomeev E, Goncharov T, Maecker H et al (2012) Cellular inhibitors of apoptosis are global regulators of NF-kappaB and MAPK activation by members of the TNF family of receptors. Sci Signal 5:ra22
11. Varfolomeev E, Vucic D (2008) (Un)expected roles of c-IAPs in apoptotic and NF-κB signaling pathways. Cell Cycle 7:1511–1521
12. Hayden MS, Ghosh S (2004) Signaling to NF-kappaB. Genes Dev 18:2195–2224
13. Scheidereit C (2006) IκB kinase complexes: gateways to NF-κB activation and transcription. Oncogene 25:6685–6705
14. Micheau O, Tschopp J (2003) Induction of TNF receptor I-mediated apoptosis via two sequential signaling complexes. Cell 114:181–190

15. Gentle IE, Silke J (2011) New perspectives in TNF-R1-induced NF-kappaB signaling. Adv Exp Med Biol 691:79–88
16. Hayden MS, Ghosh S (2014) Regulation of NF-kappaB by TNF family cytokines. Semin Immunol 26:253–266
17. Haas TL, Emmerich CH, Gerlach B et al (2009) Recruitment of the linear ubiquitin chain assembly complex stabilizes the TNF-R1 signaling complex and is required for TNF-mediated gene induction. Mol Cell 36:831–844
18. Gerlach B, Cordier SM, Schmukle AC et al (2011) Linear ubiquitination prevents inflammation and regulates immune signalling. Nature 471:591–596
19. Ikeda F, Deribe YL, Skanland SS et al (2011) SHARPIN forms a linear ubiquitin ligase complex regulating NF-kappaB activity and apoptosis. Nature 471:637–641
20. Tokunaga F, Nakagawa T, Nakahara M et al (2011) SHARPIN is a component of the NF-kappaB-activating linear ubiquitin chain assembly complex. Nature 471:633–636
21. Walczak H (2011) TNF and ubiquitin at the crossroads of gene activation, cell death, inflammation, and cancer. Immunol Rev 244:9–28
22. Vince JE, Chau D, Callus B et al (2008) TWEAK-FN14 signaling induces lysosomal degradation of a cIAP1-TRAF2 complex to sensitize tumor cells to TNFalpha. J Cell Biol 182:171–184
23. Dejardin E (2006) The alternative NF-kappaB pathway from biochemistry to biology: pitfalls and promises for future drug development. Biochem Pharmacol 72:1161–1179
24. Varfolomeev E, Blankenship JW, Wayson SM et al (2007) IAP antagonists induce autoubiquitination of c-IAPs, NF-κB activation, and TNFα-dependent apoptosis. Cell 131:669–681
25. Vallabhapurapu S, Matsuzawa A, Zhang W et al (2008) Nonredundant and complementary functions of TRAF2 and TRAF3 in a ubiquitination cascade that activates NIK-dependent alternative NF-kappaB signaling. Nat Immunol 9:1364–1370
26. Zarnegar BJ, Wang Y, Mahoney DJ et al (2008) Noncanonical NF-kappaB activation requires coordinated assembly of a regulatory complex of the adaptors cIAP1, cIAP2, TRAF2 and TRAF3 and the kinase NIK. Nat Immunol 9:1371–1378
27. Zhang J, Webster JD, Dugger DL et al (2019) Ubiquitin ligases cIAP1 and cIAP2 limit cell death to prevent inflammation. Cell Rep 27:2679–2689
28. Fulda S, Vucic D (2012) Targeting IAP proteins for therapeutic intervention in cancer. Nat Rev Drug Discov 11:109–124
29. Morrish E, Brumatti G, Silke J (2020) Future therapeutic directions for smac-mimetics. Cell 9(2):406
30. Lacasse EC, Mahoney DJ, Cheung HH et al (2008) IAP-targeted therapies for cancer. Oncogene 27:6252–6275
31. Flygare JA, Beresini M, Budha N et al (2012) Discovery of a potent small-molecule antagonist of inhibitor of apoptosis (IAP) proteins and clinical candidate for the treatment of cancer (GDC-0152). J Med Chem 55:4101–4113
32. Jensen S, Seidelin JB, Lacasse EC et al (2020) SMAC mimetics and RIPK inhibitors as therapeutics for chronic inflammatory diseases. Sci Signal 13(619):eaax8295
33. Vince JE, Wong WW, Khan N et al (2007) IAP antagonists target cIAP1 to induce TNFalpha-dependent apoptosis. Cell 131:682–693
34. De Almagro MC, Goncharov T, Newton K et al (2015) Cellular IAP proteins and LUBAC differentially regulate necrosome-associated RIP1 ubiquitination. Cell Death Dis 6:e1800
35. Dynek JN, Goncharov T, Dueber EC et al (2010) c-IAP1 and UbcH5 promote K11-linked polyubiquitination of RIP1 in TNF signalling. EMBO J 29:4198–4209
36. Goncharov T, Hedayati S, Mulvihill MM et al (2018) Disruption of XIAP-RIP2 association blocks NOD2-mediated inflammatory signaling. Mol Cell 69:551–565
37. Goncharov T, Niessen K, De Almagro MC et al (2013) OTUB1 modulates c-IAP1 stability to regulate signalling pathways. EMBO J 32:1103–1114
38. Matsumoto ML, Dong KC, Yu C et al (2012) Engineering and structural characterization of a linear polyubiquitin-specific antibody. J Mol Biol 418:134–144
39. Matsumoto ML, Wickliffe KE, Dong KC et al (2010) K11-linked polyubiquitination in cell cycle control revealed by a K11 linkage-specific antibody. Mol Cell 39:477–484
40. Newton K, Matsumoto ML, Ferrando RE et al (2012) Using linkage-specific monoclonal antibodies to analyze cellular ubiquitylation. Methods Mol Biol 832:185–196
41. Newton K, Matsumoto ML, Wertz IE et al (2008) Ubiquitin chain editing revealed by polyubiquitin linkage-specific antibodies. Cell 134:668–678

Chapter 8

Methods to Study CARD11-BCL10-MALT1 Dependent Canonical NF-κB Activation in Jurkat T Cells

Andreas Gewies, Carina Graß, and Daniel Krappmann

Abstract

Jurkat T cells have been of central importance for the discovery of signalling mediators driving NF-κB activation in response to T cell antigen receptor (TCR)/CD28 co-stimulation. The critical function of the key regulators identified in Jurkat T cells has subsequently been verified in primary murine and human T cells. CRISPR/Cas9-mediated genomic editing techniques in combination with viral reconstitution are powerful tools that now enable the investigation of the exact molecular mechanisms that govern T cell signalling, especially the impact of protein–protein interactions, protein modifications, or cancer-associated gain- or loss-of-function mutations. As exemplified by the *CARD11* gene encoding a key regulator of NF-κB signalling in T cells, we describe here the detailed workflow for the generation of CRISPR/Cas9 knockout (KO) Jurkat T cells and the subsequent reconstitution using a lentiviral transduction protocol. In addition, we explain the use of a stable NF-κB-dependent EGFP reporter system that enables a reliable quantification of NF-κB transcriptional activation in the reconstituted KO Jurkat T cells.

Key words CRISPR/Cas9, Knockout, Reconstitution, Jurkat, Viral infection, Transduction, NF-κB, EGFP, Reporter, Signal transduction, T Cell activation, CARMA1, CARD11, BCL10, MALT1, CBM complex

1 Introduction

As the core components of the CBM signalling complex, CARD11/CARMA1, BCL10 and MALT1 form a scaffold that bridges T and B lymphocyte antigen receptor (TCR and BCR) signalling to canonical activation of the transcription factor NF-κB during an adaptive immune response [1]. In addition, CBM complex formation controls activity of the MALT1 protease, which is required for optimal T and B cell activation [2]. The function of the CBM complex is tightly controlled by posttranslational modifications, such as phosphorylation and ubiquitination, and the interaction with various accessory proteins, e.g., the protein kinases PKC-θ and CK1α, the E3 ligase TRAF6 or the ubiquitin editing enzyme A20 [3–5].

Guido Franzoso and Francesca Zazzeroni (eds.), *NF-κB Transcription Factors: Methods and Protocols*, Methods in Molecular Biology, vol. 2366, https://doi.org/10.1007/978-1-0716-1669-7_8,

Many of the fundamental insights into TCR signalling came from studies carried out with Jurkat T cells (reviewed in [6]). Also for studying the function of the CBM complex for T cell signalling and activation, Jurkat T cells proved to be a valid human cellular model system. The biological relevance of key discoveries made in Jurkat T cells mirrored the effects in mice deficient in individual CBM components or in patients carrying CBM gain-of-function or loss-of-function mutations leading to primary immunodeficiency or neoplasia, respectively [1, 3, 7]. Thus, Jurkat T cells as an *in vitro* cell culture model serves an important function for the initial identification and characterization of molecular mechanisms that promote or counteract T cell activation in vivo. Especially, with the increasing possibilities of CRISPR/Cas9-directed site-specific genome editing [8], sophisticated cellular model systems are becoming even more important to decide what kind of specific manipulations will be most relevant for studying phenotypic effects in animal models. Thus, detailed in vitro studies will help to save time, enable to work cost effective and allow compliance with the Three R's (Reduction, Replacement, and Refinement) for more ethical use of animal models.

In this chapter, we describe the pipeline for fast and reliable testing of the impact of specific genetic manipulations on NF-κB signalling and activation in Jurkat T cell lines. Taking the *CARD11* gene as an example, we describe in the first part how to efficiently generate gene-specific KO Jurkat T cell clones by CRISPR/Cas9 technology. The second step involves two subsequent lentiviral transductions into the parental or KO Jurkat T cells: (1) by transducing a NF-κB-EGFP reporter gene into the CARD11 KO Jurkat T cells, we introduce a tool for a reliable quantification of NF-κB activation in single cells. (2) By viral infection of the CARD11 KO Jurkat T cells with CARD11 wild type (WT) cDNA the cells are rescued to prove that phenotypic changes are caused by the KO.

Compared to the direct generation of site-directed manipulations (e.g., single missense mutations) by CRISPR/Cas9 in Jurkat T cells, the indirect combination of KO generation followed by reconstitution harbors several advantages: First, identification and analyses of homozygous site-directed knock-in clones can be laborious. Second, off-target effects caused by CRISPR/Cas9 can be excluded by rescuing the KO Jurkat T cell clones with WT cDNAs. Third, the parallel reconstitution of WT and mutant cDNAs into the same KO Jurkat T cell clone represents a clean genetic setting to study the effects of genetic changes and minimizes the risks of measuring clonal variabilities. Fourth, by co-expression of the cell surface marker ΔCD2, which lacks any intracellular signalling domains, the lentiviral transduction system can be titrated to allow for a homogenous expression of the protein of interest at close to endogenous levels in various human and murine cell lines [9–12]. We have successfully used this approach for analyzing the effects of several protein-protein interactions and protein modifications on CBM-dependent NF-κB signalling in Jurkat T cells [11–13]. Lastly, the amendment of the basic protocol by the introduction of the NF-κB-EGFP reporter gene is not obligatory, but it allows the reliable quantification of NF-κB activation. Quantification of NF-κB activation is very helpful for analyzing augmented or partially diminished responses, as for instance observed with MALT1 phosphorylation mutants or MALT1 splicing variants [14, 15]. We have focused the chapter on the generation, lentiviral transduction, and analysis of CARD11 KO Jurkat T cells, but the protocol can easily be adapted to the same type of analysis for any other gene of interest.

2 Materials

All solutions are prepared with ultrapure water and analytical grade reagents. Access to cell culture and molecular biological equipment is required, e.g., centrifuges, sterile hoods, cell incubators, gel electrophoresis.

2.1 CRISPR/Cas9 Genomic Editing

1. Custom ordered CRISPR guide oligonucleotides (Eurofins).
2. Plasmids: psPAX2 packaging plasmid (Addgene), pMD2.G envelope plasmid (Addgene), pHAGE-ΔCD2-T2A-CDS [9], pHAGE 6×NF-κB conA-HygEGFP [14].
3. 10× Annealing Buffer: 500 mM Tris–HCl, pH 8.0, 700 mM NaCl. Add 6.06 g Tris base and 4.09 g NaCl into a beaker containing 80 ml water, adjust pH to 8.0 by carefully adding concentrated HCl upon stirring. Adjust volume to 100 ml final volume by adding water.
4. CRISPR/Cas9 plasmid px458 (full name: pSpCas9(BB)-2A-GFP) (Addgene).
5. Restriction enzymes BbsI and NdeI with compatible 10× restriction buffer (NEB).
6. Low melting agarose (Biozym).
7. TAE buffer (50× concentrated stock solution): 242 g of Tris base, 57.1 ml of glacial acetic acid, 100 ml of 0.5 M EDTA solution pH 8.0. Adjust volume to 1 l with water.
8. 10× Agarose Loading Dye: 10 mM Tris–HCl, pH 7.6, 0.25% Bromophenol Blue, 0.25% Xylene Cyanol, 50% glycerol, 100 mM EDTA; 2 μg/ml ethidium bromide.
9. 1% TAE-agarose gel: 1% agarose in TAE buffer.
10. NucleoSpin Gel and PCR Clean-up Kit (Macherey-Nagel).
11. T4 ligase and T4 ligase buffer (Thermo Fisher Scientific).

12. One Shot Stbl3 Chemically Competent *E. coli* (Invitrogen), F-mcrB mrrhsdS20(rB-, mB-) recA13 supE44 ara-14 galK2 lacY1 proA2 rpsL20(StrR) xyl-5 λ-leumtl-1.
13. LB Medium and LB Agar (Carl Roth GmbH+Co. KG).
14. 100 mg/ml Ampicillin stock in water (Carl Roth GmbH +Co. KG).
15. NucleoSpin Plasmid Mini Kit (Macherey-Nagel).
16. Jurkat T cells (DSMZ).
17. RPMI 1640 complete medium: RPMI supplemented with 10% fetal bovine serum (FBS) and 100 U/ml penicillin/streptavidin (Thermo Fisher Scientific).
18. Cell electroporator (Gene Pulser Xcell System) and 4 mm electroporation cuvettes (Gene Pulser; Bio-Rad).
19. Flow cytometers: MoFlo cell sorter (Beckman Coulter) with Summit 4.3 software and Attune Acoustic Focusing Cytometer (Thermo Fisher Scientific).
20. FACS buffer: PBS, 3% FBS.
21. Co-IP Lysis buffer: 25 mM HEPES pH 7.5, 150 mM NaCl, 1 mM glycerol, 0.2% v/v NP-40, 10 mM NaF, 8 mM β-Glycerophosphate (added fresh from frozen stock), 300 μM Sodiumvanadate (added fresh), commercial Protease inhibitor cocktail (Roche, added fresh from frozen stock) and 1 mM DTT (added fresh from frozen stock).
22. 4× SDS loading buffer: 8% SDS, 20% beta-mercaptoethanol, 40% glycerol, 0.2 M Tris, pH 6.8, and 0.05% bromophenol blue.
23. PBS (cell culture grade) w/o Ca^{2+} and w/o Mg^{2+} (Thermo Fisher Scientific).
24. Antibodies: FACS: Mouse monoclonal anti-human CD3-FITC [HIT3a], Mouse monoclonal anti-human CD28-APC [CD28.2] (BD Biosciences).
25. 2× Freezing Medium: DMSO (20%), FBS (20%), RPMI complete medium (60%).

2.2 Lentiviral Transduction

1. Lentiviral vector system: psPAX2 packaging plasmid (Addgene), pMD2.G envelope plasmid (Addgene) and the transfer plasmid pHAGE-ΔCD2-T2A-CDS [9] or pHAGE 6×NF-κB conA-HygEGFP [14].
2. HEK293T cells (DSMZ).
3. DMEM pure and DMEM complete medium supplemented with 10% FBS and 100 U/ml penicillin/streptavidin (Thermo Fisher Scientific, Waltham, MA, USA).

4. Transfection Reagent X-tremeGENE HP Transfection Reagent (Roche).
5. 0.22 μM Sterile filter.
6. 8 mg/ml Polybrene in water (Sigma-Aldrich).
7. FACS antibodies: antihuman CD2 antibody, Allophycocyanin (APC)-conjugated (Thermo Fisher Scientific).
8. Western blotting: CARD11 antibody 1D12 (NEB), anti--β-ACTIN-HRP (I-19) (Santa Cruz).

2.3 Cell Stimulation and NF-κB-EGFP Reporter

1. 1 mg/ml stock PMA (Phorbol 12-myristate 13-acetate) in DMSO (Merck Millipore).
2. 1 mg/ml stock Ionomycin in DMSO (Calbiochem).
3. 10 ng/μl TNFα in PBS (Tumor-Necrosis Factor alpha) (Biomol).
4. 1 mg/ml Stimulatory Antibodies: Mouse antihuman anti-CD3, Mouse antihuman anti-CD28, Rat anti-mouse IgG1, Rat anti-mouse IgG2a (BD Pharmingen).

3 Methods

3.1 Generation of CARD11 KO Jurkat T Cells by CRISPR/Cas9

For generating gene-specific KO Jurkat T cells, we use px458 plasmid co-expressing Cas9 and EGFP as well as the CRISPR single guide RNA (sgRNA) under control of the U6 promoter (Fig. 1a) [8]. We give general considerations for CRISPR/Cas9-targeting strategies in **Note 1**. We always use two or three alternative strategies and we test different CRISPR guide sequences in parallel to increase the chances of obtaining KO clones. So far, we always achieved a successful KO with at least one of these approaches. After limiting dilution, KO clones are identified by Western blotting to monitor ablation of the targeted protein. Optionally, the genetic modifications in the targeted locus can be verified by genomic sequencing (*see* **Note 2**). In the subsequent protocol, we describe in detail the generation and identification of CARD11 KO Jurkat T cells which has been successfully targeted by a single cut strategy in exon 3 of the *CARD11* gene [12].

1. For selection of a CRISPR target sequence region, retrieve the genomic sequence of the target gene including its exon-intron structure from the ENSEMBL genome browser and pick the stretch of sequence to be targeted. In the human *CARD11* gene, we have chosen a strategy targeting exon 3, because the first two exons largely encompass 5′ UTR regions.
2. For selection of the CRISPR guide sequence, paste the selected sequence into the CRISPRko guide sequence selection online tool 'GPP sgRNA Designer' (provided by the BROAD

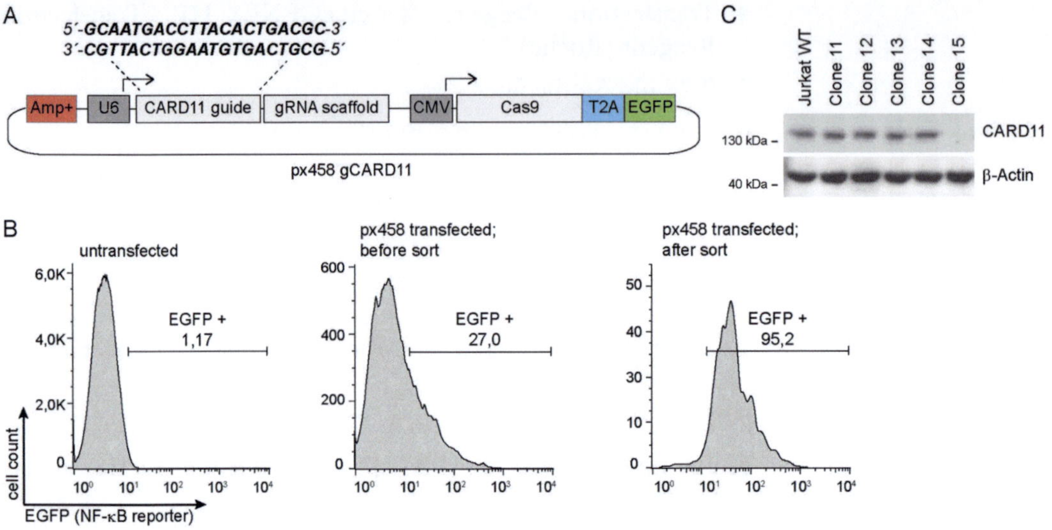

Fig. 1 Generation of CARD11 KO Jurkat T cells via the CRISPR/Cas9 system. (**a**) Vector scheme of px458 containing the CARD11 guide. (**b**) Cytometric cell sorting of Jurkat T cells electroporated with px458 gCARD11. Untransfected Jurkat T cells are used as control (left) to set the gate for cell sorting of EGFP-positive cells transfected with px458 (middle). Right histogram shows enriched EGFP-positive cells after cell sorting, which are further used for limiting cell dilution. EGFP expression was analyzed using FlowJo software (Version 10). (**c**) Western Blot analysis for the identification of potential CARD11 KO Jurkat T cell clones. Clone 15 is a CARD11 KO clone

Institute; *see* **Note 1**) and select from the result list appropriate candidate hits concerning location and on-/off-target ranking. In case of our CARD11 example, we picked the following 20 bp guide sequence: 5′- CAATGACCTTACACTGACGC - 3′. In line with the recommendation by Ran et al. [8], we added an extra G to the 5′ end of the guide sequence due to U6 RNA polymerase III promoter preferences. This is not necessary, if the picked 20 bp guide already starts with G. The final CARD11 guide sequence is: 5′- GCAATGACCTTACACT GACGC-3′ (*see* **Note 3**) (Fig. 1a).

3. For cloning the CARD11 guide sequence into the BbsI sites of px458, add to the sense and antisense oligonucleotides BbsI sticky ends at its 5′-end (5′-cacc-3′) and its 3′-end (5′-aaac-3′) to obtain a double stranded oligonucleotide with overhangs. The following oligonucleotides for targeting CARD11 were synthesized: CARD11 sense guide: 5′-caccGCAATGACCTTA CACTGACGC, CARD11 antisense guide: 5′-aaacGCGTCA GTGTAAGGTCATTGC.
4. For annealing, adjust the concentration of sense and antisense CARD11 guides to 100 μM in water.

5. Prepare the Annealing Mix by mixing 2 μl of sense guide oligo, 2 μl of antisense guide oligo, 2 μl of 10× annealing buffer and 14 μl H_2O to a total volume of 20 μl.
6. Heat the Annealing Mix 5 min at 95 °C in a heating block.
7. Switch off the block and allow to slowly cool down to room temperature over several hours.
8. Dilute annealed double stranded oligonucleotides 1:400 in water before ligation into BbsI-digested px458.
9. For restriction of px458, prepare Restriction Mix by adding px458 DNA (2 μg), BbsI enzyme (10 U), 2 μl 10× restriction buffer and water to a total volume of 20 μl.
10. Incubate the Restriction Mix at 37 °C for 2 h.
11. Add 2 μl of 10× agarose loading dye and load on a 1% TAE-agarose gel. Let it run at 100 V for 1 h.
12. Cut out the linearized vector band and elute the BbsI-linearized px458 vector band (9270 bp) (now called px458 (BB)) from the agarose gel using NucleoSpin Clean-up Kit. Eluted px458(BB) can be stored at −20 °C or directly used for ligation with guides from **step 8**.
13. For ligation of guide oligonucleotides into px458(BB), prepare Ligation Mix: px458(BB) (50 ng; **step 12**), 2 μl of annealed guide oligonucleotides (1:400 diluted; **step 8**), 2 μl of 10× T4 ligation buffer, T4 DNA ligase (20 U) and water for a total volume of 10 μl.
14. Incubate the Ligation Mix for 1 h at RT.
15. For transformation into bacteria, add 10 μl of the Ligation Mix (*see* **step 14**) to 25 μl of chemically competent Stbl3 *E. coli* thawed on ice. For mixing, gently flick the tube and keep on ice for 20 min.
16. Heat shock at 42 °C for 45 s and put the tube immediately back on ice for 2 min.
17. Add 100 μl LB medium and shake the tube for 1 h at 37 °C.
18. Plate the ligation mix onto an LB plate containing 100 μg/ml Ampicillin. Incubate the plate overnight at 37 °C.
19. For identification of px458 gCARD11 plasmid, 10–100 *E. coli* colonies should have grown on the LB agar plate (*see* **step 18**). Pick 10–20 colonies and inoculate shaking in 5 ml LB medium containing 100 μg/ml Ampicillin at 37 °C overnight.
20. Isolate plasmid DNA from bacteria using NucleoSpin Plasmid Mini Kit. Perform following enzyme restriction to identify px458 gCARD11 plasmid: Plasmid DNA (1.5 μg), BbsI (2.5 U), NdeI (2.5 U), 2 μl 10× Restriction Buffer, and H_2O to a final volume of 20 μl.

21. Incubate at 37 °C for 1 h, then add 2 μl 10× agarose loading dye and load onto a 1.5% TAE-agarose gel and analyze fragment sizes (*see* **Note 3**). If the CARD11 guide was inserted, a NdeI/NdeI restriction fragment of 437 bp is detected, if not, a BbsI/NdeI fragment of only 328 bp is generated. px458 plasmids with inserted guide are used for subsequent generation of KO cells and must be verified by sequencing using sequencing primer 5′- GACTATCATATGCTTACCGT -3′. px458-guide plasmids are re-transformed into Stbl3 E.coli for high yield maxi DNA preparation.
22. For transfection by electroporation, culture Jurkat T cells in RPMI 1640 complete medium at densities between 0.5 and 1.5×10^6 cells/ml at 37 °C and 5% CO_2 cell incubator.
23. Mix gently 8×10^6 cells in 400-μl RPMI 1640 medium with a total of 5 μg px458 plasmid DNA.
24. Transfer the mix into a 4 mm electroporation cuvette and electroporate at 220 V/1000 μF/variable resistance using a Gene pulser X electroporation device.
25. Afterwards, transfer cells to 10 ml pre-warmed RPMI 1640 complete medium and let grow for 48 h at 37 °C and 5% CO_2.
26. To enrich for transfected Jurkat T cells after electroporation and before limiting dilution, use flow cytometric cell sorting to enrich for EGFP-positive cells, because EGFP and Cas9 expression are coupled via the co-translational processing sequence T2A on the px458 plasmid (Fig. 1a). We usually use a MoFlo cell sorter, but any other cell sorter can be used that is equipped with a 488 nm laser for excitation and a 530/30 nm channel for detection of the EGFP signal. Transfected Jurkat T cells are harvested and 20,000 EGFP-positive cells are FACS sorted to enrich for EGFP-positive cells (Fig. 1b) and taken up in 5 ml RPMI 1640 complete medium, corresponding to 4000 cells/ml.
27. For isolating monoclonal Jurkat T cell clones, limiting dilution is used. Serially dilute the suspension of sorted EGFP-positive Jurkat T cells (4000 cells/ml; **step 26**) to 40 cells/ml (1:100 of 4000 cells/ml), 20 cells/ml (1:2 of 40 cells/ml) and 10 cells/ml (1:2 of 20 cells/ml) in RPMI 1640 complete medium.
28. Plate 100 μl of the respective 40, 20 and 10 cells/ml suspensions into each well of 96-well plates, corresponding to 4, 2, or 1 cells/well (*see* **Note 4**). Plate at least two complete 96-well plates per cell dilution to increase the likelihood for obtaining sufficient numbers of single cell clones growing up.
29. After 1 week at 37 °C and 5% CO_2, add 100 μl of fresh RPMI 1640 complete medium. Clonal cell colonies can be detected after 2–3 weeks of culture without any further change of

medium. When the medium turns yellowish, colonies are detected under the microscope and cell clones need to be transferred to 48-well plates with 1 ml fresh RPMI 1640 complete medium. Subsequent transfer to a volume of 2 ml in 24-well plates allows sufficient cell numbers to be obtained for analyzing loss of CARD11 in the Jurkat T cell clones by Western blot.

30. For screening Jurkat T cell clones for loss of CARD11 expression, grow 1–2 × 10^6 Jurkat T cells in 24-well plates and lyse them in 100 μl co-IP lysis buffer by incubating on an orbital shaker at 4 °C for 15 min.
31. After lysis, centrifuge the extract at 12,000 × *g* for 10 min at 4 °C.
32. Transfer supernatants into new vials, add 4× SDS loading buffer (1 volume) to supernatants (3 volumes) and incubate for 5 min at 95 °C.
33. Load 5–10 μl of lysate onto SDS-PAGE and perform Western blotting using anti-CARD11 antibody (1D12) to check, if CARD11 expression was lost in Jurkat T cell clones. Figure 1c depicts an example Western blot probing different Jurkat T cell clones after CRISPR/Cas9 targeting and limiting dilution. CARD11 expression was lost in clone #15. To verify candidate CARD11 KO Jurkat T cell clones identified by Western blot, we usually validate by sequencing the genomic region that was targeted by the CRISPR guide (*see* **Note 2**). Further, Jurkat T cells may downregulate surface expression of the CD28 co-stimulatory receptor upon long-term culturing. Therefore, we recommend confirming surface expression of CD28 and CD3 on KO Jurkat T cell clones by flow cytometry in case that anti-CD3/CD28 stimulation shall be performed (*see* **Note 5**).
34. Expand the CARD11 KO Jurkat T cell clones and grow in RPMI 1640 complete medium in a larger cell culture flask format before freezing for storage in liquid nitrogen. Use growing cells at a density of about 1 × 10^6 cells/ml for freezing (*see* **step 35**).
35. Centrifuge cells from **step 34** at 350 × *g* for 5 min, take up cells at 1 × 10^7 cells/ml in RPMI 1640 complete medium and put cells on ice for about 10 min. Mix ice-cold Jurkat T cells (1 × 10^7 cells/ml) with ice-cold 2× Freezing medium (1 volume + 1 volume) and transfer 1 ml each (5 × 10^6 cells) to several cryo-vials, which are placed in a styrofoam box for slow freezing in a −80 °C freezer.
36. After 24 h, transfer the frozen cells to liquid nitrogen for long-term storage.

37. For thawing, float a cryo-vial in a 37 °C water bath and transfer the thawed cell suspension into 10 ml PBS.
38. Centrifuge at 350 × *g* for 5 min, take up the cell pellet in 10 ml RPMI 1640 complete medium, transfer the cells to an appropriate cell culture flask and incubate at 37 °C/5% CO_2.
39. Count cells the next day and keep passaging the cells at a density between 0.5×10^6 and 1.5×10^6 cells/ml.

3.2 Lentiviral Transduction of Jurkat T Cells

Two consecutive lentiviral transductions are used to allow monitoring of NF-κB activation by an NF-κB-EGFP reporter (Subheading 3.2.1) and to rescue CARD11 KO Jurkat T cells with CARD11 (Subheading 3.2.2). Parental or CARD11 KO Jurkat T cells are transduced with replication-deficient lentiviral particles that are produced based on lentiviral vectors developed by the Trono and Mulligan labs [16, 17]. For the production of infectious, replication-deficient lentiviral particles, the psPAX2 packaging plasmid, the pMD2.G envelope plasmid and the transfer plasmid pHAGE-ΔCD2-T2A-CDS [9] are co-transfected into HEK293T viral producer cells (*see* **Note 6**). Be aware of biosafety requirements while working with infectious VSV-G-pseudo-typed lentiviral particles (*see* **Note 7**).

3.2.1 Transduction of the NF-κB-EGFP Reporter Construct into Jurkat T Cells

The NF-κB-EGFP reporter is depicted in Fig. 2a and has been described [14]. The pHAGE transfer plasmid was modified by deleting the PGK-ΔCD2-T2A-CDS cassette and inserting a cassette containing a minimal chicken conalbumin (conA) promoter and six NF-κB binding sites of the immunoglobulin κ light chain enhancer (5′-ATCTGGGGATTCCCCA-3′) for driving induction of the EGFP gene, which was fused to a Hygromycin resistance gene for optional selection. The resulting pHAGE-NF-κB-EGFP reporter plasmid (Fig. 2a) is used for lentiviral transduction of parental and CARD11 KO Jurkat T cells.

1. For production of lentiviral particles, plate 2×10^6 HEK293T cells in 10 cm cell culture dishes in 8 ml complete DMEM on the day before transfection.
2. The next day, prepare a Transfection Mix by mixing following components in a 1.5 ml reaction tube: pMD2.G (1.0 μg), psPAX2 (1.5 μg) and pHAGE-NF-κB-EGFP (2.0 μg) in 200 μl DMEM w/o supplements.
3. Add 10 μl X-tremeGENE HP Transfection Reagent.
4. Incubate Transfection Mix for 30 min at RT before adding the mix dropwise to the pre-seeded HEK293T cells.
5. Three days after transfection the virus containing supernatant is collected and sterile-filtered (0.22 μM). The viral supernatant can be used immediately or it can be stored at −80 °C.

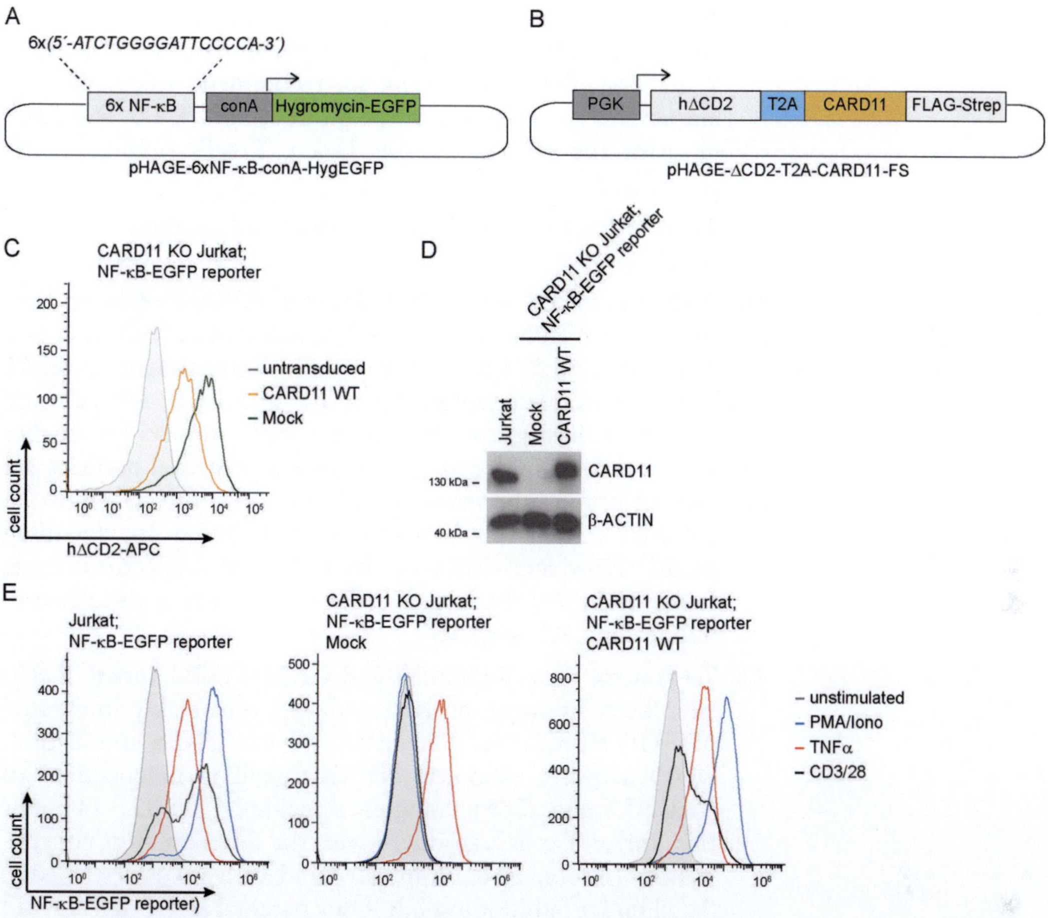

Fig. 2 Lentiviral transduction and determination of NF-κB activation by NF-κB-EGFP reporter assay. (**a**) Schematic representation of pHAGE-6×NF-κB-conA-HygEGFP reporter plasmid. Six NF-κB binding sites are fused to the conA (minimal chicken Conalbumin) promoter, which controls the expression of a hygromycin-EGFP fusion gene. (**b**) Schematic representation of the pHAGE-ΔCD2-T2A-CARD11-FlagStrep construct. The PGK (phosphoglycerate kinase) promotor controls the expression of the ΔCD2-T2A-CARD11-FlagStrep expression cassette. (**c**) Expression of the cell surface marker ΔCD2 to control transduction efficiency. An empty pHAGE-ΔCD2-T2A-FS (mock) is used as control. ΔCD2 expression was analyzed using FlowJo software (Version 10). (**d**) Western Blot analysis of reconstituted CARD11 KO Jurkat; NF-κB-EGFP reporter cell clone transduced with MOCK or CARD11 WT. Parental Jurkat were loaded as expression control for endogenous CARD11. (**e**) Induction of NF-κB-EGFP expression in parental and CARD11 KO Jurkat T cells reconstituted with Mock or CARD11 WT. Cells were unstimulated (gray), stimulated with PMA/Iono (blue), CD3/28 (black) or TNFα (red) for 5 h prior to flow cytometric analysis of EGFP expression. FACS data were analyzed using FlowJo software (Version 10)

6. For lentiviral transduction of Jurkat T cells, transfer 5×10^5 parental or CARD11 KO Jurkat T cells (*see* Subheading 3.1) into a 15 ml conical tube, spin cells down at $350 \times g$ for 5 min.
7. Discard supernatant and resuspend the cells in 500 μl RPMI 1640 complete medium.

8. Add viral supernatant from **step 5**, 2 μl of 8 mg/ml Polybrene stock and RPMI 1640 complete medium to a final volume of 2 ml. For efficient infection, we recommend using different amounts of viral supernatant, e.g., 250, 500, 1000, and 1500 μl, with the aim to get all Jurkat T cells homogenously transduced.
9. Transfer the Infection Mix to a 6-well and incubate at 37 °C/ 5% CO_2.
10. 24 h after infection wash cells 3× with PBS and resuspend the cells in 1 ml RPMI 1640 complete medium for culturing to a cell density up to 1.5×10^6 cells/ml. Keep passaging Jurkat T cells at densities between 0.5×10^6 and 1.5×10^6 cells/ml until they have grown to sufficient cell numbers for further analyses. After infection and as long as lentiviral particles are present in the supernatant of infected cells, the cells must be cultured under the biosafety level required for handling pseudo-typed lentiviruses (*see* **Note 7**). For long-term storage, freeze and thaw the transduced Jurkat T cells as described in Subheading 3.1, **step 35**.
11. To control that all parental and CARD11 KO Jurkat T cells have been homogenously transduced with stably integrated pHAGE-NF-κB-EGFP reporter, we use TNFα stimulation, which activates canonical NF-κB signalling independent of CARD11 and CBM complex signalling [11, 12, 14]. For this, Jurkat T cells transduced with the different virus concentrations (*see* **step 8**) are stimulated and analyzed as described in detail under Subheading 3.3. Only parental and CARD11 KO Jurkat T cells that display EGFP expression in more than 95% of the cells after TNFα stimulation are used for subsequent lentiviral reconstitution with CARD11 WT cDNA (*see* Fig. 2e).

3.2.2 Lentiviral Reconstitution of CARD11 KO;NF-κB-EGFP Jurkat T Cells Clones

The CARD11 cDNA is inserted in-frame downstream of the ΔCD2-T2A sequence by standard cloning techniques (*see* **Note 8**) [12]. The ΔCD2-T2A-CARD11-FlagStrep expression construct is under control of the phosphoglycerate kinase (PGK) promoter, which has been optimized for lentiviral expression (Fig. 2b) [18]. We described transduction of CARD11 KO Jurkat;NF-κB-EGFP T cells (*see* Subheading 3.2.1), but the same workflow can be used to infect parental Jurkat T cells or any other genetically modified Jurkat T cell clone.

1. Production of lentiviral particles and transduction into CARD11 KO Jurkat;NF-κB-EGFP T cells is performed exactly as described in Subheading 3.2.1, with the exception that pHAGE-NF-κB-EGFP has been replaced by pHAGE-ΔCD2-T2A-CARD11-FS (Fig. 2b). For control transduction (Mock), empty pHAGE-ΔCD2-T2A-FS is used. We recommend using

different amounts of viral supernatant, e.g., 250, 500, 1000 and 1500 μl, with the aim to get all Jurkat T cells homogenously transduced and to achieve CARD11 expression close to the levels of endogenous CARD11 in parental Jurkat T cells.

2. Control homogenous transduction of Jurkat T cells by flow cytometric analysis of ΔCD2 surface expression. Spin down 3×10^5 infected cells at $350 \times g$ for 4 min at 4 °C and then resuspended in 400 μl FACS buffer containing 1 μl of APC-conjugated anti-CD2 antibody.
3. After 15 min incubation in the dark at 4 °C, spin down the cells and wash them with 1 ml FACS buffer.
4. Resuspend the pellet in 400 μl FACS buffer. Perform the analysis of CD2 surface expression with a flow cytometer using excitation at 638 nm and 660/20 nm emission filter settings for detection of APC fluorophore.
5. Analyze ΔCD2 expression in cells by FlowJo software. Homogenous transduction of all Jurkat T cells has been achieved, if all cells are recognized by anti-CD2 antibody and shifting in the APC channel (Fig. 2c). Of note, we consistently see higher ΔCD2 surface expression in Mock transduced cells, which is due to more efficient viral packaging due to the smaller insert in the mock vector (pHAGE-ΔCD2-T2A-FS).
6. Controlling CARD11 expression is done by Western Blotting using anti-CARD11 (1D12) antibody and standard procedure (Fig. 2d). Load Extracts of Jurkat T cells to confirm exogenous CARD11 expression in transduced CARD11 KO Jurkat;NF-κB-EGFP T cells. Parental Jurkat T cells are used to control expression at endogenous levels. In case of low CARD11 expression or CARD11 overexpression, lower or higher concentrations of lentiviral particles must be used (**step 1**).

3.3 Investigating NF-κB Activity in Knockout and Reconstituted Jurkat T Cell Clones

To test for NF-κB activation, parental Jurkat T cells and CARD11 KO Jurkat T cell transduced with Mock or CARD11 WT containing the NF-κB-EGFP reporter are stimulated prior to analysis by flow cytometry for the induction of EGFP expression (*see* **Note 9**).

1. Plate 500 μl of NF-κB-EGFP reporter expressing Jurkat T cells at a density of 1×10^6 cells/ml in a 24-well format.
2. For activation of CBM-dependent NF-κB signalling, prepare a PMA/Iono (Phorbol 12-myristate 13-acetate/Ionomycin) Mastermix: add 0.15 μl of 1 mg/ml Iono stock, 0.1 μl of 1 mg/ml PMA stock and 0.75 μl PBS per stimulation.
3. Add 1 μl Mastermix to 500 μl cells to have a final concentration of 300 ng/ml Iono and 200 ng/ml PMA.
4. For TCR/CD28 co-stimulation, prepare the following Mastermix by adding per stimulation 0.5 μl murine antihuman

anti-CD3 antibody (IgG1), 1.7 μl murine antihuman anti-CD28 antibody (IgG2a), 1.7 μl anti-mouse IgG1 antibody and 1.7 μl anti-mouse IgG2a antibody for crosslinking. 5.6 μl is added to the cells to give a final concentration of 1 μg/ml (anti-CD3) and 3 μg/ml (anti-CD28, ant-IgG1 and anti-IgG2a).

5. For activation of CBM-independent NF-κB signaling, add 1 μl of 10 ng/μl TNFα stock to 500 μl cells to have a final concentration of 20 ng/ml TNFα.
6. After adding stimuli, gently mix the cells and incubated for 5 h at 37 °C/5% CO_2.
7. For FACS analysis, spin down the cells at 350 × *g* for 4 min at 4 °C and then re-suspend them in 300 μl FACS buffer. NF-κB transcriptional activation is determined by measuring EGFP-positive cells with the Attune flow cytometer at 488 nm excitation and in the 530/30 nm emission channel. EGFP expression in cells is analyzed by FlowJo software. EGFP expression is induced after PMA/Iono or anti-CD3/CD28 in parental, but lost in CARD11 KO Jurkat T cells containing NF-κB-EGFP reporter (Fig. 2e). CARD11 WT reconstitution rescues NF-κB-dependent EGFP expression in CARD11 KO Jurkat T cells. In contrast, TNFα-induced NF-κB activation is independent of CARD11 in Jurkat T cells (Fig. 2e).

In summary, by combining CRISPR/Cas9 KO generation, viral transduction of an NF-κB reporter and viral reconstitution in Jurkat T cells, it is possible to analyze the contribution of cellular mediators to NF-κB signalling upon T cell activation. Reconstitution with site-directed mutant versions of the respective cDNA will facilitate to dissect the molecular requirements and specific signalling pathways involved in T cell activation.

4 Notes

1. General considerations for the selection of the CRISPR/Cas9-targeting strategy: We identify appropriate guide sequences by using the online tool GPP sgRNA Designer provided by the BROAD Institute (https://portals.broadinstitute.org/gpp/public/analysis-tools/sgrna-design), which has been relaunched with an updated version under the name CRIS-**Pick** (https://portals.broadinstitute.org/gppx/crispick/public). The algorithm is designed for predicting maximal on-target and minimal off-target activity of the guide sequence [19]. The input sequence for analysis by the single guide selection online tool is usually a genomic region covering the first exons and introns to allow targeting of the gene by diverse strategies:
   (a) A single cut approach typically uses a guide sequence that targets an early exon in the 3′ region of the respective open reading frame (ORF). Cas9-mediated DNA double strand breaks lead to the random insertion or deletion of base pairs by non-homologous end joining (NHEJ) that can result in frameshift mutations. It is a highly effective strategy for generating KO clones, but the alternative usage of in-frame start codons downstream of the original translational start may lead to the expression of N-terminally truncated products. Therefore, antibodies directed against the C-terminus of the targeted protein should be used for Western blot analysis in order to rule out potential expression of truncated variants.
   (b) The deletion of one or several exons can be achieved by co-transfection of two px458 plasmids that express sgRNAs targeting exon-flanking intron sequences for the introduction of double strand breaks, which are subsequently annealed by NHEJ, leading to the excision of the exon-containing sequences in between. Similar to classical KO approaches, deletion of exons by two cuts in the surrounding introns is an effective approach, if the deletion of the exon(s) leads to a frameshift and no alternative splice variants exist. However, both single guides must be highly effective, because a single cut in an intron will not be sufficient for KO generation.
   (c) The introduction of frameshifts by co-transfection of two sgRNA expressing sequences targeting an exon and the adjacent intron. This approach combines the single cut strategy within an exon with the partial deletion of an exon/intron boundary, which both can lead to frameshifts as a consequence of NHEJ. Thus, this strategy is very

efficient, but the random ligation of exonic and intronic sequences may lead to artificial transcripts that are still translated.

2. For verification of CARD11 KO by sequencing, isolate genomic DNA from the Jurkat T cells. Design PCR primers that amplify a fragment of 500–1000 bp spanning the targeted locus. Purify PCR product by agarose gel electrophoresis and perform sequencing directly on the template. Direct sequence analysis of CRISPR-induced knockout clones (upon NHEJ) may be sometimes difficult, because the two alleles can carry different mutations and it may not be possible to unambiguously assign the sequences to one respective allele. In many cases, CRISPR-ID represents a good tool to separate the allelic sequences [20]. However, if unambiguous assignment of the correct sequences to the different alleles is not possible TOPO cloning of PCR fragments for subsequent sequencing of individual alleles of the targeted locus may be helpful.
3. It is not possible to directly detect the short insertion of the 20 bp guide oligonucleotide, because the BbsI cleavage sites are lost after insertion of the oligonucleotides. However, the BbsI sites are important to distinguish empty and oligonucleotide-containing px458 plasmids, because after correct insertion of the CARD11 guide, the BbsI sites are lost and cleavage of the flanking NdeI sites results in a restriction fragment of 437 bp instead of a 328 bp NdeI/BbsI fragment cut out from the empty vector. Note that the guide sequence does not include the PAM site (NGG for Streptococcus pyogenes Cas9, in this example it is 'AGG') at its 3′ end in the genomic sequence.
4. We propose to plate >1 cell/well (i.e., four or two cells per well) because many Jurkat T cells do not survive and recover after electroporation and seeding into 96-well plates at low cell density. Usually, 10–30 clonal colonies can be obtained per 96-well plate when seeding 4 cells/well. Different dilutions are seeded to increase the probability that single cell clones are derived from one dilution. Of note, after culturing the Jurkat T cell in 96-well plates for 2–3 weeks, distinct foci in the well can be observed and it can be judged under a microscope whether the cells in a well are indeed monoclonal and derived from one single cell or whether more than one cell formed multiclonal foci in one well.
5. Clonal selection of Jurkat T cells may affect CD3 or CD28 surface expression, which may be lower when compared to parental Jurkat T cells. We recommend confirming CD3 and CD28 surface expression by flow cytometry for all clones used in subsequent analysis. Resuspend Jurkat T cells (200,000 cells)

in 200 μl FACS buffer containing FITC conjugated anti-CD3 (1:100 dilution) and APC-conjugated anti-CD28 (1:50 dilution) antibodies. Incubate 20 min at RT in the dark, then wash 1× in FACS buffer and analyze with a flow cytometer (FITC channel: 488 nm excitation, 530/30 emission; APC channel: 638 nm excitation, 660/20 nm emission). Compare the mean fluorescence intensity (MFI) of CD3 and CD28 on the surface of the selected Jurkat T cells clones to MFI of parental Jurkat T cells. Expression levels in the KO clones should be comparable to parental Jurkat cells. Slight differences in the surface expression of these receptors in the KO clones compared to parental Jurkat T cells may be acceptable, if reconstitution of the clones confirms that the phenotype of the KO is caused by the specific deletion.

6. The psPAX2 packaging plasmid expresses the HIV-1 Gag, Pol, Tat and Rev proteins whereas the pMD2.G plasmid expresses the vesicular stomatatis virus G (VSV-G) envelope protein which enables the lentivirus particles to enter mammalian cells. The transfer vector pHAGE-ΔCD2-T2A-CDS [9] carries the CARD11 cDNA as coding sequence (CDS) flanked by HIV1-LTR sequences stable insertion into the genome. Its backbone is based on pHAGE-PGK-GFP-IRES-LUC-W plasmid (Addgene #46793) but instead of the GFP-IRES-LUC cassette a truncated human CD2 protein (ΔCD2) is encoded, which lacks a functional intracellular signalling domain but can be detected by extracellular staining with an anti-CD2 antibody for confirming successful transduction.
7. VSV-G-pseudo-typed lentiviral particles have a broad tropism and can potentially infect cells in the human body upon contact, inhalation or ingestion. Therefore, biosafety precautions are required also with respect to possible harmful characteristics of the transferred coding CDS, which as in the case of CARD11 mutants may code for an oncogene. Take necessary measure to protect yourself and strictly follow the guidelines and regulations in your country or region for working with lentiviral particles. Importantly, after transduction the infected cells are no longer able to produce new lentiviral particles. One week after transduction, we were not able to detect infective lentiviral particles in the supernatant of infected cells.
8. In the pHAGE plasmid, a third generation lentiviral SIN non-replicative vector, ΔCD2 and CARD11 expression are coupled by the 18 amino acids T2A sequence from *Thosea asigna*. The T2A linker co-translationally prevents the formation of its last peptide bond, allowing expression of the two genes under the control of the same promoter, here the phosphoglycerate kinase (PGK) promoter. The empty PGK-pHAGE plasmid carries a NotI-SalI-BamHI linker in

which ΔCD2 (without a STOP codon) and T2A were inserted in-frame, each with an EagI/NotI restriction site deletion strategy [9].

9. PMA is a cellular DAG (diacyl-glycerol) analogue and thus a potent activator of classical and novel PKCs (Protein kinase C), such as PKC-β and PKC-θ, respectively. In lymphocytes, CARD11 phosphorylation by PKC-β or PKC-θ triggers CBM complex formation and subsequently NF-κB activation. Ionomycin is a calcium ionophore, which triggers Ca^{2+} release from the endoplasmic reticulum and the plasma membrane into the cytosol and thereby supports T cell activation. Anti-CD3 and anti-CD28 antibodies are binding the TCR CD3 chains and CD28 co-stimulatory receptor, respectively. Addition of secondary crosslinking antibodies induces activation of TCR/CD28 by coupling the receptor on the membrane. TNFα is used as a stimulus that activates canonical NF-κB by signalling pathways independent of the CBM complex and therefore serves as a control that the common downstream events in the activation of NF-κB are intact in the Jurkat T cell clones.

References

1. Ruland J, Hartjes L (2019) CARD-BCL-10-MALT1 signalling in protective and pathological immunity. Nat Rev Immunol 19 (2):118–134. https://doi.org/10.1038/s41577-018-0087-2
2. Jaworski M, Thome M (2016) The paracaspase MALT1: biological function and potential for therapeutic inhibition. Cell Mol Life Sci 73 (3):459–473. https://doi.org/10.1007/s00018-015-2059-z
3. Meininger I, Krappmann D (2016) Lymphocyte signaling and activation by the CARMA1-BCL10-MALT1 signalosome. Biol Chem 397 (12):1315–1333. https://doi.org/10.1515/hsz-2016-0216
4. Bedsaul JR, Carter NM, Deibel KE, Hutcherson SM, Jones TA, Wang Z, Yang C, Yang YK, Pomerantz JL (2018) Mechanisms of regulated and dysregulated CARD11 signaling in adaptive immunity and disease. Front Immunol 9:2105. https://doi.org/10.3389/fimmu.2018.02105
5. Gehring T, Seeholzer T, Krappmann D (2018) BCL10 – bridging CARDs to immune activation. Front Immunol 9:1539. https://doi.org/10.3389/fimmu.2018.01539
6. Abraham RT, Weiss A (2004) Jurkat T cells and development of the T-cell receptor signalling paradigm. Nat Rev Immunol 4(4):301–308. https://doi.org/10.1038/nri1330
7. Lu HY, Bauman BM, Arjunaraja S, Dorjbal B, Milner JD, Snow AL, Turvey SE (2018) The CBM-opathies—a rapidly expanding spectrum of human inborn errors of immunity caused by mutations in the CARD11-BCL10-MALT1 complex. Front Immunol 9:2078. https://doi.org/10.3389/fimmu.2018.02078
8. Ran FA, Hsu PD, Wright J, Agarwala V, Scott DA, Zhang F (2013) Genome engineering using the CRISPR-Cas9 system. Nat Protoc 8 (11):2281–2308. https://doi.org/10.1038/nprot.2013.143
9. Hadian K, Griesbach RA, Dornauer S, Wanger TM, Nagel D, Metlitzky M, Beisker W, Schmidt-Supprian M, Krappmann D (2011) NF-kappaB essential modulator (NEMO) interaction with linear and lys-63 ubiquitin chains contributes to NF-kappaB activation. J Biol Chem 286(29):26107–26117. https://doi.org/10.1074/jbc.M111.233163
10. Schimmack G, Schorpp K, Kutzner K, Gehring T, Brenke JK, Hadian K, Krappmann D (2017) YOD1/TRAF6 association balances p62-dependent IL-1 signaling to NF-kappaB. elife 6. https://doi.org/10.7554/eLife.22416
11. Meininger I, Griesbach RA, Hu D, Gehring T, Seeholzer T, Bertossi A, Kranich J,

Oeckinghaus A, Eitelhuber AC, Greczmiel U, Gewies A, Schmidt-Supprian M, Ruland J, Brocker T, Heissmeyer V, Heyd F, Krappmann D (2016) Alternative splicing of MALT1 controls signalling and activation of CD4(+) T cells. Nat Commun 7:11292. https://doi.org/10.1038/ncomms11292

12. Seeholzer T, Kurz S, Schlauderer F, Woods S, Gehring T, Widmann S, Lammens K, Krappmann D (2018) BCL10-CARD11 fusion mimics an active CARD11 seed that triggers constitutive BCL10 oligomerization and lymphocyte activation. Front Immunol 9:2695. https://doi.org/10.3389/fimmu.2018.02695
13. Schlauderer F, Seeholzer T, Desfosses A, Gehring T, Strauss M, Hopfner KP, Gutsche I, Krappmann D, Lammens K (2018) Molecular architecture and regulation of BCL10-MALT1 filaments. Nat Commun 9(1):4041. https://doi.org/10.1038/s41467-018-06573-8
14. Gehring T, Erdmann T, Rahm M, Grass C, Flatley A, O'Neill TJ, Woods S, Meininger I, Karayel O, Kutzner K, Grau M, Shinohara H, Lammens K, Feederle R, Hauck SM, Lenz G, Krappmann D (2019) MALT1 phosphorylation controls activation of T lymphocytes and survival of ABC-DLBCL tumor cells. Cell Rep 29(4):873–888. e810. https://doi.org/10.1016/j.celrep.2019.09.040
15. Kutukculer N, Seeholzer T, O'Neill TJ, Grass C, Aykut A, Karaca NE, Durmaz A, Cogulu O, Aksu G, Gehring T, Gewies A, Krappmann D (2021) Human immune disorder associated with homozygous hypomorphic mutation affecting MALT1B splice variant. J Allergy Clin Immunol 147:775–778 e778. https://doi.org/10.1016/j.jaci.2020.07.034
16. Murphy GJ, Mostoslavsky G, Kotton DN, Mulligan RC (2006) Exogenous control of mammalian gene expression via modulation of translational termination. Nat Med 12(9):1093–1099. https://doi.org/10.1038/nm1376
17. Salmon P, Trono D (2007) Production and titration of lentiviral vectors. Curr Protoc Hum Genet 12:10. https://doi.org/10.1002/0471142905.hg1210s54
18. Ben-Dor I, Itsykson P, Goldenberg D, Galun E, Reubinoff BE (2006) Lentiviral vectors harboring a dual-gene system allow high and homogeneous transgene expression in selected polyclonal human embryonic stem cells. Mol Ther 14(2):255–267. https://doi.org/10.1016/j.ymthe.2006.02.010
19. Doench JG, Fusi N, Sullender M, Hegde M, Vaimberg EW, Donovan KF, Smith I, Tothova Z, Wilen C, Orchard R, Virgin HW, Listgarten J, Root DE (2016) Optimized sgRNA design to maximize activity and minimize off-target effects of CRISPR-Cas9. Nat Biotechnol 34(2):184–191. https://doi.org/10.1038/nbt.3437
20. Dehairs J, Talebi A, Cherifi Y, Swinnen JV (2016) CRISP-ID: decoding CRISPR mediated indels by Sanger sequencing. Sci Rep 6:28973. https://doi.org/10.1038/srep28973

Chapter 9

Analysis of Calcium Control of Canonical NF-κB Signaling in B Lymphocytes

Corbett T. Berry, Michael J. May, and Bruce D. Freedman

Abstract

The central role of calcium (Ca^{2+}) signaling in lymphocyte development and acquisition of functional immunity and tolerance is well established. Ca^{2+} signals are initiated upon antigen binding to cognate receptors on lymphocytes that trigger store operated Ca^{2+} entry (SOCE). The underlying mechanism of SOCE in lymphocytes involves TCR and BCR mediated activation of Stromal Interaction Molecule 1 and 2 (STIM1/2) embedded in the ER membrane. Once activated, STIM proteins oligomerize and re-localize to ER domains juxtaposed to the plasma membrane where they activate Orai channels to allow Ca^{2+} to enter the cell across the plasma membrane. Importantly, STIM/Orai-dependent Ca^{2+} signals guide antigen induced lymphocyte development and function principally by regulating the activity of transcription factors.

The most widely studied of these transcription factors is the Nuclear Factor of Activated T cells (NFAT). NFAT is expressed ubiquitously and the mechanism by which Ca^{2+} regulates NFAT activation and signaling is well known. By contrast, a mechanistic understanding of how Ca^{2+} signals also shape the activation and specificity of NF-κB to control the expression of pro-inflammatory genes has lagged. Here we discuss the methodology used to investigate Ca^{2+} dependent mechanisms of NF-κB activation in lymphocytes. Our approach focuses on three main areas of signal transduction and signaling: (1) antigen receptor engagement and Ca^{2+} dependent initiation of NF-kB signaling, (2) Ca^{2+} dependent induction of NF-κB heterodimer activation and nuclear localization, and (3) and how Ca^{2+} regulates NF-κB dependent expression of target genes and proteins.

Key words NF-κB, p65, c-Rel, Calcium, IκBα, STIM, Orai, B Lymphocyte, BCR

1 Introduction

The central role for calcium (Ca^{2+}) signaling in lymphocytes has been appreciated for 30 years. However, our understanding of the regulation and specific mechanisms of Ca^{2+} action have advanced more rapidly since the identification of the Stromal Interaction Molecule 1 and 2 proteins (STIM1/2) and Orai channels. STIM and Orai are expressed in most cell types and although the mechanisms and biological consequences of this store operated Ca^{2+} entry

Guido Franzoso and Francesca Zazzeroni (eds.), *NF-κB Transcription Factors: Methods and Protocols*, Methods in Molecular Biology, vol. 2366, https://doi.org/10.1007/978-1-0716-1669-7_9,

(SOCE) pathway vary widely, in virtually all instances STIMs play a pivotal role [1–3]. Triggering STIM1 and STIM2 begins with Ca^{2+} dissociation from their EF hand domains within the endoplasmic reticulum (ER) lumen. Stimulus-induced decreases in ER [Ca^{2+}] promote its dissociation from STIM EF hands to trigger conformational activation, oligomerization, and relocalization of STIMs [4]. This resulting physical interaction with Orai channels initiates their activation providing a pathway for Ca^{2+} entry across the plasma membrane. In lymphocytes STIM1 plays a dominant role in antigen receptor (TCR and BCR)-induced Orai1 activation while STIM2 controls cytoplasmic Ca^{2+} levels under resting conditions [1, 2, 5, 6].

More than a decade before the details of SOCE were identified, the mechanism by which Ca^{2+} controls NFAT activation was established [7]. Despite knowing that Ca^{2+} also tunes NF-κB activity [8–15], only recently has our mechanistic understanding, including the basis for the distinct Ca^{2+} sensitivity of NF-κB been revealed in detail [16, 17]. In the case of canonical NF-κB signaling, inhibitory kappa-B (IκB) proteins sequester hetero-/homo-dimers of p65, c-Rel, and p50 proteins in the cytoplasm [18, 19]. Antigen receptor coupled phospholipases (PLCγ-1 and PLCγ-2 in T cells and B cells respectively) hydrolyze plasma membrane phosphatidylinositol 4,5-bisphosphate (PIP_2) to form diacylglycerol (DAG) and Inositol 1,4,5 trisphosphate (IP_3). IP_3 activates IP_3 receptor/channels in the ER membrane and the resulting loss of ER Ca^{2+} activates STIM1 and STIM2, which initiate Orai-mediated Ca^{2+} entry. DAG, in some instances in conjunction with Ca^{2+}, activates one of several protein kinase C (PKC) isoforms that phosphorylate CARMA1, promoting its association with BCL10 and MALT1 to form the CBM complex [20, 21]. The CBM complex in turn activates the IκB kinase (IKK) complex comprised of IKKα, IKKβ, and NEMO [22]. IKKβ mediated phosphorylation of IκB bound to p65/p50 and c-Rel/p50 targets it for proteasomal degradation, freeing p65/p50 and c-Rel/p50 dimers, which migrate to the nucleus where they can initiate gene expression. More recently, we established that phosphorylation controls p65 nuclear localization and transcriptional activation in T lymphocytes [23] and established the Ca^{2+} sensitivity of p65 and c-Rel activation in B lymphocytes [16]. Although much remains to be discovered, we and others have outlined a framework whereby multiple Ca^{2+} regulated checkpoints control NF-κB activation and function in T and B cells.

In this chapter we describe approaches we have developed to interrogate the activity of sequential checkpoints in canonical NF-kB activation in B lymphocytes. We are focused on the control of these steps by BCR-induced Ca^{2+} signals but the framework we outline could also be used to examine the regulation of control

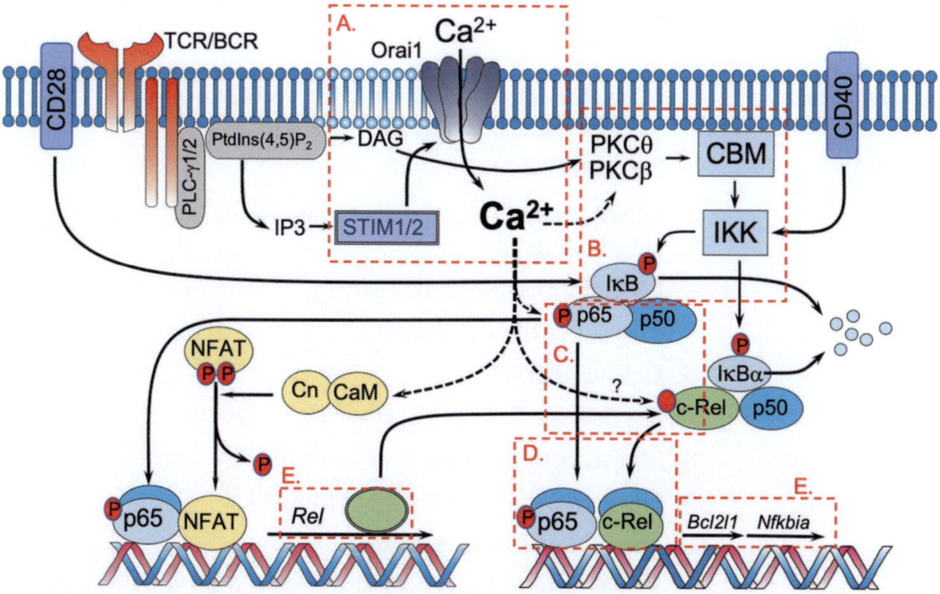

Fig. 1 Mechanisms of canonical NF-kB activation by antigen receptor induced calcium signals. This chapter focuses on how to dissect sequential steps in canonical NF-κB signaling in lymphocytes. A critical requirement for studying these actions of Ca^{2+} is a general understanding of the mechanisms of antigen induced Ca^{2+} mobilization and how to manipulate these signals. In our studies, we use two approaches to selectively modulate Ca^{2+} entry without disrupting other consequences of antigen receptor engagement. In response to antigen receptor engagement, the predominant entry pathway for extracellular Ca^{2+} are ion channels in the plasma membrane encoded by Orai1 (**a**). Orai1 channels are gated/activated by ER membrane resident STIM1 and STIM 2 proteins whose activity is triggered by IP3 mediated depletion of Ca^{2+} from the ER. By deleting STIM1 and STIM2 in B lymphocytes (or T lymphocytes) one can dissect how Ca^{2+} entry regulates NF-κB activation and actions. A second strategy used to assess the role of Ca^{2+} entry is to stimulate B lymphocytes in Ca^{2+} deficient medium. In this scenario, in the absence of extracellular Ca^{2+}, Orai channels are activated but no extracellular Ca^{2+} is available to enter the cell. Using these approaches to manipulate Ca^{2+} entry, one can then examine the impact of these signals on IKK complex activation (**a**, not a focus of this chapter) and its impact on IκBα degradation (**b**). IκBα phosphorylation initiates IκBα proteasomal degradation. IκBα phosphorylation, degradation, and re-expression can be assessed by immunoblot analysis. IκBα degradation is a prerequisite for the nuclear localization of p65, c-Rel, and p50 to the nucleus to initiate gene expression. One critical action of Ca^{2+} we identified is phosphorylation (**c**) of p65 to regulate its nuclear localization (**d**). Upon entering the nucleus, p65 and c-Rel initiate the expression of immune-regulatory genes (**e**) including the re-expression of IκBα (*NFKB1A*) following its degradation, expression of the pro-survival genes including BcL-xL (*BCL2L1*) and c-Rel (*Rel*)

points in canonical NF-kB signaling in any cells type in response to any stimulus. Key mechanisms by which BCR induced Ca^{2+} signals regulate proximal and distal checkpoints in canonical NF-κB activation are illustrated in Fig. 1. Each of the boxes indicates a step in this cascade for which we outline the experimental approach and methods that can be used to query its activity.

2 Materials

2.1 B Lymphocyte Isolation and Stimulation

1. Mice: C57Bl/6 mice, STIM1$^{f/f}$ STIM2$^{f/f}$Mb1^{Cre+}, STIM1$^{f/f}$ STIM2$^{f/f}$Mb1^{Cre-} (*see* **Notes 1** and **2**).
2. 70% ethanol.
3. CD23 coated microbeads (Miltenyi Biotec) (*see* **Note 3** for T lymphocytes).
4. LS Columns (Miltenyi Biotec).
5. Magnet (Miltenyi Biotec).
6. Cell strainers.
7. Surgical tools: scissors, tweezers.
8. 50 mL conical tubes.
9. Sterile 3 mL syringes (1 per mouse).
10. Falcon Petri Dishes 35 mm.
11. FACS buffer: 1× DPBS, 0.5% BSA, 0.25 mM EDTA.
12. Complete RPMI medium (cRPMI): RPMI 1640 medium supplemented with 10% fetal bovine serum (FBS), 2 mM L-glutamine, penicillin (50 U/mL), 50 U/mL streptomycin, 1 mM pyruvate, 100 mM MEM Non-essential amino acids solution, 50 mM 2-mercapto ethanol, 10 mM HEPES.
13. Complete Ca^{2+} "free" RPMI (cRPMI CF): Complete RPMI with 0.5 mM EGTA.
14. Complete DMEM medium (cDMEM): DMEM medium, supplemented with 10% FBS, 2 mM L-glutamine, penicillin (50 U/mL), 50 U/mL streptomycin, 1 mM pyruvate, 100 mM MEM Non-essential amino acids solution, 50 mM 2-mercapto ethanol, 10 mM HEPES.
15. Calcium free DMEM (cDMEM CF): Calcium free DMEM solution with 0.5 mM EGTA along with added components as above.
16. Normal experimental solution (NES): 145 mM NaCl, 4.5 mM KCl, 2 mM $CaCl_2$, 1 mM $MgCl_2$, 10 mM glucose, 10 mM HEPES, 2 mM glutamine, 0.5% FBS, glutamax.
17. Calcium free experimental solution (NES CF): 145 mM NaCl, 4.5 mM KCl, 1 mM $MgCl_2$, 0.25 mM EGTA, 10 mM glucose, 10 mM HEPES, 2 mM glutamine, 0.5% FBS, glutamax.
18. Sterile Bottle Top Filters.
19. AffiniPure $F(ab')_2$ Fragment Goat Anti-Mouse IgM, μ chain specific (Jackson ImmunoResearch).

2.2 SDS-Polyacrylamide Gel Electrophoresis (SDS-PAGE)

1. Separating buffer (4×): 1.5 M Tris–HCl, pH 8.8, 0.4% (w/v) SDS. Stored at room temperature.
2. Stacking buffer (4×): 0.5 M Tris–HCl, pH 6.8, 0.4% (w/v), SDS. Stored at room temperature.
3. Thirty percent acrylamide/bis solution (37.5:1with 2.6% C) stored at 4 °C.
4. *N,N,N,N′*-tetramethyl ethylenediamine (TEMED): Stored at room temperature.
5. Ammonium persulfate (APS): Prepare in 10% (w/v) and store at 4 °C.
6. Methanol.
7. Running buffer: 25 mM Tris, 250 mM, glycine, 0.1% (w/v) SDS. Prepare a 5× stock stored at room temperature.
8. Spacer plates, short plates, and combs.
9. Pre-stained protein molecular weight standard markers.

2.3 Sample Preparation and Blotting for IκBα Phosphorylation and Degradation

1. NP40 lysis buffer: 50 mM Tris–HCl, pH 7.5, 20 mM EDTA, 1% NP-40.
2. Protease inhibitors cocktail (PIC): 1 mM Sodium Orthovanadate, 1 mM PMSF, 10 mg/mL Leupeptin, 5 mg/mL Aprotinin.
3. UV-visible Spectrophotometer.
4. SDS–polyacrylamide gel electrophoresis (4–15%, Bio-Rad).
5. Transfer buffer: 25 mM Tris, 192 mM glycine, 20% (v/v) methanol.
6. Ice pack.
7. Methanol at room temperature.
8. Polyvinylidene difluoride (PVDF) membranes.
9. Thick cellulose chromatography filter paper.
10. Tris-buffered saline with Tween (TBS-T): 25 mM Tris–HCl, pH 8.0, 140 mM NaCl, 3 mM KCl, 0.05% Tween-30.
11. Blocking buffer: 5% (w/v) nonfat dried milk in TBS-T.
12. Stripping buffer: 100 mM 2-mercaptoethanol, 2% SDS, 62.5 mM Tris–HCl.
13. Primary anti-human antibodies: Phospho-IκBα (Ser32) (14D4) Rabbit mAb, IκBα (L35A5) Mouse mAb (Amino-terminal Antigen) (Cell Signaling), anti-alpha-tubulin (Sigma Aldrich).
14. Secondary antibodies: Protein A HRP (Cell signaling), Goat anti-Rabbit IgG (H + L) HRP.

15. ECL Western Blotting Substrate (Pierce).
16. ChemiDoc or similar gel documentation system.
17. NuPAGE LDS sample buffer (4× stock) and NuPAGE Reducing Agent (10× stock) (Invitrogen).

2.4 Confocal and Immunocytochemical Analysis of p65 and c-Rel Localization

1. DMEM medium containing 2 mM Ca^{2+} or Ca^{2+} free equivalent solution (*see* Subheading 2.1, **item 12**).
2. Cell-Tak (Corning).
3. 15 mm glass coverslips.
4. 0.2% Triton-X-100.
5. 4% paraformaldehyde.
6. 1-5% BSA in DPBS.
7. 1 mg/mL anti-p65 primary antibody (Santa Cruz Biotechnology).
8. Anti-c-Rel-PE primary antibody (Miltenyi biotec).
9. Alex Fluor 488 goat anti-rabbit secondary antibody (Thermo Fischer).
10. 4 mg/mL Hoechst 33342 (Life Technologies).
11. Fluoromount (Fisher Scientific).
12. Leica confocal microscope (Leica Microsystems, Wetzlar, Germany) or similar confocal microscope.
13. ImageJ (NIH, https://imagej.net/ImageJ).

2.5 Quantitative RT-PCR Analysis of NF-κB Target Gene Expression

1. RNeasy Plus Mini or Micro Kit (Qiagen).
2. High capacity cDNA reverse transcription kit (Applied Biosystems).
3. Power SYBR Green PCR Master Mix (Applied Biosciences).
4. 384 well plate.
5. Plate covers.
6. 7500 Fast Real-Time PCR System (Applied Biosciences).
7. Primers for Ca^{2+} and NF-κB dependent gene expression are: *Rel* (Forward: GGCCTTTTTCTCCTTTGGCG, Reverse: GATCCACAAAAGTGTCCCAGC), *Bcl2l1* (Forward: TGACAACCGTGCCCCAAATA, Reverse: TTGGCGGTGTACATCAGCTT), *Irf4* (Forward: CTTCAAGGCTTGGGCATTGTT, Reverse: TGGCCATCTGTGTGTCATCC), *Ubc* (Reference gene, Forward: GCCCAGTGTTACCACCAAGA , Reverse: CCCATCACACCCAAGAACA).

3 Methods

3.1 B Lymphocyte Isolation and Stimulation (See Note 2 About T Cells)

1. Uncap a 50 mL conical tube for each spleen harvested during dissection.
2. Prepare a 30 mm petri dish with 3 mL of sterile filtered FACS buffer for each spleen to be dissected. Keep the dish covered.
3. Euthanize mice using an IACUC approved method such as CO_2 asphyxiation.
4. Spray the site of incision on the left abdomen with ethanol, tent the skin and snip in a rostral to caudal orientation, and use 70% ethanol sterilized instruments to isolate and remove the spleen.
5. Resect the spleen taking care not to damage the splenic capsule. Remove vessels from the hilum. Place each spleen into a separate petri dish. Move petri dishes to a laminar flow hood for processing and continue through step 15 with sterile technique.
6. Unwrap and place a 40μM pore size cell strainer into each petri dish and move the spleen into the cell strainer.
7. Remove the sterile thumb-rest portion of a plunger from a 3 cc syringe and use it to gently press the spleen through the bottom of the cell strainer to dissociate tissue into single cell suspension. The parenchyma and capsule will remain in the cell strainer. Repeat for each spleen with a new plunger.
8. Lift the cell strainer holding its tab and pipette an additional 10 mL of FACS buffer through it into the petri dish.
9. Centrifuge samples at 100 × *g* for 5 min at room temperature (*see* **Note 4**).
10. Remove red blood cells (RBCs) from the cell preparations (*see* **Note 5**).
11. Purify CD23 B cells using CD23⁺ positive selection microbeads (Miltenyi biotec). Save a 30μL sample of isolated cells to check the purity before and after isolation by flow cytometry (*see* **Notes 6** and **7**).
12. Resuspend isolated B cells at ten million cells/mL in cRPMI or cDMEM (*see* **Note 8**).
13. Centrifuge at 300 × *g* for 5 min.
14. Pour off supernatant and resuspend B lymphocytes in pellet by flicking the tube and resuspend cells in cRPMI at ten million cells/mL (*see* **Notes 8** and **9**).
15. Stimulate B lymphocytes immediately after isolation by adding an equal volume of medium containing anti-IgM (*see* **Note 10**).

16. Add anti-IgM at 2× the desired final stimulation concentration (20μg/mL) that will be added to an equal volume of cells for rapid mixing/stimulation (*see* **Note 11**).
17. At appropriate stimulation time points, recover cells and process for further analysis.

3.2 SDS-PAGE

1. These directions are for casting and running 10% SDS gels using the Bio-Rad Mini Protean Electrophoresis apparatus. Directions for other systems are similar and an alternative to making gels would be purchasing precast gels (*see* **Note 12**).
2. Clean the back spacer plates and front short plates before use. Plates should be wiped down with 70% ethanol to avoid gel sticking unevenly to the plates.
3. Assemble the glass plates in the gel-casting apparatus placing a short front plate with a back spacer plate lined up evenly.
4. In a 50 mL conical tube, prepare a 10% resolving gel by mixing 6.25 mL water, 3.75 mL 4× separating buffer, 5 mL 30% acrylamide/bis solution, 50μL APS solution, and 20μL TEMED. Mix by inversion.
5. Pipette the solution into the gel chamber leaving 0.5″ for the stacking gel. Layer 1 mL of methanol onto the resolving gel while the gel solidifies. Allow the gel to polymerize for about 30 min.
6. Prepare the stacking gel in a 50 mL conical by mixing 3.05 mL of water, 1.25 mL of 4× stacking buffer, 0.65 mL of 30% acrylamide/bis solution, and 25μL of APS solution.
7. When the resolving gel is polymerized, place a piece of paper towel along the top of each gel chamber and invert the casting mold to absorb the methanol out of the gel chamber in preparation of the stacking gel.
8. Add 10μL of TEMED to the stacking gel solution and mix by inversion.
9. Pipette stacking gel on top of the resolving gel and insert a 10-well comb.. The stacking will polymerize in 5–10 min.
10. Assemble the electrophoresis cassette in the tank when gel has fully polymerized. Two gels will be required to create an inner chamber, using a dam plate if only one gel is required. Fill inner chamber entirely and 1/3 of the outer chamber with 1× running buffer.
11. Remove the comb by pulling straight up on both sides of the comb to keep wells intact.
12. Using a gel-loading tip, add 10μL of protein molecular weight standard to the first well. If possible, skip one well and add up to 25μL samples to subsequent wells with gel-loading tips from Subheading 3.3, **step 16**.

13. Connect the power supply and run at 150–180 V until the samples have separated over the entire length of the gel (visualized by watching the separation of the pre-stained markers). The gel is run at room temperature.

3.3 Analysis of Phospho-IκBα and IκB (Fig. 1b, See Note 13)

1. Use three million purified B lymphocytes per condition. Cells can be stimulated in sterile 0.7 mL microcentrifuge tubes or in 12 well plates. Microcentrifuge tubes are the simplest for downstream processing and are optimal for minimizing cell loss during collection.
2. Label eight microcentrifuge tubes with genotype, timepoint, and external Ca^{2+} condition. We commonly employ two experimental designs to address the calcium entry dependence of NF-kB activation (*see* **Note 14**).
3. Pipette 300μL of cell suspension (i.e., three million cells in cRPMI or in cRPMI CF, *see* **Note 15**) into to each tube.
4. Set a timer for each individual time point.
5. Pipette 300μL of 2× stimulation media into each tube.
6. Place in a 5% CO_2 incubator with the lids open.
7. Sample collection.
   (a) For stimulation in microcentrifuge tubes: At each timepoint, remove the corresponding samples from the incubator. Replace the cap and centrifuge to max speed to collect the cells. Aspirate the supernatant taking care not to disturb the cell pellet at the bottom of the tube.
   (b) For stimulation in 12 well plates: At each time point, remove the plate from the incubator, gently pipette to resuspend the cells. Transfer to a microcentrifuge tube and centrifuge up to max speed to collect the cells.
8. Aspirate the supernatant taking care not to disturb the cell pellet at the bottom of the tube or plate. Take care not to leave any liquid behind. At this point, samples can be placed directly in a −80 °C freezer or immediately processed as described below.
9. Pipette 40μL of cold NP40 lysis buffer with PIC directly onto the cell pellet.
10. Vortex for 60 s and place on ice for 30 min.
11. Centrifuge at 20,000 × *g* for 20 min.
12. Transfer supernatant to a new microcentrifuge tube.
13. Use 1μL of protein lysate to check the protein concentration in cell lysates via Bradford assay.
14. Transfer the volume equivalent of 50μg of protein to a new microcentrifuge tube. The sample volume must be <32.5μL to allow for loading into 50μL wells.

15. Prepare samples for SDS-PAGE in sample buffer. Pipette the appropriate volume of NuPAGE LDS sample buffer (4× stock) and NuPAGE Reducing Agent (10× stock) to obtain a 1× dilution.
16. Heat samples at 95 °C for 10 min.
17. Resolve proteins by gel electrophoresis as directed below. Connect the power supply and run at 150–180 V until the samples have separated over the entire length of the gel (visualized by watching the separation of the pre-stained markers, *see* **Note 16**). The gel is run at room temperature.

3.4 Blotting and Immunostaining

1. These directions for protein transfer are for the Life Technologies Electrophoretic Transfer Cell wet transfer apparatus.
2. Fill the transfer chamber with prefilled complete transfer buffer (containing 20% methanol) and place an ice pack in the outer chamber.
3. Cut two pieces of chromatography filter paper and a single piece of PVDF membrane large enough to cover the gel. Saturate the PVDF membrane with 100% methanol and immediately immerse membrane in transfer buffer. Do not allow membrane to dry.
4. Fill a dish large enough to hold the transfer cassette with complete transfer buffer. Immerse the membrane, filter paper, and foam pads in transfer buffer.
5. Remove the gel from the electrophoresis apparatus and trim away the stacking gel using a clean razor blade.
6. Assemble the gel transfer "sandwich" (keeping all materials immersed in transfer buffer) in the following order: sponge, filter paper, gel, PVDF membrane, filter paper, and sponge (*see* **Note 17**).
7. Remove any air bubbles from the assembly using a roller or plastic 5–10 mL serological pipette.
8. Place the sandwich in a cassette and insert the cassette into the inner chamber of the transfer apparatus with the gel side of the sandwich closest to the negative (black) electrode and the membrane side of the sandwich closest to the positive (red) electrode.
9. Run the transfer at 125 V for 1 h 20 m at 4 °C.
10. When transfer is complete the markers should be visible on the membrane. Disassemble the cassette and sandwich starting at one corner and place the PVDF membrane in 5% solution (w/v) of BSA in TBS-T buffer (Subheading 2.4, **item 10**) with constant gentle agitation on a rocker for 30 minutes at room temperature.

11. Incubate the membrane with anti-phospho-(Ser32/36)-IκBα (dilution 1:1000 in TBS-T with 2% BSA) for 3 h at room temperature or overnight at 4 °C with gentle agitation (*see* **Note 18**).
12. Wash the membrane with TBS-T buffer three times for 5 min.
13. Apply a secondary HRP-conjugated antibody (dilution 1:4000 in TBS-T with 2% BSA) for 1 h at room temperature with gentle agitation (*see* **Note 19**).
14. Wash three times for 5 min with TBS-T buffer.
15. During the final wash, mix 1 mL of each of the ECL reagents for chemiluminescence detection. Immediately after discarding the TBS-T from the final wash, pipette the mixed luminal reagent onto the blot. Incubate at room temperature for 3 min.
16. Pick up the membrane using forceps and gently dab off the luminal using Kim-Wipes. and record chemiluminescence with a ChemiDoc or similar gel documentation system.
17. Strip the membrane by stripping buffer for 30 min at 50 °C then wash for 10 min in TBS-T buffer and repeat wash two times in order to re-probe.
18. Repeat blocking in 4% BSA in TBS-T for 15 min at room temperature.
19. Re-probe the membrane with anti-IκBα antibody (dilution 1:1000 in PBS-T with 2% BSA).
20. To detect the degradation IκBα, analyze relative amount of the protein in the input samples by probing with a-tubulin (1:10,000). We do not strip the blots for this analysis since alpha-tubulin (50 KDa) does not migrate with IκBα (*see* **Note 20** for alternative approaches to analyze proximal steps of activation).
21. Additional Ca^{2+} and NF-κB dependent targets including c-Rel, Irf4, IκBα, and Bcl-xL can be evaluated using immunoblotting or flow cytometry.

3.5 Immuno-Fluorescence Analysis of p65 and c-Rel Localization (Fig. 1d)

1. Following IκBα phosphorylation and degradation, NF-κB proteins are released from the cytoplasm and localize to the nucleus (*see* **Note 21**). To assess localization, we use the procedures outlined in Subheadings 3.1 and 3.2 to isolate and activate cells (also *see* **Note 22** for typical experimental design). We typically stimulate lymphocytes in 96 well flat bottom plates and then collect and adhere them to 15 mm glass coverslips for efficient downstream processing.
2. Coat 15 mm glass #1.5 coverslips with Cell-Tak.
3. Wash coverslips in sterile water and 70% ethanol to remove particulates.

4. Wipe dry with Kim-Wipe. Pipette 5μL of Cell-Tak onto the center of the coverslip.
5. Spread the liquid on the coverslip using a plastic pipette tip bent at a 45–90° angle.
6. Continue spreading for 3–5 min or until liquid has evaporated.
7. Submerge the coverslip in 70% ethanol and rinse with nanopore filtered water.
8. Allow the coverslip to airdry then place coverslips on parafilm on a plastic or metal tray (*see* **Note 23**).
9. Pipette 100μL of cell suspension (i.e., one million cells) into each well of a 96-well plate.
10. Pipette 100μL of 2× stimulation media (Ca^{2+} containing and/or Ca^2 free) or normal media (Ca^{2+} containing and/or Ca^2 free) into each well.
11. Place in a 5% CO_2 incubator at 37 °C for 60 min.
12. Remove from incubator. Gently pipette to suspend cells.
13. Collect the cell suspension and transfer to a 0.7 mL microcentrifuge tube.
14. Centrifuge at 100 × *g* for 5 min.
15. Discard the supernatant.
16. Gently resuspend the cell pellet in 30μL of cRPMI or calcium free cRPMI. Pipette slowly to avoid generating bubbles or damaging cells (*see* **Note 21** for experimental design alternatives).
17. Dispense the cell suspension onto the center of a Cell-Tak coated coverslip.
18. Place in a 5% CO_2 incubator at 37 °C for 10–15 min to allow for effective adhesion.
19. Wash three times with DPBS using a plastic polyethylene bulb pipette. Dispense an adequate volume to cover the entire coverslip. For the rest of this procedure, when washing or adding an additional solution, carefully vacuum suction off the residual fluid using a pipette tip or glass pipette.
20. Add 4% paraformaldehyde in DPBS to each coverslip. Incubate for 10 min at RT.
21. Wash three times with DPBS as described above.
22. Add 0.2% Triton-X-100 to permeabilize cells for 5 min at RT.
23. Wash three times with DPBS for 5 min.
24. Add 5% BSA in DPBS and incubate at RT for 1 h.
25. Prepare primary antibody solution. Calculate the total volume of primary antibody staining solution needed to stain each

sample with 50μL of solution. Make an additional 20% to cover for volume loss due to pipetting. Dilute primary antibodies including anti-p65 (1:100) and anti-c-Rel-PE antibodies (1:60) in 1% BSA in DPBS. Minimize exposure to light (also *see* **Note 7** regarding Fc receptor binding).

26. Dispense a 50μL droplet onto fresh parafilm in a tray. Using forceps, carefully invert the coverslip onto the antibody solution so that the cell side is now down. Cover with tin foil to block photobleaching by ambient light. Incubate at room temperature for 1 h or overnight at 4 °C.
27. Use forceps to invert each coverslip onto fresh parafilm so that the cells face upward.
28. Wash each coverslip three times with 1% BSA in DPBS for 5 min per wash.
29. Prepare secondary antibody solution. Calculate the total volume of secondary antibody staining solution necessary to stain all samples with 80μL per coverslip. Make an additional 20% to cover for volume loss with pipetting. Dilute secondary antibodies including Alexa fluor 488 goat anti-rabbit 1:250 in DPBS with 1% BSA.
30. Pipette 80μL of secondary antibody solution directly onto each coverslip (do not invert). Cover with foil and incubate for 45 min at RT.
31. Wash three times with 1% BSA in PBS for 5 min each.
32. Add Hoechst 33342 to label nuclei. Cover with foil and incubate at RT for 20 min.
33. Wash each coverslip three times for 5 min with 1% BSA in PBS.
34. Three 10–15 mm coverslips can be mounted on a single glass slide. Label each slide with relevant information (e.g., date, cell type, stimulation, duration).
35. Pipette 5–10μL of Fluoromount onto the slide. Using forceps, carefully grasp each coverslip and briefly dip it into the beaker containing water.
36. Touch the edge of a Kimwipe to the coverslip and attempt to remove as much liquid as possible.
37. Invert the coverslip (cell side down) onto the drop of Fluoromount. Repeat for additional coverslips. Allow to dry overnight at RT.
38. Evaluate the localization of NF-κB images on a confocal microscope with channels for Hoechst, c-Rel (PE), p65 (Fluor 488), and bright field or DIC.
39. Measure nuclear intensities of p65 and c-Rel (*see* **Note 24**).

3.6 Measurement of NF-κB Driven Gene Expression (Fig. 1e)

1. Use the procedure outlined in Subheadings 3.1 and 3.2 along with a sample density of three million cells/sample for analysis of gene expression. Cells can be stimulated in sterile 0.7 mL Eppendorf tubes or in flat bottom 12-well plates.
2. Label wells with genotype, stimulus, and external Ca^{2+} condition. See below for two commonly used experimental designs similar to those described above (*see* **Note 16**).
3. Pipette 300μL of cell suspension (i.e., three million cells) into each microcentrifuge tube.
4. Pipette 300μL of 2× stimulation media (Ca^{2+} containing and/or Ca^2 free) or normal media (Ca^{2+} containing and/or Ca^2 free) into each well.
5. Place in a 5% CO_2 incubator at 37 °C for 60 min.
6. Remove from incubator.
7. Centrifuge at 100 × *g* for 5 min.
8. Discard the supernatant.
9. Isolate RNA using the RNeasy Plus Mini kit.
10. Check RNA concentration and purity.
11. Generate cDNA using High capacity cDNA reverse transcription kit. We use 100–500 ηg of RNA as input. It is critical to use the same quantity of input RNA for all samples.
12. Dilute cDNA in nuclease free water. For 100–200 ηg, we dilute 1:10. For 500 ηg, we dilute 1:20.
13. Prepare 384 well plate for qRT-PCR. Determining the number of samples, replicates (at least three per condition), and targets. Among the validated NF-κB targets we typically assess are *Nfkbia* (encodes IκBα protein), *Rel* (encodes c-Rel protein), *Bcl2l1* (encodes Bcl-xL protein), and *Irf4* (encodes Irf4 protein). We use 10μL reactions (5μL of Power SYBR Green PCR Master Mix, 4μL cDNA, and 1μL of primer to each well).
14. Thaw SYBR Green Master Mix (2×) on ice.
15. Add diluted cDNA into PCR strip tubes.
16. Dilute stock primers to 10μM in nuclease free water. Add equal volume forward and reverse primer into a nuclease free microcentrifuge tube for a final concentration of 5μM.
17. Prepare primers with the master mix. Label a clean microcentrifuge with each target. Incorporate an error factor (*see* **Note 25**). Pipette each volume of primer and master mix into a clean microcentrifuge tube.
18. Vortex and place on ice.
19. Load 6μL of primer/master mix into PCR plate.

20. Cover with a plate sealer, vortex, and centrifuge the plate at 200 × *g* for 1 min.
21. Load onto qPCR machine. Cycle parameters should be optimized for each gene target and instrument.
22. Analyze with the instrument software with the appropriate reference gene and sample identified.

4 Notes

1. $STIM1^{f/f}$ $STIM2^{f/f}$ were obtained from Anjana Rao (La Jolla Immunology Institute). We crossed these to Mb1 Cre^+ mice to generate a B cell-specific deletion of STIM1 and STIM2. If deletion is desired in T cells, then these mice can be crossed with commercially available Lck Cre^+ or Vav Cre^+ mice or crossed with any tissue specific Cre^+ mouse line.
2. Mice (8 weeks or older) are maintained in accordance with guidelines of the Institutional Animal Care and Use Committee (IACUC) and recommendations in the Guide and Use of Laboratory Animals of the National Institutes of Health (NIH).
3. Studies of NF-κB activation in T cells follow the same general approach T lymphocytes can be isolated from spleens using a Miltenyi negative or positive selection kits (Miltenyi biotec).
4. It is critical to never allow freshly isolated lymphocytes to sit in a pellet. It is also critical to avoid vortexing. Cell pellets should be resuspended by "flicking" the bottom of tubes after centrifugation. Furthermore, during isolation, only polypropylene tubes (15 and 50 mL conical) should be used.
5. Procedures to purify splenic lymphocyte subsets all include a step to remove red blood cells. The typical way to do this is to lyse RBCs with ACK Lysis solution (Ammonia chloride potassium) for 10 minutes at room temperature. We do NOT recommend using this procedure with B or T lymphocytes because it induces cell swelling. RBCs to not have volume regulatory mechanisms and they rupture whereas lymphocytes can regulate their volume and resist hypotonic stress and survive. However, swelling activates a number of signaling pathways in lymphocytes and mobilizes calcium in B lymphocytes specifically [24]. Consequently, this approach should be avoided. An alternative approach that is incorporated into the Miltenyi B lymphocyte negative selection kits involves immune-magnetically depleting RBCs with an antibody (Ter111 clone) that recognizes an RBC specific antigen. Alternatively, the positive selection approach we use allows us to "fish" B cells away from T lymphocytes, RBCs, and other cells

based upon their expression of a unique membrane surface antigen, in this case CD23. Thus, RBCs and other CD23 negative cells are eluted from the purification columns.

6. When working with primary B lymphocytes, it is critical to maintain the cells in a healthy and viable state. Unstimulated primary murine B lymphocytes undergo apoptosis relatively quickly in culture following isolation so it is imperative that they be handled gently, that the isolation procedure be performed as quickly as possible, and that they be activated (when relevant) as quickly as possible following isolation and purification.
7. B lymphocytes express receptors for the conserved Fc domains of antibody heavy chains. Consequently, when using antibodies for cell purification or activation it is critical that these receptors be blocked to prevent non-specific binding to and/or activation of these receptors. Consequently, whenever antibodies are used to purify, stain, or activate B lymphocytes, Fc receptors should be blocked. To do this, cells must first be treated with a Fc blocking antibody (BD Biosciences).
8. The composition of medium used to culture primary B cells is critical and the osmolarity should be at least 290 milliosmolar. While murine B lymphocytes can be cultured in either RPMI or DMEM medium, an important detail often overlooked is that DMEM has a higher bicarbonate concentration than RPMI and is optimized for 10% CO_2. If B lymphocytes in DMEM are placed in 5% CO_2 incubators the pH of the medium will be basic (>7.4). Lymphocytes are intolerant of basic conditions.
9. B cells should be suspended in medium containing heat inactivated serum to kill any complement in the serum. Also, B lymphocytes should not be suspended in serum- or protein-free solution or in PBS for extended periods (>5–10 min) as the low osmolarity promotes cell death. The inclusion of the reducing agent 2-mercaptoethanol is critically important when culturing murine B lymphocytes.
10. While freshly isolated B lymphocytes are polyclonal, meaning their antigen receptors (BCRs) are each specific for a distinct antigen, all naïve B lymphocytes express IgM on their surface. Consequently, we can use a polyclonal antibody that recognizes a conserved motif on IgM to crosslink these receptors and mimic activation by cognate antigen.
11. We use Anti-IgM $F(ab')_2$ fragments to stimulate naïve B cells. We use these fragments because they do not have Fc domains and will not bind to Fc receptors. However, they are still able to cross link the BCR. Pipette the appropriate volume of anti-IgM stock into complete media and/or Ca^{2+} deficient media to obtain a final concentration of 20μg/mL anti-IgM. For

example, to make 2.5 mL of 2× stimulation media from a stock concentration of anti-IgM at 1.3 mg/mL, pipette 77μL antibody into 2423μL of complete RPMI.

12. Precast gels can be obtained from several commercial sources. These area relatively expensive alternatives to casting gels but generally provide improved reproducibility and may be preferable for labs that do not run more than a few gels per week.
13. NF-κB proteins are maintained in an inactive state through binding to the inhibitory protein IκBα. IκBα resides in a trimeric complex with the canonical NF-κB proteins p65 or c-Rel and p50. Within minutes following antigen receptor engagement on both B and T cells, IκBα is phosphorylated on Ser32 by the activated IKK complex and this phosphorylation triggers IκBα proteasomal degradation. So, a simple strategy to assess activation of canonical NF-κB is to measure IκBα Ser32 phosphorylation and subsequently the level of steady-state IκBα protein expression. Interestingly, following its degradation, IκBα expression is controlled by NF-κB activity in the nucleus. Thus, another indicator of functional signaling following its initial degradation is IκBα re-expression to pre-stimulation levels (30–60 min after activation).
14. The Ca^{2+} entry dependence of NF-κB activation is examined by (a) stimulating cells in Ca^{2+} free medium (*see* **Note 9**) or (b) by stimulating B cells in normal medium but by using B lymphocytes isolated from *Stim1*$^{fl/fl}$ *Stim2*$^{fl/fl}$ × *Mb1*^{cre} (STIM DKO) that exhibit defective STIM/Orai-dependent Ca^{2+} entry.

Approach #1 (WT B lymphocytes cultured in the presence and absence of extracellular Ca^{2+})	
WT, 0′, $-Ca^{2+}$	WT, 0′, $+Ca^{2+}$
WT, 15′, $-Ca^{2+}$	WT, 15′, $+Ca^{2+}$
WT, 30′, $-Ca^{2+}$	WT, 30′, $+Ca^{2+}$
WT, 60′, $-Ca^{2+}$	WT, 50′, $+Ca^{2+}$

Approach #2 (WT and STIM DKO B lymphocytes cultured in Ca^{2+} containing media)	
WT, 0′	STIM DKO, 0′
WT, 15′	STIM DKO, 15′$^{+}$
WT, 30′	STIM DKO, 30′
WT, 60′	STIM DKO, 60′

15. There are several sources of ionized Ca^{2+} in cell culture medium. DMEM contains 1.8 mM Ca^{2+}, RPMI contains 0.42 mM Ca^{2+}, and FBS contains 3–4 mM Ca^{2+}. We add 0.5 mM EGTA to cRPMI and Ca^{2+} free cDMEM to buffer the free [Ca^{2+}].
16. Run the gel until the dye front emerges into the running buffer or smallest MW marker can be seen near the bottom of the gel. This will provide enough separation to distinguish NF-κB proteins following chemiluminescence detection.
17. Chromatography filter paper cut to pieces of 3″ × 4″ provide sufficient coverage of the gel and fit in the transfer cassette without overhanging. A membrane size of 2.5″ × 3.5″ will cover the separating portion of the gel.
18. Primary antibodies diluted in 1% BSA in TBS-T containing 0.1% (w/v) sodium azide can reliably be reused up to three times.
19. Sodium azide inhibits the enzymatic activity of HRP and should not be used for the preparation of secondary antibodies in blocking buffer. Secondary antibodies should be prepared freshly and not stored for later or multiple uses since there is no sodium azide.
20. Another intermediate step of NF-κB activation upstream of IκBα is IKK complex activation, which can be assessed by immunoblot analysis of its phosphorylation. Downstream of IκBα degradation one can assess posttranslational modifications, including phosphorylation on p65 p-S534 (S536 in humans). These modifications license nuclear localization and/or transcriptional activation of these proteins. Together, these sequential events can also be used to evaluate NF-κB p65 and c-Rel activation and licensing [23]. While not described in detail here, immunoblot analysis of IKK and p65 phosphorylation follows the same protocol as outlined below for IκBα. Greater cell numbers and protein quantities may be necessary to detect phospho-p65 and -IKK.
21. IκBα proteasomal degradation is required for the release of p65 and c-Rel, but their nuclear localization and transcriptional activation is controlled separately. We previously showed that 65 pS536 phosphorylation licenses p65 for nuclear localization in T lymphocytes [17]. Thus, to directly assess this competence, one can directly quantify the extent of nuclear localization for p65 and c-Rel. This is assessed by immunostaining these proteins within the cell and analyzing their localization by confocal microscopy.

22. Culture in 96-well flat bottom plates instead of round bottom or microcentrifuge tubes to avoid cell clumping. Larger volume flat bottom plates can also be used as long as a similar cell density is used. Label wells with genotype, stimulus, and external Ca^{2+} condition. Two commonly used experimental designs similar to those described above.

Approach #1 (WT B lymphocytes cultured in the presence and absence of extracellular Ca^{2+})	
WT, unstimulated, $-Ca^{2+}$	WT, unstimulated, $+Ca^{2+}$
WT, anti-IgM, $-Ca^{2+}$	WT, anti-IgM, $+Ca^{2+}$

Approach #2 (WT and STIM DKO B lymphocytes cultured in 0.42 mM extracellular Ca^{2+})	
WT, unstimulated, $+Ca^{2+}$	STIM DKO, unstimulated, $+Ca^{2+}$
WT, anti-IgM, $+Ca^{2+}$	STIM DKO, anti-IgM, $+Ca^{2+}$

23. Coverslip staining and incubation containers can easily be assembled using an empty plastic pipette box and parafilm. Open the pipette box and stretch the parafilm over the top. For extended incubation times, we tape a filter paper wetted with ddH_2O to the top of the lid for humidification.
24. We use ImageJ to for these analyses. Utilize the Hoechst channel to generate a nuclear ROI. Use this ROI to then create an overlay and measure average p65 and c-Rel intensity. Some image acquisition and analysis software have cell scoring capabilities and can be used to calculate nuclear to cytoplasmic protein ratios. However, newly activated primary B cells have a very thin rim of cytoplasm and extreme care must be taken to not incorporate this into the nuclear ROI. Nuclear intensities are usually compared using non-parametric statistics and displayed as histograms using R or Matlab.
25. We use 0.8 when using a multichannel pipette. For example, for ten samples with triplicates, we calculate the error factor as 3.8×10 samples $= 38$. Multiply this factor by 1 and 5 to calculate the volume for primer and mastermix, respectively.

References

1. Liou J, Kim ML, Heo WD, Jones JT, Myers JW, Ferrell JE Jr, Meyer T (2005) STIM is a Ca2+ sensor essential for Ca2+-store-depletion-triggered Ca2+ influx. Curr Biol 15:1235–1241
2. Roos J, DiGregorio PJ, Yeromin AV, Ohlsen K, Lioudyno M, Zhang S, Safrina O, Kozak JA, Wagner SL, Cahalan MD, Velicelebi G, Stauderman KA (2005) STIM1, an essential and conserved component of store-operated Ca2+ channel function. J Cell Biol 169:435–445
3. Zhang SL, Yu Y, Roos J, Kozak JA, Deerinck TJ, Ellisman MH, Stauderman KA, Cahalan MD (2005) STIM1 is a Ca2+ sensor that

activates CRAC channels and migrates from the Ca2+ store to the plasma membrane. Nature 437:902–905

4. Stathopulos PB, Li GY, Plevin MJ, Ames JB, Ikura M (2006) Stored Ca2+ depletion-induced oligomerization of stromal interaction molecule 1 (STIM1) via the EF-SAM region: an initiation mechanism for capacitive Ca2+ entry. J Biol Chem 281:35855–35862
5. Vig M, Peinelt C, Beck A, Koomoa DL, Rabah D, Koblan-Huberson M, Kraft S, Turner H, Fleig A, Penner R, Kinet JP (2006) CRACM1 is a plasma membrane protein essential for store-operated Ca2+ entry. Science 312:1220–1223
6. Feske S, Gwack Y, Prakriya M, Srikanth S, Puppel SH, Tanasa B, Hogan PG, Lewis RS, Daly M, Rao A (2006) A mutation in Orai1 causes immune deficiency by abrogating CRAC channel function. Nature 441:179–185
7. Flanagan WM, Corthesy B, Bram RJ, Crabtree GR (1991) Nuclear association of a T-cell transcription factor blocked by FK-506 and cyclosporin A. Nature 352:803–807
8. Dolmetsch RE, Xu K, Lewis RS (1998) Calcium oscillations increase the efficiency and specificity of gene expression. Nature 392:933–936
9. Healy JI, Dolmetsch RE, Lewis RS, Goodnow CC (1998) Quantitative and qualitative control of antigen receptor signalling in tolerant B lymphocytes. Novartis Found Symp 215:137–144
10. Dolmetsch RE, Lewis RS, Goodnow CC, Healy JI (1997) Differential activation of transcription factors induced by Ca2+ response amplitude and duration. Nature 386:855–858
11. Healy JI, Dolmetsch RE, Timmerman LA, Cyster JG, Thomas ML, Crabtree GR, Lewis RS, Goodnow CC (1997) Different nuclear signals are activated by the B cell receptor during positive versus negative signaling. Immunity 6:419–428
12. Trushin SA, Pennington KN, Carmona EM, Asin S, Savoy DN, Billadeau DD, Paya CV (2003) Protein kinase Calpha (PKCalpha) acts upstream of PKCtheta to activate IkappaB kinase and NF-kappaB in T lymphocytes. Mol Cell Biol 23:7068–7081
13. Trushin SA, Pennington KN, Algeciras-Schimnich A, Paya CV (1999) Protein kinase C and calcineurin synergize to activate IkappaB kinase and NF-kappaB in T lymphocytes. J Biol Chem 274:22923–22931
14. Steffan NM, Bren GD, Frantz B, Tocci MJ, O'Neill EA, Paya CV (1995) Regulation of IkB alpha phosphorylation by PKC- and Ca2+-dependent signal transduction pathways. J Immunol 155:4685–4691
15. Frantz B, Nordby EC, Bren G, Steffan N, Paya CV, Kincaid RL, Tocci MJ, O'Keefe SJ, O'Neill EA (1994) Calcineurin acts in synergy with PMA to inactivate I kappa B/MAD3, an inhibitor of NF-kappa B. EMBO J 13:861–870
16. Berry CT, Liu X, Myles A, Nandi S, Chen YH, Hershberg U, Brodsky IE, Cancro MP, Lengner CJ, May MJ, Freedman BD (2020) BCR-induced Ca2+ signals dynamically tune survival, metabolic reprogramming, and proliferation of naive B cells. Cell Rep 31:107474
17. Berry CT, May MJ, Freedman BD (2018) STIM- and Orai-mediated calcium entry controls NF-kappaB activity and function in lymphocytes. Cell Calcium 74:131–143
18. Hayden MS, West AP, Ghosh S (2006) NF-kappaB and the immune response. Oncogene 25:6758–6780
19. Ghosh S, May MJ, Kopp EB (1998) NF-kappa B and Rel proteins: evolutionarily conserved mediators of immune responses. Annu Rev Immunol 16:225–260
20. Altman A, Villalba M (2003) Protein kinase C-theta (PKCtheta): it's all about location, location, location. Immunol Rev 192:53–63
21. Villalba M, Coudronniere N, Deckert M, Teixeiro E, Mas P, Altman A (2000) A novel functional interaction between Vav and PKCtheta is required for TCR-induced T cell activation. Immunity 12:151–160
22. Hayden MS, Ghosh S (2008) Shared principles in NF-kappaB signaling. Cell 132:344–362
23. Liu X, Berry CT, Ruthel G, Madara JJ, MacGillivray K, Gray CM, Madge LA, McCorkell KA, Beiting DP, Hershberg U, May MJ, Freedman BD (2016) T cell receptor-induced NF-kappaB signaling and transcriptional activation are regulated by STIM1- and Orai1-mediated calcium entry. J Biol Chem 291 (16):8440–8452
24. Zhu P, Liu X, Labelle EF, Freedman BD (2005) Mechanisms of hypotonicity-induced calcium signaling and integrin activation by arachidonic acid-derived inflammatory mediators in B cells. J Immunol 175:4981–4989

Chapter 10

A Kinase Assay for Measuring the Activity of the NIK-IKK1 Complex Induced via the Noncanonical NF-κB Pathway

Tapas Mukherjee, Yashika Ratra, Balaji Banoth, Alvina Deka, Smarajit Polley, and Soumen Basak

Abstract

Nuclear factor-kappa B (NF-κB) inducing kinase (NIK), a key component of the noncanonical NF-κB pathway, directs a range of physiological processes, such as lymphoid organogenesis, immune cell differentiation, and immune responses. Aberrant noncanonical NF-κB signaling also causes human ailments, including autoimmune and neoplastic diseases. As such, NIK is constitutively degraded in resting cells, and accumulates upon noncanonical NF-κB signaling. NIK then associates with and phosphorylates IkappaB kinase 1 (IKK1, alternately IKKα). Subsequently, the NIK-IKK1 complex mediates the phosphorylation of p100 that triggers partial proteolysis of p100 into p52. Typically, accumulation of NIK or processing of p100 is estimated by immunoblot analyses, and these indirect measurements are used as a surrogate for cellular NIK activity. However, studies involving knockout and cancerous cells indicated that the activity of NIK-IKK1 might not always correlate with the abundance of NIK or with the relative level of p52 and p100. In this report, we describe a specific and sensitive assay for direct evaluation of cellular NIK-IKK1 activity. Here, NIK immunoprecipitates are examined for the presence of IKK1-dependent kinase activity toward p100. The NIK-IKK1 assay captured selectively noncanonical NF-κB activation in the context of multiple cell activating stimuli and cell types, including patient-derived myeloma cells. We suggest that our assay may help advance our understanding of the role of NIK in health and diseases.

Key words NF-kappa B (NF-κB), NF-κB inducing kinase (NIK), Noncanonical NF-κB, *Nfkb2/* p100, Immunoprecipitation, Kinase assay

1 Introduction

NF-κB transcription factors play critical roles in the development and functioning of the immune system. They are activated via two distinct pathways, namely the canonical and noncanonical pathways. Pro-inflammatory substances, such as tumor necrosis factor (TNF), trigger canonical NF-κB signaling [1]. In this pathway, a

Tapas Mukherjee and Yashika Ratra contributed equally to this work.

Guido Franzoso and Francesca Zazzeroni (eds.), *NF-κB Transcription Factors: Methods and Protocols*, Methods in Molecular Biology, vol. 2366, https://doi.org/10.1007/978-1-0716-1669-7_10,

complex consisting of NF-κB essential modulator (NEMO) and IKK2 (also known as IKKβ) phosphorylates the inhibitory IκB proteins, the major isoform being IκBα, bound to the RelA:p50 heterodimers in the cytoplasm. Signal-induced phosphorylation leads to proteasomal degradation of IκBs and release of RelA:p50 into the nucleus, where they activate immune response genes. On the other hand, immune-differentiating cues, such as that signal through lymphotoxin β receptor (LTβR) and B-cell activating factor receptor (BAFF-R), induce NIK-dependent noncanonical NF-κB signaling, which activates RelB:p52 heterodimers [2]. In resting cells, an E3 ubiquitin-ligase complex comprising of TNF receptor-associated factor 3 (TRAF3), TRAF2, cellular inhibitor of apoptosis 1 (cIAP1) and cIAP2 tags NIK for constitutive proteasomal degradation; whereas, p100 encoded by *Nfkb2* retains RelB in the cytoplasm in a multimeric complex [3–5]. Noncanonical signaling disrupts the ubiquitin-ligase complex causing cellular accumulation of NIK, which then phosphorylates and thereby activates IKK1. In addition, NIK recruits IKK1 to p100 promoting IKK1-mediated phosphorylation of p100 at the C-terminal serine 866 and 870 residues [6–8]. Subsequent to phosphorylation involving the NIK-IKK1 kinase complex, proteasome-mediated partial proteolysis removes the C-terminal inhibitory domain of p100 that generates the mature p52 NF-κB subunit and liberates RelB:p52 into the nucleus. As a constituent of the noncanonical pathway, NIK directs RelB:p52-mediated gene expressions [2].

NIK activated downstream of LTβR in lymphoid stromal cells regulates secondary lymph node development and germinal center formation. Recent studies substantiated a cell-autonomous role of NIK in the maturation and survival of B cells [9]. Furthermore, NIK modulates effector and memory T cell responses and instruct maintenance of regulatory T cells [10, 11]. Examination of DC-specific NIK knockout mice revealed that although dispensable for DC development, NIK is required for antigen cross-presentation [12]. In macrophages, NIK-dependent noncanonical NF-κB signaling suppresses anti-viral type-1 interferon responses [13, 14]. In addition, NIK was shown to reinforce innate immune response by mediating cell-type-specific crosstalk between canonical and noncanonical NF-κB signaling [15, 16]. Deregulated NIK activity has also been implicated in human ailments [17]. Previous studies identified gain-of-function mutations in genes encoding LTβR and NIK as well as inactivating mutations in genes encoding NIK inhibitors TRAF2, TRAF3, and cIAPs in hematologic malignancies, including multiple myeloma, mantle cell lymphoma and T cell leukemia [18–20] as well as in solid tumors [21–23]. More so, the role of NIK was suggested in the immune pathologies associated with diabetes-induced obesity and systemic lupus erythematosus [24, 25]. Because of its involvement in human diseases, NIK is currently considered as a therapeutic target.

Despite the importance of NIK, an assay that directly measures the activity of the NIK-IKK1 complex induced during noncanonical NF-κB signaling is lacking. Typically, processing of p100 or accumulation of NIK is estimated by immunoblot analyses; the level of p52 in relation to p100 or the abundance of NIK is then used as a surrogate for cellular NIK-IKK1 activity. While p52 and p100 could be conveniently detected, detection of NIK in immunoblots possesses considerable challenge in the majority of cell types owing to the low cellular abundance of this kinase. Importantly, recent studies indicated that these indirect measurements might not always reliably capture the NIK-IKK1 activity. For instance, genetic deficiency of p50 led to constitutive processing of p100 into p52 in *Nfkb1*$^{-/-}$ mouse embryonic fibroblasts (MEFs) even in the absence of NIK-activating signals presumably because of increased RelA binding to p52 [26]. In cancerous cells derived from myeloma patients, activating mutations in the NIK-pathway instead caused complete degradation of p100 in a Fbxw7α-dependent manner [27, 28]. Recent reports further suggested that RelB as well as IκBδ, the oligomeric form of p100, inhibited p52 generation from p100 by NIK-mediated signaling [29, 30]. On the other hand, IKK1 deficiency led to the accumulation of NIK in knockout cells; but despite the augmented cellular abundance, NIK was unable to trigger noncanonical NF-κB signaling in the absence of IKK1 [31]. These studies warranted an assay for directly measuring the activity of NIK-IKK1 complex induced during noncanonical NF-κB signaling.

Here, we describe a specific and sensitive immunoprecipitation-based kinase assay for determining directly the activity of cellular NIK-IKK1 complex. The NIK-IKK1 kinase assay captured selectively noncanonical NF-κB pathway activation in response to multiple stimuli and in diverse cell types.

2 Materials

2.1 Cell Culture

2.1.1 Primary Cells and Cell Lines

1. Mouse embryonic fibroblasts (MEFs): Isolate primary MEFs from E12.5–E14.5 embryos and immortalized them as described earlier [32].
2. Lymphoid stromal-derived mouse cell lines: BLS4 and BLS12.
3. Primary bone marrow derived macrophages (BMDMs).
4. Splenic primary B cells.
5. Human myeloma cell lines (HMCLs): KMS28PE, KMS20, OciMy1, and OciMy5. KMS28PE and KMS20 are characterized by genetic aberration in locus encoding cIAPs that inhibit noncanonical NF-κB signaling. OciMy1 cells have an inactivating mutation in *TRAF3*, which encodes a repressor of the

noncanonical NF-κB pathway. As a control, use OciMy5 cells that harbor an amplification in *NFKB1* encoding the canonical NF-κB component p50/p105.

2.1.2 Cell Culture Media and Supplements

1. Dulbecco modified essential media (DMEM, Corning).
2. Roswell Park Memorial Institute (RPMI) 1640 medium (Gibco).
3. Bovine calf serum (BCS, Hyclone).
4. Fetal bovine serum (FBS, Gibco).
5. M-CSF (eBioscience).
6. Penicillin-Streptomycin (Corning).
7. L-Glutamine (Corning).

2.2 Reagents and Standard Biochemical Assays

2.2.1 Ligands and Kinase Substrates

1. Recombinant mouse TNF (Roche).
2. LTβR agonist (αLTβR) (Biogen Idec).
3. Recombinant mouse BAFF (Peprotech).
4. NIK small molecule inhibitors (SMIs) (Genentech, San Francisco, USA).
5. Recombinant $p100_{(406-899)}$, $p100_{(406-899\ SS/AA)}$ and GST-$I\kappa B\alpha_{(1-54)}$ (Biobharati Lifesciences).

2.2.2 Chemicals and Buffers

1. Phosphate Buffer Saline (PBS), pH 7.4.
2. 0.5 M Ethylenediaminetetraacetic acid (EDTA), pH 8.8.
3. 1 M HEPES-KOH, pH 7.9.
4. 1 M HEPES pH, 7.7.
5. 5 M NaCl.
6. 1 M $MgCl_2$.
7. 10% NP-40.
8. Tween-20.
9. 1 M Sodium fluoride (NaF).
10. 2 M β-glycerophosphate.
11. 1 M Sodium orthovanadate (Na_3VO_4).
12. 1× Protease Inhibitor Cocktail (PIC, Sigma).
13. 1 M Dithiothreitol.
14. Protein G 4 Fast Flow Sepharose beads (GE Amersham).
15. Resolving gel buffer: 1.5 M Tris–HCl, pH 8.8.
16. Stacking gel stock: 0.5 M Tris–HCl, pH 6.8.
17. ProtoGel: 30% (w/v) acrylamide: 0.8% (w/v): methylene bisacrylamide solution, 37.5:1 (National Diagnostics).
18. 10% (w/v) Ammonium persulfate.

19. *N,N,N′,N′*-Tetramethylethylenediamine (TEMED).
20. Pierce™ ECL Plus Western Blotting Substrate (ThermoFisher Scientific).
21. Sodium dodecyl sulfate.

2.2.3 Antibodies

1. Anti-IκBα (sc-371) (Santa Cruz Biotechnology).
2. Anti-IKKα (sc-7184) (Santa Cruz Biotechnology).
3. Anti-Actin (sc-1615) (Santa Cruz Biotechnology).
4. p52/p100 (#4882) (Cell Signaling Technology).
5. NIK (#4994) (Cell Signaling Technology).
6. IKKα (#2682) (Cell Signaling Technology).
7. NEMO (BB-AB0035) (Biobharati Lifesciences).
8. β-Tubulin (BB-AB0150) (Biobharati Lifesciences).

2.2.4 Biochemical Assays Buffers

1. Cell Extraction Buffer (CEB): 10 mM HEPES-KOH, pH 7.9, 250 mM NaCl, 1 mM EDTA, 0.5% NP-40, 0.2% Tween-20, 10 mM NaF, 20 mM β-glycerophosphate, 100μM Na_3VO_4, 1× Protease Inhibitor Cocktail (PIC) and 2 mM Dithiothreitol (DTT).
2. Wash Buffer: 20 mM HEPES, pH 7.7, 20 mM β-glycerophosphate, 100 mM NaCl, 100μM Na_3VO_4, 10 mM $MgCl_2$, 10 mM NaF, 1× PIC and 2 mM DTT.
3. 10× Kinase Assay Buffer: 20 mM HEPES pH 7.7, 20 mM β-glycerophosphate, 100 mM NaCl, 100μM Na_3VO_4, 10 mM $MgCl_2$, 10 mM NaF, 1× PIC, 2 mM DTT and 20 mM cold ATP.
4. Kinase Reaction Mix: 2μl of 10× Kinase Assay Buffer, 0.5μg (0.5μl of 1.0μg/μl stock) $p100_{(406–899)}$ or $p100_{(406–899\ SS/AA)}$, 10μCi γ-[P^{32}] ATP* (~1μl of 10μCi/μl stock, ~2×10^5 cpm in the scintillation counter) and 17μl H_2O.
5. 1× Running buffer: 250 mM Tris, 1.92 M glycine and 1% SDS.
6. 1× Transfer buffer: 250 mM Tris, 1.92 M glycine and 10% methanol.
7. Tris Buffer Saline with 0.1% Tween-20 (TBST): 200 mM Tris, 3 M NaCl, pH 8.0 and 0.05% Tween-20.

2.2.5 Equipments

1. End to end rotor.
2. Cyclomixer.
3. Microcentrifuge (Eppendorf Centrifuge 5415R).
4. Dry bath (30 °C).
5. Mini PROTEAN tetra System and Western semi-dry transfer apparatus (Bio-Rad).

6. Gel Drier (Bio-Rad, Model 583).
7. 3 mm Whatman chromatography paper.
8. PVDF blotting membrane.
9. Rocker.
10. Phosphoscreen (GE healthcare).
11. Typhoon 9400 PhosphorImager (GE Amersham, UK).

3 Methods

A detailed schematic flowchart of the NIK-IKK1 kinase activity assay has been represented in Fig. 1 and the step-by-step procedure is described below.

3.1 Cell Culture

1. For adherent cells: Culture primary or immortalized MEFs in DMEM supplemented with 10% BCS. Likewise, maintain BLS4 and BLS12 cell lines in DMEM containing 10% FBS. Passage cells every third day at a confluency of 80–90%. Generate BMDMs from mouse bone marrow cells by culturing the cells in RPMI containing 10% FBS and 50 ng/ml of M-CSF for five days. Supplement all culture media with 1% penicillin-streptomycin and 1% of L-Glutamine. Incubate cells at 37 °C in a humidified atmosphere with 5% CO_2.
2. For suspension cells: Culture isolated B cells from mouse spleen in RPMI containing 10% FBS. Maintain human myeloma cell lines (HMCLs) in RPMI containing 10% FBS. Supplement all culture media with 1% penicillin-streptomycin and 1% of L-Glutamine. Incubate cells at 37 °C in a humidified atmosphere with 5% CO_2.

3.2 Treatment with Ligands

1. For adherent cells: Maintain primary or immortalized MEFs according to NIH 3T3 protocol. Stimulate semiconfluent cultures in 6 or 10 cm dish with 2μg/ml of αLTβR for 12 h or treat cells as indicated in the respective figure legends (Figs. 2 and 3 and Fig. 4e). Subsequently, harvest cells and prepare cell lysates. For BLS4 and BLS12 cells or BMDMs, stimulate cells in 6 cm dishes for 24 h using 0.3μg/ml of αLTβR (Fig. 4a, b).
2. For suspension cells: To explore noncanonical NF-κB signaling in B cells, treat primary mature B cells ($B220^+$) with 100 ng/ml of BAFF for 8 h (Fig. 4c). For human myeloma cell lines (HMCLs); KMS28PE, KMS20 and OciMy1 harboring mutations in the noncanonical NF-κB pathway (Fig. 4d), keep cells untreated. Culture all suspension cells in T-25/T-75 flask.

3.3 Cell Lysate Preparation

Collection and processing of cells should be performed as described below.

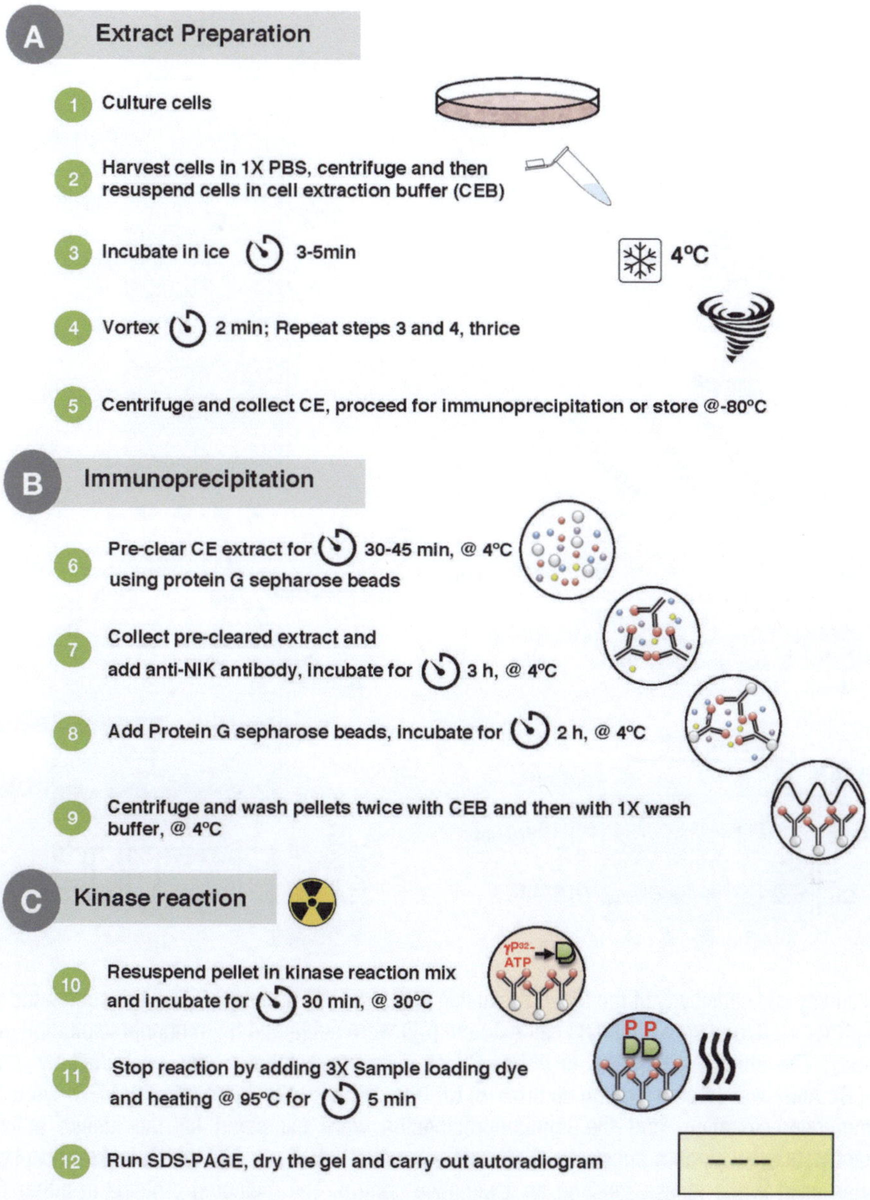

Fig. 1 Schematic flowchart illustrating an immunoprecipitation-based NIK-IKK1 kinase assay. As outlined, NIK-IKK1 kinase assay comprises of three major steps. (**a**) *Cell extract preparation*: In this step, cells are harvested, centrifuged, suspended in cell extraction buffer (CEB) and cell extracts are prepared. (**b**) *Immunoprecipitation*: Cell extracts are pre-cleared, and subjected to immunoprecipitation using anti-NIK antibody. Immunoprecipitates are resuspended in 1× kinase wash buffer. (**c**) *Kinase Reaction*: Immunoprecipitates are examined for the presence of kinase activity using recombinant $p100_{(406–899)}$ as a substrate and γ^{32}P-ATP. Sample are resolved in SDS-PAGE gels and the presence of radiolabelled $p100_{(406–899)}$ is detected by autoradiography

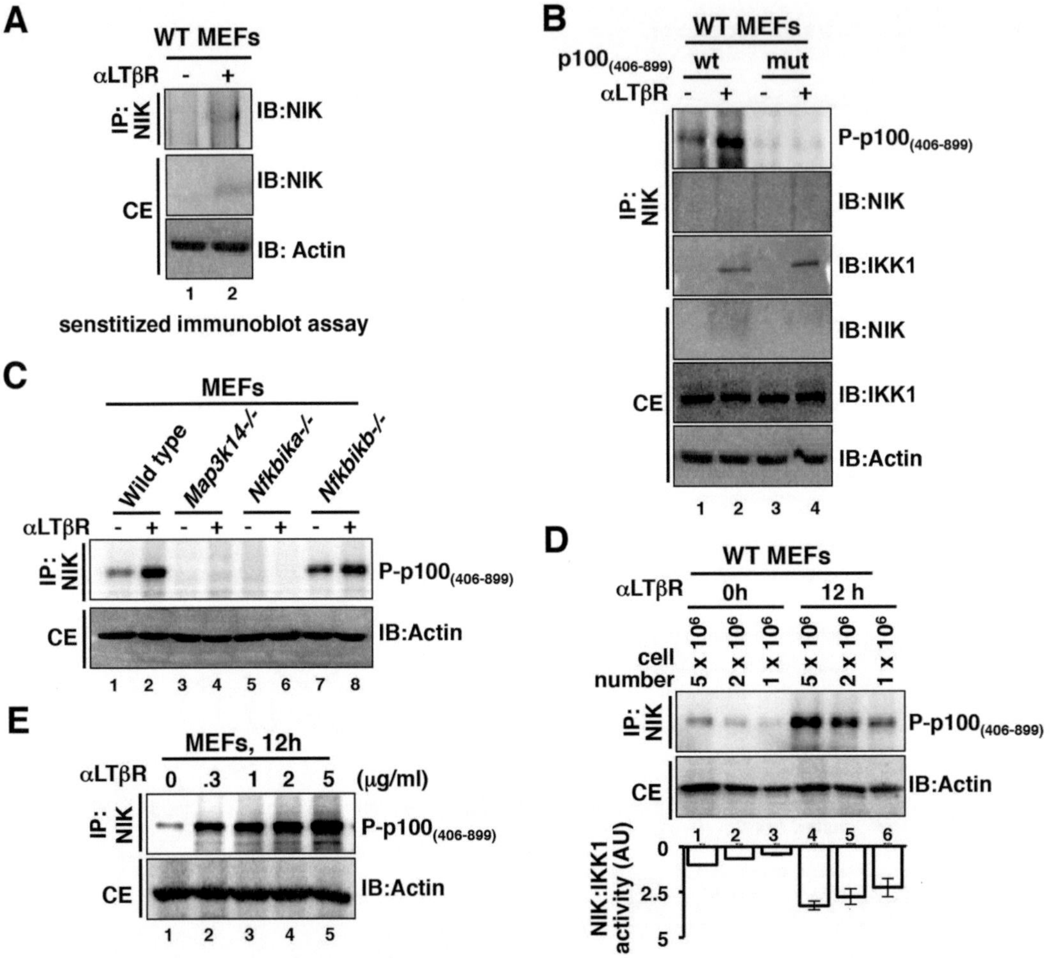

Fig. 2 Specificity and sensitivity of the NIK-IKK1 kinase assay. (**a**) WT primary MEFs were stimulated with 2μg/ml of αLTβR for 12 h, and corresponding cell extracts (CE) were subjected to immunoprecipitation using anti-NIK antibody. The abundance of NIK in either CE or immunoprecipitates was examined by immunoblot analyses (IB). Actin served as a loading control. (**b**) CE from αLTβR-stimulated primary MEFs were subjected to NIK immunoprecipitation, and the immunoprecipitates were examined for the kinase activity using recombinant $p100_{(406–899)}$ as a substrate. A mutant version of $p100_{(406–899)}$, which lacks phosphoacceptor serines, was used in the kinase reaction as a negative control. The indicated proteins in either CE or NIK immunoprecipitates were also examined by IB. Actin in the input extracts served as a loading control. (**c**) WT, NIK-deficient (*Map3k14*$^{-/-}$), IKK1-deficient (*Nfkbika*$^{-/-}$) and IKK2-deficient (*Nfkbikb*$^{-/-}$) immortalized MEFs were stimulated for 12 h with 2μg/ml of αLTβR before being subjected to the NIK-IKK1 kinase assay. (**d**) Indicated numbers of WT MEFs were stimulated with 2μg/ml of αLTβR before being subjected to the kinase assay. Bottom, bar graphs represent quantified data from three different experiments. (**e**) WT primary MEFs stimulated with different concentrations of αLTβR for 12 h were subjected to the kinase assay. Data information: In all panels, figures are representative of three independent biological experiments. For detecting NIK in (**a**), we loaded ~225μg of total CE or immunoprecipitates derived from ~400μg equivalent of total protein per lane. Immunoblot conditions were also sensitized (*see* **Note 2**). For all other figure panels involving the kinase assay, we used routine cell extraction protocol; accordingly, immunoblot analyses of NIK in (**b**) was performed using ~15μg of total CE or immunoprecipitates from ~50μg equivalent of total protein per lane

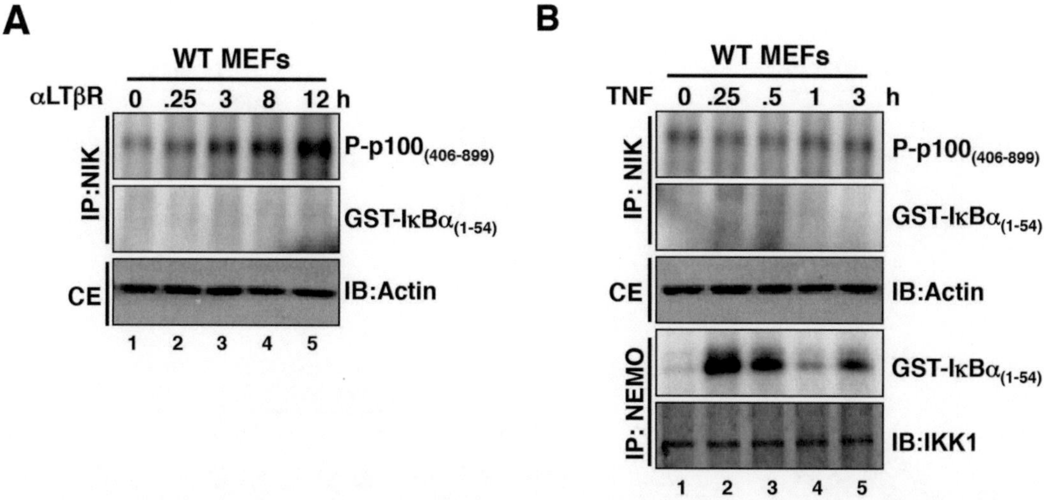

Fig. 3 Evaluating the selectivity of the NIK-IKK1 kinase assay toward noncanonical NF-κB signaling. (**a**, **b**) WT MEFs were stimulated with 0.3μg/ml of αLTβR (**a**) or treated with 1 ng/ml of TNF (**b**) for the indicated time periods. Cells were harvested, and extracts were subjected to immunoprecipitation using anti-NIK antibody. Immunoprecipitates were examined in the kinase reaction using either recombinant $p100_{(406-899)}$ or GST-IκB$\alpha_{(1-54)}$ as a substrate. Actin present in the input extracts served as a loading control. Furthermore, NEMO immunoprecipitates obtained from TNF-treated MEFs were evaluated for the IκBα kinase activity (bottom, **b**). Data information: In all panels, figures are representative of three independent biological experiments

3.3.1 For Adherent Cells (MEFs, BLS4, BLS12, and BMDMs)

1. Aspirate off media from the 6 cm plate, and add 1.0 ml of ice-cold 1× PBS containing 1 mM EDTA over the cells.
2. Gently scrap cells using a rubber policeman and collect it in a 1.7 ml microcentrifuge tube.
3. Centrifuge cells at 371 × *g* for 3 min. Aspirate off the supernatant carefully without disturbing the cell pellet. Unless otherwise mentioned, harvest ~5 × 10^6 of αLTβR-stimulated MEFs (*see* **Notes 1** and **2**) or BLS4 or BLS12 cells for cell lysate preparation. Similarly, harvest ~2 × 10^6 BMDMs for preparing cell lysates.

3.3.2 For Suspension Culture (B Cells or HMCLs)

1. Collect cells in a 15 ml falcon tube and centrifuge at 208 × *g* for 3 min.
2. Aspirate off media from falcon tubes and wash cells slowly by resuspending them in 1.0 ml of ice-cold 1× PBS containing 1 mM EDTA.
3. Transfer cells into a 1.7 ml microcentrifuge tube and centrifuge at 208 × *g* for 3–5 min. Aspirate off the supernatant carefully without disturbing the cell pellet. Harvest ~2 × 10^6 cells of mature B cells and HMCLs for cell lysate preparation.

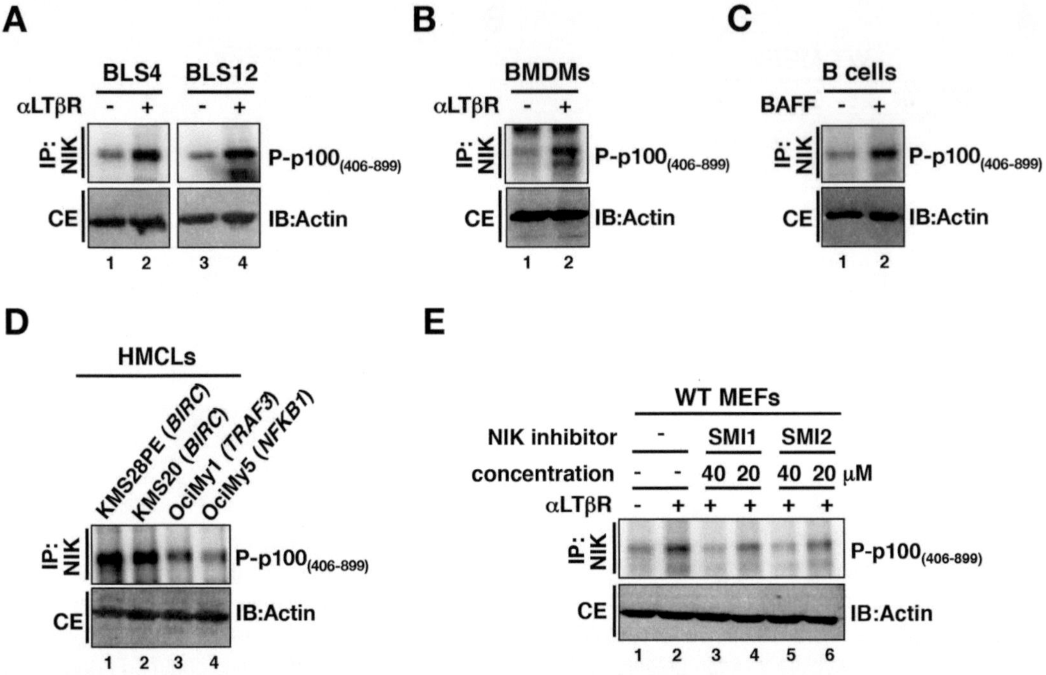

Fig. 4 Versatile applications of the NIK-IKK1 kinase assay. (**a**, **b**) The NIK-IKK1 kinase assay involving lymph node stromal-derived BLS4 and BLS12 mouse cell lines (**a**) or bone marrow derived macrophages (BMDMs) (**b**) stimulated with 0.3μg/ml of αLTβR for 24 h. (**c**) Similarly, primary B220+ B cells isolated from the mouse spleen were stimulated with 100 ng/ml of BAFF for 8 hours before being subjected to the NIK-IKK1 kinase assay. (**d**) The NIK-IKK1 kinase assay revealing the noncanonical NF-κB pathway activity in the indicated human myeloma cell lines (HMCLs). These HMCLs harbor indicated mutations (see the parentheses) in the NF-κB pathway regulators. Subsequently, cells were harvested and subjected to the NIK-IKK1 kinase assay. (**e**) WT MEFs were stimulated with 0.3μg/ml of αLTβR in the presence or absence of indicated concentrations of NIK inhibitors, SMI1 and SMI2. Data information: In all panels, figures are representative of three independent biological experiments

3.4 Preparing Cell Lysate for Immunoprecipitation

1. Resuspend cell pellet in 120μl of ice-cold CEB by gently pipetting up and down (*see* **Notes 3** and **4**).
2. Incubate the cell lysate on ice for 3 min and then vortex for 2 min at setting 7–8. Repeat this step thrice.
3. Centrifuge cell lysate at maximum speed for 5 min at 4 °C, and transfer the supernatant into a fresh 1.7 ml microcentrifuge tube. Next, either subject the cell lysate for NIK immunoprecipitation, as described in Subheading 3.5 or store cell lysate at −80 °C for future use.

3.5 Immunoprecipitation of Cellular NIK

1. Normalize for the total protein content of cell lysates derived from different sets (e.g., untreated or treated) using the Lowry method (Bio-Rad, according to the manufacturer's protocol).

2. Mix ~120μl of cell lysate (~200μg) with 5μl of Protein G 4 Fast Flow bead (~50% slurry suspended in CEB) and incubate for 30 min at 4 °C using an end-to-end rotator (*see* **Note 5**).
3. Centrifuge cell lysate at 2000 × *g* for 2 min at 4 °C.
4. Transfer 100μl of the pre-cleared cell lysate into a fresh 1.7 ml microcentrifuge tube. Keep 20μl of the cell lysate aside for subsequent western blot analyses as the input of immunoprecipitation reaction.
5. Next, add 5μl of anti-NIK antibody to the cell lysate and incubate in the end-to-end rotator for 3 h at 4 °C.
6. After 3 h, add 13μl of Protein G 4 Fast Flow beads to the extract, and incubate the mixture for another 2 h at 4 °C in the end-to-end rotator (*see* **Note 5**).
7. Centrifuge antibody (Ab)-bead complexes at 2000 × *g* for 2 min. Remove the supernatant was carefully.
8. Resuspend the immunopellet in 500μl of ice-cold CEB and subject to centrifugation again. Repeat **steps** 7 and **8** twice.
9. Subsequently, resuspend the immunopellet in 500μl of ice-cold 1× Wash Buffer (*see* **Note 6**) and centrifuge again at 2000 × *g* for 2 min. Utilize this washed immunopellet in the kinase assay, as described below.

3.6 In Vitro Kinase Assay

1. Resuspend the washed immunopellet in 20μl of Kinase Reaction Mix by gently pipetting up and down (*see* **Notes** 7 and **8**). Examine the immunoprecipitates from different cell types for the presence of kinase activity using recombinant $p100_{(406–899)}$ protein, which contains signal-responsive serine residues at position 866 and 870, as a substrate (Fig. 2b, top row, lane 1–2). Further, to determine the specificity of the assay protocol, use $p100_{(406–899\ SS/AA)}$ protein as a negative control substrate, where the serine residues of p100 are mutated to alanine (Fig. 2b, top row, lane 3–4).
2. Incubate the reaction mixture for 30 min at 30 °C (*see* **Note 9**).
3. Stop the kinase reaction by adding 13μl 3× SDS loading buffer and subsequently, heating the samples at 95 °C for 5 min.

3.7 SDS-PAGE and Autoradiography

1. Prepare 10% SDS-PAGE gel.
2. Load 10μl of the reaction mixture on a 10% SDS-PAGE gel (*see* **Note 2**), and run the gel at 80–120 V (*see* **Note 10**). Load protein ladder in a well.
3. Transfer the gel over a pair of 3 mm Whatman chromatography sheet of the dimension 6 cm × 8.5 cm and cover it with a saran wrap. Dry the gel in Bio-Rad Gel Drier for 1 h 15 min in the gradient temperature mode (1 °C rise/min).

4. Expose the dried gel to the phosphor-screen in a phosphor-imaging cassette for 12–14 h. Scan the exposed screen using a Typhoon 9400 Variable Mode Scanner (Amersham Biosciences, GE Healthcare). Phosphorylated $p100_{(406-899)}$ will be detected in the autoradiogram as a band migrating approximately as a ~50 kDa protein (*see* **Note 11**).

3.8 Immunoblotting

1. For immunoblotting, load pre-cleared cell lysates (*see* Subheading 3.5, **step 4**) or immunoprecipitation samples in 10% SDS-PAGE gel and run at 80–120 V (*see* **Note 2**). Subsequently, carry out western blot into PVDF membrane using semi-dry transfer apparatus (*see* **Note 12**).
2. After the transfer, incubate the blot (PVDF membrane) for 45 min at room temperature (RT, 22–25 °C) with blocking reagent (5% non-fat milk made in 1× TBST) and then incubate with primary antibody diluted in 5% non-fat milk, overnight at 4 °C.
3. Wash the blot with 1× TBST thrice, for 10 min each.
4. Next, prepare the secondary antibody by diluting it in 2.5% non-fat milk. Add secondary antibody to the blot and incubated for 1 h at RT.
5. Wash the blots with TBST thrice, for 10 min each.
6. Remove TBST from the blot and properly add 1 ml Pierce™ ECL Plus Western Blotting Substrate throughout the blot.
7. Subsequently, scan the blot in Typhoon 9400 Variable Mode Scanner (Amersham Biosciences, GE Healthcare). Detection of actin from the pre-cleared cell lysate should be used for confirming consistent loading.

3.9 Summary

Here, we developed an immunoprecipitation-based assay that directly captured the NIK-IKK1 kinase activity (Fig. 1). Our biochemical analyses established that the NIK-IKK1 assay faithfully reflected the activation state of the noncanonical NF-κB pathway. In our assay, NIK immunoprecipitates obtained from cells stimulated with αLTβR specifically phosphorylated the C-terminal serine 866 and serine 870 residues of p100 (Fig. 2b).

The cellular level of NIK is rather low. NIK could be detected in cell lysates and in NIK-immunoprecipiates only when concentrated cell extracts were used and immunoblot conditions were further sensitized. In fact, we utilized ~225μg of total CE or immunoprecipitates derived from ~400μg equivalent of total protein per lane that led to detection of a rather weak band corresponding to NIK in immunoblots (Fig. 2a) and (*see* **Note 2**). On the other hand, this routine kinase assay protocol involved loading of ~15μg of total CE protein or immunoprecipitates derived from ~50μg equivalent of total protein per lane. Although, these conditions did not reveal

NIK in immunoblots, we could conveniently detect the activity of NIK-IKK1 in our radioactive kinase assay (Fig. 2b). Indeed, this sensitive kinase assay could readily measure the NIK-IKK1 activity even in a relatively small number of cells (10^6 MEFs, Fig. 2d). Further, our assay allowed superior quantitative analyses as compared to immunoblot-based studies. Indeed, we noted satisfactory data reproducibility among experimental replicates (Fig. 2d, *see* error bars associated with the technical replicates in the bar graph) and at least a half-log dynamical range. Additionally, our assay exhibited dose-response effect and mirrored minute changes in the dose of the noncanonical stimulus αLTβR used for treating MEFs (Fig. 2e). We used a range of doses of αLTβR from 0.3 to 5μg/ml for cell treatments.

Our analyses involving NIK-deficient, IKK1-deficient and IKK2-deficient cell lines not only substantiated the specificity of the assay protocol but also indicated that the NIK immunoprecipitates required IKK1 for phosphorylating p100 (Fig. 2c). Of note, a previous study demonstrated that IKK1 immunoprecipitates obtained from TNF-treated cells, akin to NEMO immunoprecipitates, efficiently phosphorylated recombinant IκBα [33]. In their study, Dejardin et al. [33] used an anti-IKK1 antibody that led to immunoprecipitation of both NEMO-associated and NEMO-independent IKK1 complexes. They found that IKK1 immunoprecipitates from LTβR-stimulated MEFs phosphorylate both recombinant IκBα as well as recombinant p100, while those from TNF-treated cells phosphorylate only IκBα. As such, TNF signaling does not accumulate NIK. Therefore, these analyses substantiated that p100 phosphorylation by IKK1 requires NIK. However, their assay was unable to discern if the NEMO-associated IKK1 complex was solely responsible for IκBα phosphorylation or NEMO-independent IKK1 complexes, particularly those isolated from LTβR-stimulated cells, also phosphorylated IκBα. Interestingly, NIK immunoprecipitates, even those obtained from TNF-treated cells, were devoid of the IκBα kinase activity (Fig. 3a, b). Hence, our assay selectively detected noncanonical NF-κB signaling induced by LTβR and was insensitive to canonical NF-κB pathway activation by TNF (Fig. 3a, b). Taken together, our assay suggests that IKK1 exists at least in two biochemically distinguishable complexes, which have separate biological functions. As a part of the NEMO-IKK complex, IKK1 phosphorylates IκBα in the canonical NF-κB pathway, whereas NIK-associated IKK1 drives noncanonical NF-κB signaling by specifically phosphorylating p100.

Furthermore, our assay was robust enabling the assessment of NF-κB signaling in the context of multiple noncanonical stimuli and a wide range of cell types. We could estimate the activity of NIK-IKK1 in lymphoid stromal-derived BLS4 and BLS12 cell lines, and in αLTβR-stimulated BMDMs or BAFF-treated B cells (Fig. 4a–c). We also detected mutational activation of NIK-IKK1 in cancerous myeloma cell lines (Fig. 4d). Furthermore, our analyses suggested that our NIK-IKK1 assay could be used for screening and validation of chemical inhibitors of NIK (Fig. 4e). Because NIK is considered as a potential therapeutic target, our assay bear significance for future biomedical research.

4 Notes

1. Immortalized MEFs were used for generating Fig. 2c; all other MEF data in this draft were generated using primary cells.
2. For detecting NIK by immunoblot analyses in either CE or immunoprecipitates (Fig. 2a), we harvested ~2 × 10^7 cells and prepared a ~5-fold concentrated extract with the total protein concentration of ~7.5 mg/ml. We loaded ~225μg of total protein (30μl) per lane for analyzing CE and examined immunoprecipitates derived from ~400μg equivalent of total protein per lane. Furthermore, immunoblot conditions were also sensitized - a 1:250 dilution of anti-NIK antibody and a 1:500 dilution of secondary antibody was used. Otherwise, we used routine cell extraction protocol for assessing the NIK-IKK1 kinase activity as described under Subheading 3 (Figs. 2b–d and 3). Accordingly, immunoblot analyses of NIK described in Fig. 2b were performed using ~15μg of total CE protein or immunoprecipitates derived from ~50μg equivalent of total protein per lane. Moreover, immunoblot analyses involved a more routinely 1:500 dilution of anti-NIK antibody and a 1:1000 dilution of secondary antibody. Although inadequate for detecting NIK in immunoblots, these amounts were sufficient for conveniently detecting the activity of NIK-IKK1 in our radioactive kinase assay.
3. While preparing cell lysates, NaF, β-glycerophosphate, Na_3VO_4, PIC and DTT should be freshly added into CEB. CEB should be prepared in excess as it would be also required in various other steps (*see* Subheadings 3.5, **steps 2** and **8**).
4. All procedures involving cell lysate processing and preparation should be carried out in ice i.e., 4 °C. Cell pellets should be completely dislodged, and cells should be properly resuspended for efficient cell lysis.
5. Ethanol should be removed from Protein G 4 Fast Flow Sepharose bead (~50% slurry) and resuspended with CEB (50% v/v). Protein G 4 Fast Flow Sepharose bead slurry should be mixed well each time before pipetting into the cell lysate in the pre-clearing step (*see* Subheading 3.5, **step 2**) or into the immunoprecipitation mixture (*see* Subheading 3.5, **step 6**). While adding the Protein G 4 Fast Flow Sepharose bead slurry,

the end of 200μl pipette tips should be snipped to ensure accurate delivery of the beads and to prevent plugging of tips by the beads.

6. 1× Wash Buffer should be freshly prepared from stock solutions. All washing steps should be carried out at 4 °C. In between washes, about 20μl of supernatant should be left to avoid aspirating the beads. After the final wash (*see* Subheading 3.5, **step 9**), the supernatant should be carefully aspirated as much as possible. During this step the residual 20μl supernatant should be gently aspirated using a narrow-bore pipette tip.
7. Kinase reaction mix should be prepared from freshly prepared 10× Kinase Assay Buffer. Cold ATP should be thawed in ice only (4 °C). γ-[P^{32}] ATP should be stored in −20 °C and was thawed at RT while preparing Kinase Reaction Mix.
8. It should be ensured that the immunopellet is properly mixed with the kinase reaction mixture by gently tapping the microcentrifuge tube and taking care that the beads are not splashed onto the walls of the tube.
9. As the immunopellet settles down during the kinase reaction step, the immunopellet-kinase reaction mix should be gently mixed briefly (for 2–3 s) with a pipette tip every 10 min.
10. Only a small fraction of γ-[P^{32}] is utilized in the kinase reaction for the phosphorylation of $p100_{(406–899)}$. Importantly, the bulk of the radioactivity remains as γ-[P^{32}] ATP and resolves at or near the dye front following SDS-PAGE. Hence, the electrophoresis run should be terminated before tracking dye front leaves the gel (~1–2 cm from the bottom), to prevent and minimize radioactive contamination in the running buffer. In this way, more than 95% of the radioactive waste can be isolated in the trimmed gel and easily collected in a suitable radioactive waste container. γ-[P^{32}] ATP should be handled with all precautions in accordance with the prescribed institutional regulation for the use of radioactivity.
11. Determining the appropriate exposure time is important and depends on the quality of γ-[P^{32}] ATP. Intense and sharp bands from γ-[P^{32}] phosphorylated $p100_{(406–899)}$ substrate is generally obtained in the first and second half-life of γ-[P^{32}] ATP.
12. PVDF membrane should be activated using methanol prior to western blot transfer.

Acknowledgments

We thank Dr. Nico Ghilardi, Genentech Inc., Dr. Michael Kuehl, NCI and Dr. Tomoya Katakai, Kansai Medical University for research reagents. Research in the PI's laboratory has been funded by SERB, Department of Science and Technology, Govt. of India (EMR/2015/000658), the Wellcome Trust DBT India Alliance (500094/Z/09/Z) and NII-Core. TM thanks UGC, YR thanks DBT, and AD thanks DST-INSPIRE for research fellowships. We thank Vijendra Kumar, SIL for technical help and other laboratory members for valuable discussions.

Author Contributions: TM and YR conducted cell-based analyses with the assistance from AD and BB, and the guidance from SP and SB TM, YR, and SB designed the experiments, curated and validated the data. TM and YR wrote the manuscript with SB. The authors declare no conflicts of interest.

References

1. Mitchell S, Vargas J, Hoffmann A (2016) Signaling via the NFκB system. Wiley Interdiscip Rev Syst Biol Med 8:227–241
2. Sun S-C (2017) The non-canonical NF-κB pathway in immunity and inflammation. Nat Rev Immunol 17:545–558
3. Tao Z, Fusco A, Huang D-B et al (2014) p100/IκBδ sequesters and inhibits NF-κB through kappaBsome formation. Proc Natl Acad Sci 111:15946–15951
4. Vallabhapurapu S, Matsuzawa A, Zhang WZ et al (2008) Nonredundant and complementary functions of TRAF2 and TRAF3 in a ubiquitination cascade that activates NIK-dependent alternative NF-κB signaling. Nat Immunol 9:1364–1370
5. Zarnegar BJ, Wang Y, Mahoney DJ et al (2008) Noncanonical NF-κB activation requires coordinated assembly of a regulatory complex of the adaptors cIAP1, cIAP2, TRAF2 and TRAF3 and the kinase NIK. Nat Immunol 9:1371–1378
6. Xiao G, Fong A, Sun SC (2004) Induction of p100 processing by NF-κB-inducing kinase involves docking IκB kinase α (IKKα) to p100 and IKKα-mediated phosphorylation. J Biol Chem 279:30099–30105
7. Polley S, Passos DO, Bin HD et al (2016) Structural basis for the activation of IKK1/α. Cell Rep 17:1907–1914
8. Liang C, Zhang M, Sun SC (2006) β-TrCP binding and processing of NF-κB2/p100 involve its phosphorylation at serines 866 and 870. Cell Signal 18:1309–1317
9. Hahn M, Macht A, Waisman A et al (2016) NF-κB-inducing kinase is essential for B-cell maintenance in mice. Eur J Immunol 46:732–741
10. Li Y, Wang H, Zhou X et al (2016) Cell intrinsic role of NF-κB-inducing kinase in regulating T cell-mediated immune and autoimmune responses. Sci Rep 6:22115
11. Murray SE, Polesso F, Rowe AM et al (2011) NF-κB-inducing kinase plays an essential T cell-intrinsic role in graft-versus-host disease and lethal autoimmunity in mice. J Clin Invest 121:4775–4786
12. Katakam AK, Brightbill H, Franci C et al (2015) Dendritic cells require NIK for CD40-dependent cross-priming of CD8 + T cells. Proc Natl Acad Sci 112:14664–14669
13. Jin J, Hu H, Li HS et al (2014) Noncanonical NF-κB pathway controls the production of type I interferons in antiviral innate immunity. Immunity 40:342–354
14. Parvatiyar K, Pindado J, Dev A et al (2018) A TRAF3-NIK module differentially regulates DNA vs RNA pathways in innate immune signaling. Nat Commun 9:2770
15. Banoth B, Chatterjee B, Vijayaragavan B et al (2015) Stimulus-selective crosstalk via the NF-κB signaling system reinforces innate immune response to alleviate gut infection. elife 4:1–25
16. Chatterjee B, Banoth B, Mukherjee T et al (2016) Late-phase synthesis of IκBα insulates the TLR4-activated canonical NF-κB pathway from noncanonical NF-κB signaling in macrophages. Sci Signal 9(457):ra120
17. Maubach G, Feige MH, Lim MCC et al (2019) NF-kappaB-inducing kinase in cancer. Biochim Biophys Acta Rev Cancer 1871:40–49
18. Roy P, Sarkar U, Basak S (2018) The NF-κB activating pathways in multiple myeloma. Biomedicines 6:59

19. Rahal R, Frick M, Romero R et al (2014) Pharmacological and genomic profiling identifies NF-κB-targeted treatment strategies for mantle cell lymphoma. Nat Med 20:87–92
20. Saitoh Y, Yamamoto N, Dewan MZ et al (2008) Overexpressed NF-κB-inducing kinase contributes to the tumorigenesis of adult T-cell leukemia and Hodgkin Reed-Sternberg cells. Blood 111:5118–5129
21. Uno M, Saitoh Y, Mochida K et al (2014) NF-κB inducing kinase, a central signaling component of the non-canonical pathway of NF-κB, contributes to ovarian cancer progression. PLoS One 9:e88347
22. Vazquez-Santillan K, Melendez-Zajgla J, Jimenez-Hernandez LE et al (2016) NF-kappa B-inducing kinase regulates stem cell phenotype in breast cancer. Sci Rep 6:1–17
23. Duran LD, Jung J-U et al (2016) NIK regulates MT1-MMP activity and promotes glioma cell invasion independently of the canonical NF-κB pathway. Oncogenesis 5:e231–e212
24. Sheng L, Zhou Y, Chen Z et al (2012) NF-κB-inducing kinase (NIK) promotes hyperglycemia and glucose intolerance in obesity by augmenting glucagon action. Nat Med 18:943–949
25. Brightbill HD, Suto E, Blaquiere N et al (2018) NF-κB inducing kinase is a therapeutic target for systemic lupus erythematosus. Nat Commun 9:1–14
26. Basak S, Shih VF-S, Hoffmann A (2008) Generation and activation of multiple dimeric transcription factors within the NF-κB signaling system. Mol Cell Biol 28:3139–3150
27. Busino L, Millman SE, Scotto L et al (2012) Fbxw7α-and GSK3-mediated degradation of p100 is a pro-survival mechanism in multiple myeloma. Nat Cell Biol 14:375–385
28. Roy P, Mukherjee T, Chatterjee B et al (2017) Non-canonical NFκB mutations reinforce pro-survival TNF response in multiple myeloma through an autoregulatory RelB:p50 NFκB pathway. Oncogene 36:1417–1429
29. Wang VY-F, Fusco AJ, Tao Z et al (2016) The NF-κB subunit RelB controls p100 processing by competing with the kinases NIK and IKK1 for binding to p100. Sci Signal 9:ra96–ra96
30. Mitchell S, Hoffmann A (2019) Substrate complex competition is a regulatory motif that allows NFκB RelA to license but not amplify NFκB RelB. Proc Natl Acad Sci 116:10592–10597
31. Razani B, Zarnegar B, Ytterberg AJ et al (2010) Negative feedback in noncanonical NF-κB signaling modulates NIK stability through IKKα-mediated phosphorylation. Sci Signal 3:1–10
32. Mukherjee T, Chatterjee B, Dhar A et al (2017) A TNF-p100 pathway subverts noncanonical NFκB signaling in inflamed secondary lymphoid organs. EMBO J 36:3501–3516
33. Dejardin E, Droin NM, Delhase M et al (2002) The lymphotoxin-beta receptor induces different patterns of gene expression via two NF-kappaB pathways. Immunity 17:525–535

Chapter 11

Analyze the SUMOylation of IKKγ/NEMO During Genotoxic Stress

Zhao-Hui Wu and Shigeki Miyamoto

Abstract

SUMOylation is an important posttranslational modification of substrate proteins that regulates their functions in a variety of cellular processes including epigenetic and transcriptional regulation of gene expression, genomic stability, DNA repair, subcellular translocation, and protein turnover. The critical roles of SUMOylation in regulating NF-κB signaling is exemplified by the findings that it regulates IκBα stability, transactivity of RelA and RelB, as well as initiating the export of nuclear DNA damage signal to cytoplasmic IKK complex through NEMO SUMOylation. Detection of SUMOylated protein is technically challenging due to only a small fraction of substrate proteins is SUMOylated and this process is also reversible by highly active SUMO-deconjugating enzymes. In this protocol, we outline a method for detecting SUMOylation of NEMO in mammalian cells treated by genotoxic agents.

Key words NEMO, SUMOylation, Genotoxic stress, Immunoprecipitation, Immunoblot

1 Introduction

Protein posttranslational modifications play crucial roles in controlling the activity and stability of a variety of core components of the nuclear factor kappa B (NF-κB) signaling pathways [1]. SUMOylation is a form of posttranslational modification that covalently attaches a relatively small protein (~20 kDa) called SUMO (small ubiquitin-like modifier) to target proteins [2, 3]. Like ubiquitylation, SUMOylation is a three-step process which is catalyzed by SUMO-activating enzyme (E1), SUMO-conjugating enzyme (E2), and SUMO ligase (E3) [4, 5]. There is only one known SUMO E1, heterodimeric SAE1/SAE2, and one known SUMO E2, UBC9, in humans thus far. Several SUMO E3 ligases have been reported to facilitate SUMOylation in a substrate-specific manner, which includes RanBP2 (Ran-binding protein 2) [6], Pc2 (polycomb group protein) [7], and the protein inhibitors of activated STAT (PIAS) family members [8]. Five SUMO

Guido Franzoso and Francesca Zazzeroni (eds.), *NF-κB Transcription Factors: Methods and Protocols*, Methods in Molecular Biology, vol. 2366, https://doi.org/10.1007/978-1-0716-1669-7_11,

paralogs, SUMO1/2/3/4/5, have been identified in the human genome so far, which can modify substrate proteins either as single moiety conjugation or with homo/ heterotypic polySUMO chains [5]. Protein SUMOylation is involved in many cellular processes, which may alter the organization of nuclear structures and chromatin, regulate gene transcription, and direct nucleo-cytoplasmic translocation.

SUMOylation of several proteins involved in NF-κB signaling, such as IκBα [9], p100 [10], RelB [11], RelA/p65 [12], and NEMO [13] has been reported to play pivotal roles in controlling the activation and resolution of NF-κB signaling. IκBα is the first NF-κB signaling-related protein whose function was reported to be altered by SUMOylation [9]. Conjugation of SUMO on the Lys 21 of IκBα blocked the ubiquitylation of IκBα on the same lysine residue, which is one of two critical lysine residues (Lys21/Lys22) mediating its ubiquitin-proteasome-dependent degradation [14, 15]. Consequently, IκBα SUMOylation leads to inhibition of proteasomal degradation of IκBα and suppression of NF-κB activation. In contrast, SUMOylation of p100 is required for p100 phosphorylation and processing into p52 in response to various stimuli, supporting its critical role in enhancing alternative NF-κB signaling [10]. However, SUMOylation of the key alternative NF-κB pathway transcription factor RelB converts it from transcriptional activator into a repressor [11], which may contribute to the fine-tuning and reprogramming of RelB-dependent transcriptome. SUMOylation of RelA/p65 was shown to be induced by TNF-α and hypoxia in hepatocellular carcinoma cells, which enhances its nuclear localization and NF-κB activity [12]. Our previous studies on genotoxic NF-κB signaling demonstrated an essential role of the SUMOylation of NEMO, the IKK regulatory subunit, in transmitting nuclear DNA damage signal to cytoplasmic NF-κB activating machinery, thereby leading to NF-κB activation upon DNA damage [13, 16].

SUMOylation is a reversible modification that can be removed by a class of SUMO proteases, including SENP1/2/3/5/6/7 [17]. A combination of only a small fraction of substrate proteins being SUMOylated and the presence of highly active SUMO-deconjugation enzymes renders the detection of protein SUMOylation under physiological conditions rather challenging [18]. Using human and mouse cells, we have demonstrated that NEMO is conjugated by SUMO1 at Lys277 and Lys309 in response to genotoxic treatments, which is essential for the accumulation of NEMO in the nucleus. The SUMOylation of NEMO enables its subsequent modifications, including phosphorylation and mono-ubiquitylation, and export from the nucleus in complex with DNA damage response apical kinase ATM, which leads to the activation of cytoplasmic IKK complex in response to the nuclear DNA damage signal [13]. This chapter describes a detailed protocol for the enrichment and analysis of NEMO SUMOylation in mammalian cells upon genotoxic stress. This is achieved by preserving protein SUMOylation by cell lysis in the presence of an inhibitor of SUMO-deconjugating enzymes, denaturation of cell lysates to further inactivate SUMO proteases and to dissociate any NEMO-associated proteins, isolating NEMO by immunoprecipitation, and detecting SUMOylated NEMO by immunoblotting with a SUMO1 antibody (Fig. 1).

2 Materials

1. Cell lysis buffer: 20 mM Tris–HCl, pH 7.0, 250 mM NaCl, 3 mM EDTA, 3 mM EGTA, 0.5% NP-40, 2 mM DTT, 0.5 mM PMSF, 20 mM β-glycerol phosphate, 1 mM sodium orthovanadate, 1 μg/ml leupeptin, 1 μg/ml aprotinin, 10 mM p-nitrophenyl phosphate, 10 mM sodium fluoride (*see* **Note 1**).
2. Phosphate-Buffered Saline (PBS): 137 mM NaCl, 2.7 mM KCl, 4.3 mM Na_2HPO_4, and 1.5 mM KH_2PO_4.
3. 10% SDS.
4. N-ethylmaleimide.
5. Primary antibody suitable for immunoprecipitation of NEMO, such as anti-NEMO/IKKγ (BD Pharmingen) (or other protein of interest).
6. (*Optional*) Antibodies against epitope tags if analyzing transfected NEMO with tags.
7. Nonspecific IgG control (R&D).
8. Antibodies against SUMO-1 (GMP-1, Invitrogen).
9. Protein G-sepharose or protein A-sepharose beads, pre-equilibrated to the cell lysis buffer.
10. 2× SDS sample buffer: 125 mM Tris–HCl, pH 6.7, 4% SDS, 0.2 M DTT, 20% glycerol, and 0.01% bromophenol blue.
11. 30% Acrylamide/Bis Solution, 29:1.
12. Ammonium Persulfate.
13. Tris-buffered saline (TBS): 150 mM NaCl, 25 mM Tris-Base, pH 7.4.
14. PVDF membrane.
15. 5% nonfat milk in TBS, 0.1% Tween-20.
16. HRP-conjugated secondary antibodies.
17. Enhanced Chemioluminescence (ECL).
18. X-ray films.

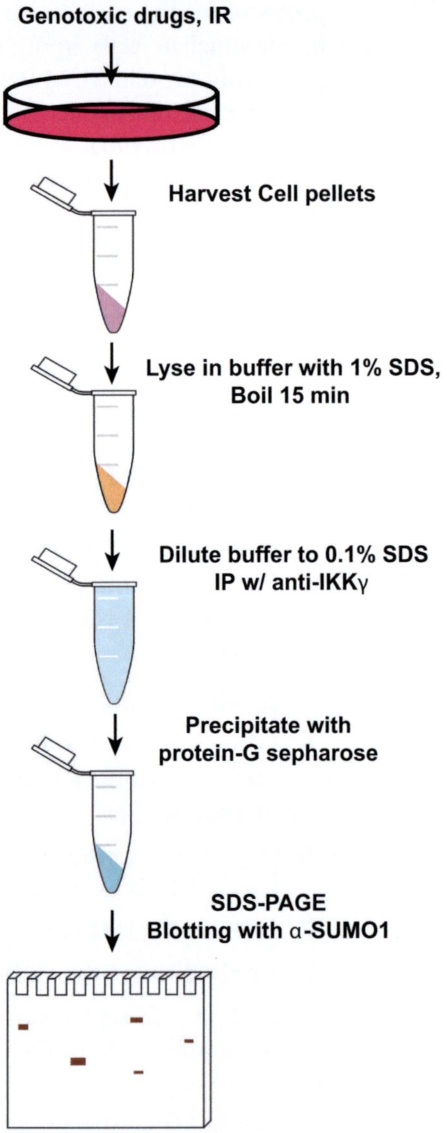

Fig. 1 Schematic protocol for the enrichment and analysis of NEMO SUMOylation in mammalian cells upon genotoxic stress

3 Methods

1. Culture cells at sub-confluent density and treat them with genotoxic agents (e.g., 10 μM Etoposide) for a predetermined time (*see* **Note 2**). A relatively larger cell number is preferred for SUMOylation analysis. Typically, 1–3 × 10^7 cells are used for each sample.

2. Place the plate of cells on ice right before harvesting, remove media by aspiration, rinse cells with 5 ml of PBS, then remove PBS.
3. Add 1 ml of ice-cold PBS to the plate, scrape cells off plates with a cell scraper, transfer to a 1.5 ml microcentrifuge tube, centrifuge at 1500 × *g* for 1 min at 4 °C and remove the PBS. Rinse the cells by resuspending the cell pellets with 1 ml ice-cold PBS and re-pellet cells by centrifugation as above. If suspension cells are used for analysis, centrifuge cell suspension in conical tubes at 800 × *g* for 5 min at 4 °C, and remove the media by aspiration. Rinse the cells by resuspending the cell pellets with 1 ml ice-cold PBS, transfer the cell suspension to 1.5 ml microcentrifuge tube, and re-pellet cells by centrifugation as above.
4. Add 20 μl PBS to each microcentrifuge tube and loosen the cell pellet by tapping the bottom of the tube.
5. Add 180 μl cell lysis buffer supplemented with 20 mM N-ethylmaleimide (*see* **Note 3**) to each tube and resuspend cells by tapping, sit on ice for 5 min (*see* **Note 4**).
6. Add 20 μl 10% SDS into each tube and mix by tapping (~1% SDS final concentration in lysates).
7. Boil cell lysates for 15 min (*see* **Note 5**).
8. Centrifuge at 12,000 × *g* for 10 min at 4 °C.
9. Carefully transfer the supernatant to a fresh 2 ml microcentrifuge tube.
10. Remove 10 μl lysate and save in a fresh 1.5 ml microcentrifuge tube as input control samples (5% of total). Keep this input sample at −20 °C until analysis.
11. Add 1.8 ml cell lysis buffer (without SDS) to each 2 ml microcentrifuge tube and mix with cell lysates (~0.1% SDS final concentration in lysates) (*see* **Note 6**).
12. Add an appropriate amount of NEMO antibody sufficient to quantitatively precipitate NEMO (e.g., 2 μg of BD anti-NEMO antibody for 1 × 10^7 HEK293 cells) (or antibody against the protein of interest) to each tube.
13. Rotate tubes for 2 h at 4 °C.
14. While samples are incubated with NEMO antibody, wash Protein G-sepharose beads (20 μl/sample) with cell lysis buffer, centrifuge, and resuspend Protein G-sepharose with an equal volume of cell lysis buffer.
15. Add 20 μl pre-equilibrated Protein G-sepharose (1:1 slurry) in each microcentrifuge tube.
16. Rotate tubes for overnight (12–14 h) at 4 °C.

17. Centrifuge the tube at 12,000 × *g* for 5 min at 4 °C and remove supernatants.
18. Wash the Protein G-sepharose with NEMO immunoprecipitates with cell lysis buffer by adding 500 μl ice-cold cell lysis buffer in each tube, resuspending Protein G-sepharose beads by inverting the tube, then pelleting the Protein G-sepharose beads by centrifugation at 12,000 × *g* for 1 min at 4 °C. Remove the supernatant by pipetting and repeat washing with fresh ice-cold cell lysis buffer for three more times as above to remove any residual nonspecific proteins.
19. Add 10 μl 2× SDS sample buffer to each tube, resuspend, and boil for 5 min.
20. Load the immunoprecipitated samples and input controls on precasted SDS-PAGE gel (10%), run and transfer the protein samples onto PVDF membrane.
21. Rinse the transferred PVDF membrane with PBS.
22. Incubate the membrane in TBS with 5% nonfat milk and 0.1% Tween-20 for 10 min at room temperature to pre-block the membrane.
23. Incubate the membrane with anti-SUMO-1 (GMP-1, 1:1000 in TBS with 5% nonfat milk and 0.1% Tween-20) on rotating platform overnight at 4 °C.
24. Wash the membrane with PBS for three times.
25. Incubate the membrane with HRP-conjugated secondary antibody for 2 h at room temperature.
26. Wash the membrane with PBS for three times.
27. Incubate the membrane with ECL and expose to X-ray film (*see* **Notes 7** and **8**).

4 Notes

1. To make the cell lysis working solution, all the protease inhibitors should be added freshly into the cell lysis buffer stock solution (20 mM Tris–Hcl, pH 7.0, 250 mM NaCl, 3 mM EDTA, 3 mM EGTA, 0.5% NP-40).
2. Each protein of interest may have different kinetics of SUMOylation in response to a specific treatment. The optimal time point for detecting target protein SUMOylation should be determined by pilot experiments. For example, our experiments indicate NEMO SUMOylation in response to genotoxic treatment often peaks around 60 min after treatment (when time points such as 45, 60, and 90 min are taken), which is thus often used for our studies [13].

3. N-ethylmaleimide should be freshly dissolved in DMSO before added into cell lysis buffer.
4. Make sure the cells are completely resuspended in the buffer before adding SDS. Otherwise, cell pellets become "clumpy" without SDS reaching all the proteins in the cell lysate and most proteins are lost in the pellets during subsequent centrifugation following boiling.
5. This step ensures the inactivation of deSUMOylases/other proteases and the preservation of protein SUMOylation. It also removes proteins noncovalently associated with the protein being immunoprecipitated, which enhances the specificity of the SUMOylation detection.
6. Reducing SDS concentration to 0.1% in cell lysates is essential for the following immunoprecipitation.
7. Generally, the fraction of substrate protein being SUMOylated is rather small. Extended exposure time or hypersensitive ECL reagents may be needed for detecting the SUMOylated protein species.
8. Detecting SUMOylation by endogenous SUMO1 could be challenging, which may be improved by altering the expression of the proteins modulating the SUMOylation level of the substrate of interest. As we found NEMO SUMOylation is facilitated by SUMO E3 ligase PIASy [19] and suppressed by SUMO protease SENP2 [20], overexpressing PIASy or silencing SENP2 in cells substantially improved our capacity to detect NEMO SUMOylation. Other molecules that could enhance NEMO SUMOylation, such as PIDD [21] and Tax [22], were also shown to enhance detection of NEMO SUMOylation when overexpressed in the cells.

References

1. Perkins ND (2006) Post-translational modifications regulating the activity and function of the nuclear factor kappa B pathway. Oncogene 25(51):6717–6730. https://doi.org/10.1038/sj.onc.1209937
2. Mahajan R, Delphin C, Guan T, Gerace L, Melchior F (1997) A small ubiquitin-related polypeptide involved in targeting RanGAP1 to nuclear pore complex protein RanBP2. Cell 88(1):97–107. https://doi.org/10.1016/s0092-8674(00)81862-0
3. Matunis MJ, Coutavas E, Blobel G (1996) A novel ubiquitin-like modification modulates the partitioning of the Ran-GTPase-activating protein RanGAP1 between the cytosol and the nuclear pore complex. J Cell Biol 135(6 Pt 1):1457–1470. https://doi.org/10.1083/jcb.135.6.1457
4. Mabb AM, Miyamoto S (2007) SUMO and NF-kappaB ties. Cell Mol Life Sci 64 (15):1979–1996. https://doi.org/10.1007/s00018-007-7005-2
5. Celen AB, Sahin U (2020) Sumoylation on its 25th anniversary: mechanisms, pathology, and emerging concepts. FEBS J 287 (15):3110–3140. https://doi.org/10.1111/febs.15319
6. Pichler A, Gast A, Seeler JS, Dejean A, Melchior F (2002) The nucleoporin RanBP2 has SUMO1 E3 ligase activity. Cell 108 (1):109–120. https://doi.org/10.1016/s0092-8674(01)00633-x

7. Kagey MH, Melhuish TA, Wotton D (2003) The polycomb protein Pc2 is a SUMO E3. Cell 113(1):127–137. https://doi.org/10.1016/s0092-8674(03)00159-4
8. Schmidt D, Muller S (2002) Members of the PIAS family act as SUMO ligases for c-Jun and p53 and repress p53 activity. Proc Natl Acad Sci U S A 99(5):2872–2877. https://doi.org/10.1073/pnas.052559499
9. Desterro JM, Rodriguez MS, Hay RT (1998) SUMO-1 modification of IkappaBalpha inhibits NF-kappaB activation. Mol Cell 2 (2):233–239. https://doi.org/10.1016/s1097-2765(00)80133-1
10. Vatsyayan J, Qing G, Xiao G, Hu J (2008) SUMO1 modification of NF-kappaB2/p100 is essential for stimuli-induced p100 phosphorylation and processing. EMBO Rep 9 (9):885–890. https://doi.org/10.1038/embor.2008.122
11. Leidner J, Voogdt C, Niedenthal R, Möller P, Marienfeld U, Marienfeld RB (2014) SUMOylation attenuates the transcriptional activity of the NF-κB subunit RelB. J Cell Biochem 115 (8):1430–1440. https://doi.org/10.1002/jcb.24794
12. Liu J, Tao X, Zhang J, Wang P, Sha M, Ma Y, Geng X, Feng L, Shen Y, Yu Y, Wang S, Fang S, Shen Y (2016) Small ubiquitin-related modifier 1 is involved in hepatocellular carcinoma progression via mediating p65 nuclear translocation. Oncotarget 7(16):22206–22218. https://doi.org/10.18632/oncotarget.8066
13. Huang TT, Wuerzberger-Davis SM, Wu ZH, Miyamoto S (2003) Sequential modification of NEMO/IKKgamma by SUMO-1 and ubiquitin mediates NF-kappaB activation by genotoxic stress. Cell 115(5):565–576
14. Chen ZJ, Parent L, Maniatis T (1996) Site-specific phosphorylation of IkappaBalpha by a novel ubiquitination-dependent protein kinase activity. Cell 84(6):853–862
15. Hayden MS, Ghosh S (2004) Signaling to NF-kappaB. Genes Dev 18(18):2195–2224. https://doi.org/10.1101/gad.1228704
16. Wu ZH, Shi Y, Tibbetts RS, Miyamoto S (2006) Molecular linkage between the kinase ATM and NF-kappaB signaling in response to genotoxic stimuli. Science 311 (5764):1141–1146. https://doi.org/10.1126/science.1121513
17. Kunz K, Piller T, Muller S (2018) SUMO-specific proteases and isopeptidases of the SENP family at a glance. J Cell Sci 131(6). https://doi.org/10.1242/jcs.211904
18. Hay RT (2007) SUMO-specific proteases: a twist in the tail. Trends Cell Biol 17 (8):370–376. https://doi.org/10.1016/j.tcb.2007.08.002
19. Mabb AM, Wuerzberger-Davis SM, Miyamoto S (2006) PIASy mediates NEMO sumoylation and NF-kappaB activation in response to genotoxic stress. Nat Cell Biol 8(9):986–993. https://doi.org/10.1038/ncb1458
20. Lee MH, Mabb AM, Gill GB, Yeh ET, Miyamoto S (2011) NF-kappaB induction of the SUMO protease SENP2: a negative feedback loop to attenuate cell survival response to genotoxic stress. Mol Cell 43(2):180–191. https://doi.org/10.1016/j.molcel.2011.06.017
21. Janssens S, Tinel A, Lippens S, Tschopp J (2005) PIDD mediates NF-kappaB activation in response to DNA damage. Cell 123 (6):1079–1092. https://doi.org/10.1016/j.cell.2005.09.036
22. Kfoury Y, Setterblad N, El-Sabban M, Zamborlini A, Dassouki Z, El Hajj H, Hermine O, Pique C, de Thé H, Saïb A, Bazarbachi A (2011) Tax ubiquitylation and SUMOylation control the dynamic shuttling of Tax and NEMO between Ubc9 nuclear bodies and the centrosome. Blood 117 (1):190–199. https://doi.org/10.1182/blood-2010-05-285742

Part III

Methods for Analyzing NF-κB Activation in Physiology and Disease

Chapter 12

Analysis of the Contribution of NF-κB in the Regulation of Chemotherapy-Induced Cell Senescence by Establishing a Tetracycline-Regulated Cell System

Francesco Pacifico, Elvira Crescenzi, and Antonio Leonardi

Abstract

Therapy-induced senescence (TIS or therapy-induced premature senescence) is a key cellular program triggered in the course of cancer radiotherapy and chemotherapy with genotoxic drugs, both in cancer cells and in normal cells, whose activation critically affects the outcome of cancer therapy. Drug-induced senescent cells undergo a permanent cell cycle arrest, acquire distinctive morphological and biochemical alterations, and an enhanced secretory ability, referred to as senescence-associated secretory phenotype (SASP). The transcription factor NF-κB acts as a master regulator of the SASP, driving the expression of senescence-associated secretome components.

Here we describe protocols for the establishment of a tetracycline-regulated cell system for the investigation of the role of NF-κB in TIS. We also describe protocols routinely used in our laboratory, to investigate TIS in this Tet-On inducible expression system. Finally, we describe techniques for the validation of TIS induction.

Key words NF-κB, IκBαM, Tet-On system, Therapy-induced senescence, SASP, SA-beta-gal, BrdU, DNA damage foci, Conditioned medium

1 Introduction

Therapy-induced senescence (TIS) is a state of permanent cell cycle arrest that ensues both in normal and neoplastic cells following conventional cancer treatments, such as chemo- or radiotherapy [1–3]. Senescent cells gradually acquire an enhanced secretory ability (SASP or senescence-associated secretory phenotype), meaning they secrete cytokines, chemokines, and growth factors [4, 5], mostly under NF-κB transcriptional control [6, 7]. It is now appreciated that the activation of a premature senescence program critically contribute to the outcome of cancer therapy [8–10], although the biological effects of senescence are complex, and

Guido Franzoso and Francesca Zazzeroni (eds.), *NF-κB Transcription Factors: Methods and Protocols*, Methods in Molecular Biology, vol. 2366, https://doi.org/10.1007/978-1-0716-1669-7_12,

both protumoral and antitumoral activities of senescent cells have been demonstrated [11, 12].

Although there is no single specific marker for the identification of senescent cells [13], TIS cells acquire several distinctive features that can be used for their identification and characterization. In this chapter, we describe protocols for the detection of three characteristic features of TIS cells, i.e., an increased activity of lysosomal beta-galactosidase; a permanent cell cycle arrest; and the presence of irreparable DNA damage which results in persistent DNA damage foci.

The upregulation of lysosomal beta-galactosidase in senescent human cells, the so-called senescence-associated β-gal or SA-β-gal, has been described in 1995 [14] and still represents the most widely used marker for the detection of senescence [15, 16].

Long-term, irreversible loss of proliferative ability is a key feature of senescence, which allows for the distinction between senescent and quiescent cells. Here we describe a classical method for the detection of DNA replication, using 5-bromo-2′-deoxyuridine (BrdU) incorporation and simultaneous determination of DNA content. BrdU is an analogue of thymidine and it is incorporated into newly synthesized DNA. Hence, when exposed to a pulse of BrdU, only cycling cells that actively synthesize DNA are detected using an anti-BrdU antibody.

Finally, although some forms of senescence are induced in the absence of DNA damage [17, 18], TIS cells display irreparable DNA damage and a persistent activation of DNA damage response [19–21]. Hence, TIS cells are characterized by an accumulation of DNA damage response proteins in the nuclear area surrounding irreparable DNA lesions, to form DNA damage foci that can be detected and quantified by immunofluorescence microscopy.

The tetracycline-inducible Tet-Off and Tet-On systems are used to regulate ectopic expression of different genes in eukaryotic cells [22–24]. These systems take advantage of the basic principles regulating the activity of the tetracycline resistance operon in bacteria (Fig.1).

Binding of tetracycline gene repressor (tetR) to tetracycline operator (tetO) sequence blocks transcription of tetracycline resistance (tetA) gene (Fig. 1a, c). The addition of doxycycline (Dox), a tetracycline analog, triggers a conformational switch in tetR that dissociates from tetO sequence thereby allowing tetA transcription (Fig. 1a, b).

The Tet-Off system leads to silencing of gene expression following treatment with Doxycycline (Dox), as a consequence of the tetracycline-controlled transcriptional activator (tTA) inability to activate transcription from transgene minimal promoter (mP) (Fig. 2b). In the absence of Dox, tTA, that arises from the fusion between tetR and herpes simplex virus VP16 sequences (Fig. 2a), binds to the tetracycline responsive element (TRE) of the transgene mP and drives its transcription (Fig. 2c).

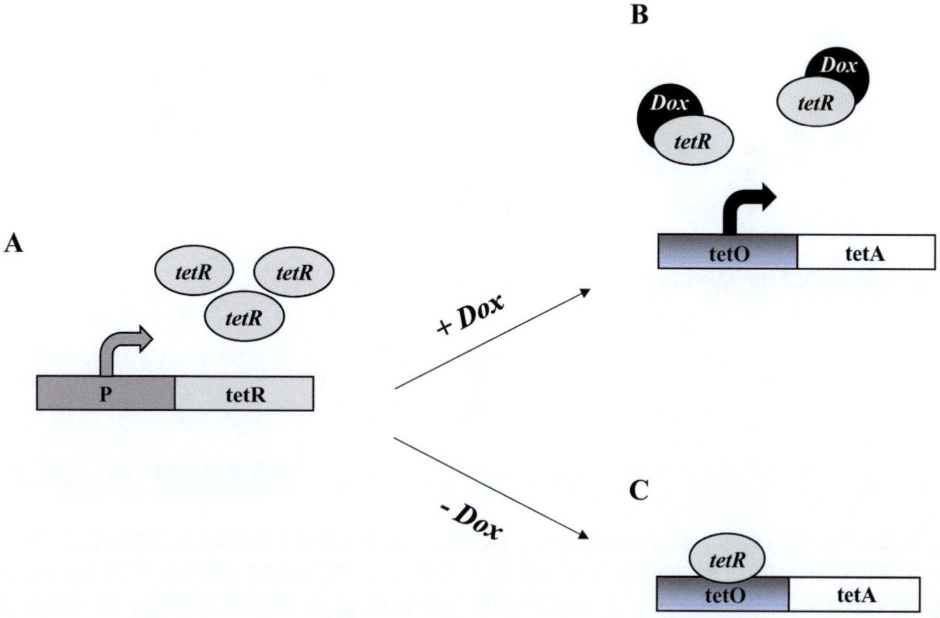

Fig. 1 Tetracycline repressor (tetR) (A) binds to tetracycline operator (tetO) in the absence of doxycycline (Dox) (C) thus blocking tetracycline resistance (tetA) gene transcription, that on the contrary is activated by the presence of Dox that prevents tetR binding to tetO (B)

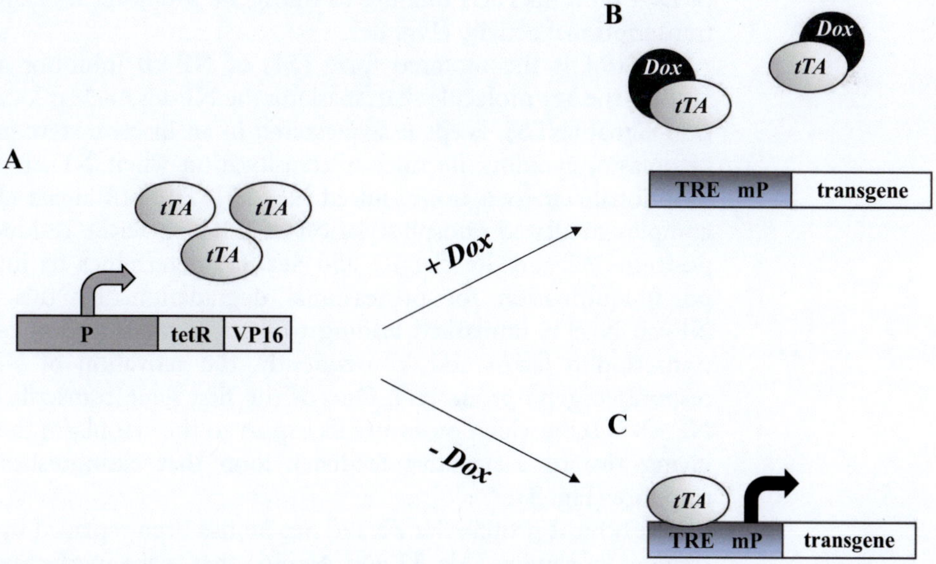

Fig. 2 In the Tet-Off system the tetracycline-controlled transcriptional activator (tTA) (A) is not able to activate transcription from transgene minimal promoter (mP) following treatment with doxycycline (Dox) (B), while it binds to the tetracycline responsive element (TRE) of the transgene mP in the absence of Dox (C)

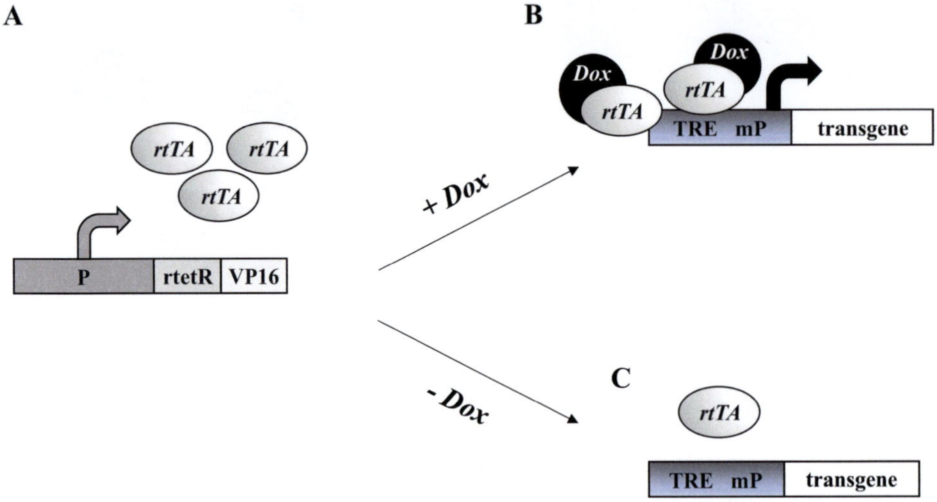

Fig. 3 In the Tet-On system doxycycline (Dox) promotes tetracycline responsive element (TRE)/minimal promoter (mP) binding (B) of the reverse tetracycline-controlled transcriptional activator (rtTA) (A), thus driving transgene expression. The absence of Dox inhibits rtTA binding to TRE/mP blocking its transcriptional activity (C)

On the contrary, in the Tet-On system Dox furthers TRE/mP binding of a mutated form of tTA, called reverse tTA (rtTA) (Fig. 3a), that leads to transgene expression (Fig. 3b). The absence of Dox prevents rtTA binding to transgene promoter blocking its transcriptional activity (Fig. 3c).

IκBαM is the mutated form (M) of NF-κB inhibitor alpha (IκBα), the key molecule that, masking the NF-κB nuclear localization signal (NLS), keeps it sequestered in an inactive state in the cytoplasm, avoiding its nuclear translocation when NF-κB transcriptional activity is not required [25–27]. The IκB kinase (IKK) complex-mediated phosphorylation of IκBα on serine residues in positions 32 and 36 (Ser 32 and Ser 36) determines its further polyubiquitination for proteasomal degradation. By this way, NF-κB NLS is unmasked leading to nuclear translocation of the transcription factor and, consequently, the activation of NF-κB responsive gene promoters. One of the first gene transcribed by NF-κB is IκBα which promotes its export to the cytoplasm thereby giving rise to a negative feedback loop that extinguishes the response (Fig. 4).

In IκBαM protein Ser 32 and Ser 36 has been replaced by two residues of alanine (Ala 32 and Ala 36) that make ineffective the IKK complex-mediated phosphorylation. Consequently, polyubiquitination does not occur, IκBα is not degraded by the proteasome and NF-κB remains sequestered in the cytoplasm. For this reason, this mutated form of IκBα, also called super-repressor, is often used to shut down constitutive or inducible NF-κB activity in many cell types.

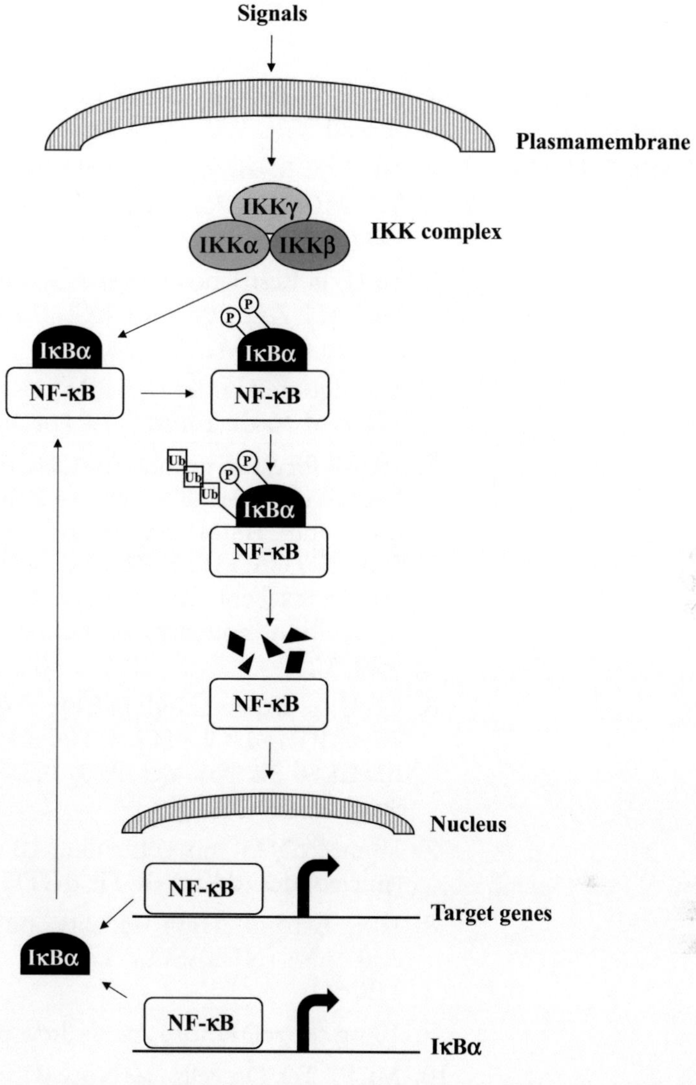

Fig. 4 NF-κB activation is driven by ubiquitin-mediated IκBα degradation following IKK complex-mediated phosphorylation (*see* Subheading 1 for details)

In this chapter, we describe the establishment of a tetracycline-regulated cell system for the investigation of the role of NF-κB in TIS. We present protocols used in our lab to induce TIS in cancer cell lines in vitro, and methods to validate the development of a senescent phenotype, the cell cycle arrest and the presence of persistent DNA damage foci. Finally, we describe the procedure to collect conditioned media for analysis of SASP factors released by cultured TIS cells.

2 Materials

2.1 Establishment of a Tetracycline-Regulated Cell System

1. 0.5µg/µl Tet-On cloning vector pTRE2hyg (*see* **Note 1**). Store at −20 °C.
2. 10 U/µl Restriction enzymes BamHI in store buffer: 10 mM Tris–HCl pH 7.4, 300 mM KCl, 5 mM $MgCl_2$, 0.1 mM EDTA, 1 mM DTT, 0.5 mg/ml BSA, 50% glycerol.
3. 10 U/µl Restriction enzymes SalI in store buffer: 10 mM Tris–HCl pH 7.4, 300 mM KCl, 0.1 mM EDTA, 1 mM DTT, 0.5 mg/ml BSA, 50% glycerol.
4. 1× Digestion buffer D: 6 mM Tris–HCl pH 7.9, 6 mM $MgCl_2$, 150 mM NaCl, 1 mM DTT (*see* **Note 2**). Store at −20 °C.
5. 10µM forward primer (5′-CG**GGATCC** ACCATGGACTA CAAGGACGAC-3′ including BamHI restriction site, bolded text) and 10µM reverse primer (5′-GC**GTCGAC** TCA TAACGTCAGACGCTGG -3′ including SalI restriction site, bolded text) containing two bases flanking restriction enzyme recognition sequences to increase cutting efficiency. Store at −20 °C.
6. 3.5 U/µl Expand High Fidelity Taq Polymerase in store buffer: 20 mM Tris–HCl, pH 7.5, 100 mM KCl, 1 mM DTT, 0.1 mM EDTA, 0.5% Nonidet P40, 0.5% Tween 20, 50% glycerol. Store at −20 °C.
7. 10 mM dNTP mix containing 10 mM each of the four deoxynucleotides (dATP, dCTP, dGTP, TTP). Store at −20 °C.
8. 10× Expand High Fidelity buffer: 500 mM Tris–HCl, 220 mM $(NH_4)_2SO_4$, 15 mM $MgCl_2$, pH 8.9. Store at −20 °C.
9. 10 ng/µl pcDNA3FLAG-IκBαM plasmid.
10. MCF7 Tet-On cells (*see* **Note 3**).
11. 20µg of total proteins from pTRE2hyg-transfected MCF7 Tet-On cells.
12. 20µg of total proteins from pTRE2hyg FLAG-IκBαM-transfected MCF7 Tet-On cells.
13. 5% nonfat dry milk.
14. Anti-FLAG antibodies (*see* **Note 4**).
15. Anti-Actin antibodies.
16. Horseradish peroxidase-conjugated secondary antibodies.
17. Enhanced Chemiluminescent western blotting substrate.
18. Dual-Luciferase reporter assay system (*see* **Note 5**).
19. 2000 U/ml TNFα: Dilute stock solution 2×10^6 U/ml in cell culture medium. Store at −80 °C.

20. 2 mg/ml Doxycycline (Dox) stock solution in water (*see* **Note 6**): Mix by vortexing, filter-sterilize using a membrane filter (0.22μm), aliquot and store at −20 °C.
21. Tris-buffered saline (TBS): 0.15 M NaCl, 0.02 M Tris–HCl, pH 7.4.
22. Tris-buffered saline-Tween (TBST): TBS containing 0.1% Tween-20.

2.2 Induction of Premature Senescence

1. 1 mg/ml Doxorubicin Hydrochloride (Adriamycin) stock solution in distilled water (*see* **Note 7**). Mix by vortexing, filter-sterilize using a membrane filter (0.22μm), aliquot and store at −20 °C. The stock solution can be kept for several months. Once thawed, the tube should be kept in the dark (work with the hood light off when treating cells).
2. 1 mg/ml *cis*-diammineplatinum (II) dichloride (Cisplatin, CDDP) stock solution in deionized water or saline (0.9% sodium chloride) (*see* **Notes 8** and **9**). Mix by vortexing, filter-sterilize using a membrane filter (0.22μm), aliquot and store at −20 °C. The stock solution can be kept for several months.
3. 50 mg/ml 5-Fluorouracil (5-FU) stock solution in Phosphate-Buffered Saline (PBS). Heat the mixture to 37 °C (*see* **Note 10**). Filter-sterilize using a membrane filter (0.22μm), aliquot and store at −20 °C. The stock solution is stable for several months.
4. 30% (w/w) hydrogen peroxide solution in H_2O (*see* **Note 11**).

2.3 Senescence-Associated β-Galactosidase (SA-β-gal) Staining

1. Three percent formaldehyde fixation buffer (*see* **Note 12**): Dilute formaldehyde 37% in PBS or in culture medium (81μl/ml) in a fume hood.
2. SA-β-gal staining solution: 1 mg/ml 5-Bromo-4-chloro-3-indolyl β-D-galactopyranoside (X-Gal), 40 mM citric acid, 12 mM sodium phosphate, 5 mM potassium ferrocyanide, 5 mM potassium ferricyanide, 150 mM sodium chloride, 2 mM magnesium chloride, pH 6.0 (*see* **Note 13**).

2.4 BrdU Staining of Cells for Flow Cytometric Analysis

1. Cold ethanol (*see* **Note 14**).
2. 10 mg/ml (32.5 mM) BrdU solution in PBS, pH 7.2. Mix by vortexing, filter-sterilize using a membrane filter (0.22μm), aliquot and store at −80 °C (*see* **Note 15**).
3. 4 M HCl solution (*see* **Note 16**).
4. Tris-buffered saline (TBS): 0.15 M NaCl, 0.02 M Tris–HCl, pH 7.4.
5. Tris-buffered saline-Tween (TBST): TBS containing 0.1% Tween-20.

6. 1 mg/ml Propidium Iodide Stock Solution in PBS. Store at 4 °C. Protect the tube from light (foil-wrapped) (*see* **Note 17**).
7. 5μg/ml Propidium Iodide staining solution in PBS containing 10μg/ml RNase (DNase-free). Protect the tube from light (foil-wrapped).

2.5 Detection of DNA Damage Foci by Immunofluorescence Microscopy

1. 37% HCl (*see* **Note 18**).
2. Methanol (*see* **Note 19**).
3. Acetone.
4. TBST.
5. TBST containing 2% bovine serum albumin (BSA).
6. TBST containing 10% fetal bovine serum (FBS).
7. DAPI (4′,6-Diamidino-2-phenylindole dihydrochloride) 1 mg/ml solution in distilled water (*see* **Note 20**).

2.6 Conditioned Media Preparation

1. Cell culture growth media supplemented with 1% fetal bovine serum (FBS) (*see* **Note 21**).
2. Opti-MEM.

3 Methods

3.1 Establishment of a Tetracycline-Regulated Cell System

1. PCR reaction mix (50μl final volume) for FLAG-IκBαM cDNA amplification:
   - 5μl 10× Expand High Fidelity buffer (1× final).
   - 1μl 10 mM dNTP mix (200μM final).
   - 1μl 10μM forward primer (200 nM final).
   - 1μl 10μM reverse primer (200 nM final).
   - 2μl 10 ng/μl pcDNA3FLAG-IκBαM plasmid (20 ng final).
   - 0.5μl 3.5 U/μl Expand High Fidelity Taq Polymerase (1.75 U final).
   - 39.5μl sterile H_2O.
2. PCR thermal cycler (BioRad MJmini) protocol:
   - 94 °C—2 min—1 cycle—Initial Denaturation.
   - 94 °C—30 s—Denaturation.
   - 55 °C—30 s—Annealing } 35 cycles.
   - 72 °C—1 min—Extension.
   - 72 °C—5 min—Final Extension.
   - 4 °C—unlimited time—End.
3. Western blot analysis of MCF7 IκBαM inducible clones: use 20μg of total proteins from pTRE2hyg empty vector and

pTRE2hyg FLAG-IκBαM-transfected MCF7 Tet-On cells, incubated in the absence or in the presence of 1μg/ml Dox for 24 h, for standard 10% SDS-PAGE gel electrophoresis and then blot onto nitrocellulose membrane.

Block filters for 1 h 30 min at room temperature with 5% nonfat dry milk in TBST buffer and incubate with 1:2000 dilution of anti-FLAG or anti-actin antibodies for 1 h 30 min at room temperature. After TBST washing, incubate blots for 1 h with horseradish peroxidase-conjugated secondary antibodies diluted 1:5000 in TBST buffer and then reveal by ECL (Fig. 5).

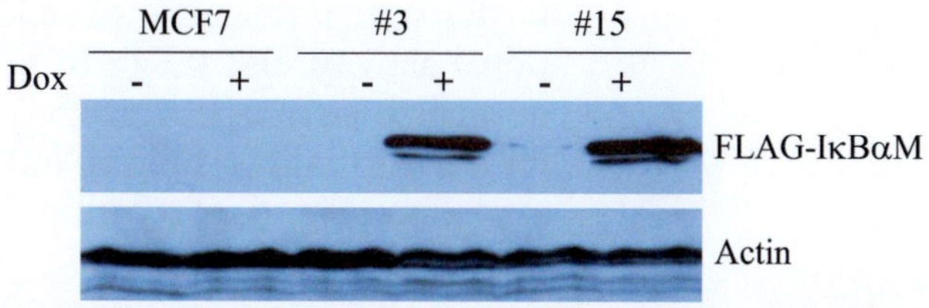

Fig. 5 Western blot analysis of MCF7 IκBαM inducible clones: doxycycline (Dox) activates FLAG-IκBαM expression in pTRE2hyg FLAG-IκBαM-transfected MCF7 cells (clones #3 and #15), but not in control vector transfected cells (MCF7). Actin was used to normalize for protein content between cell lines

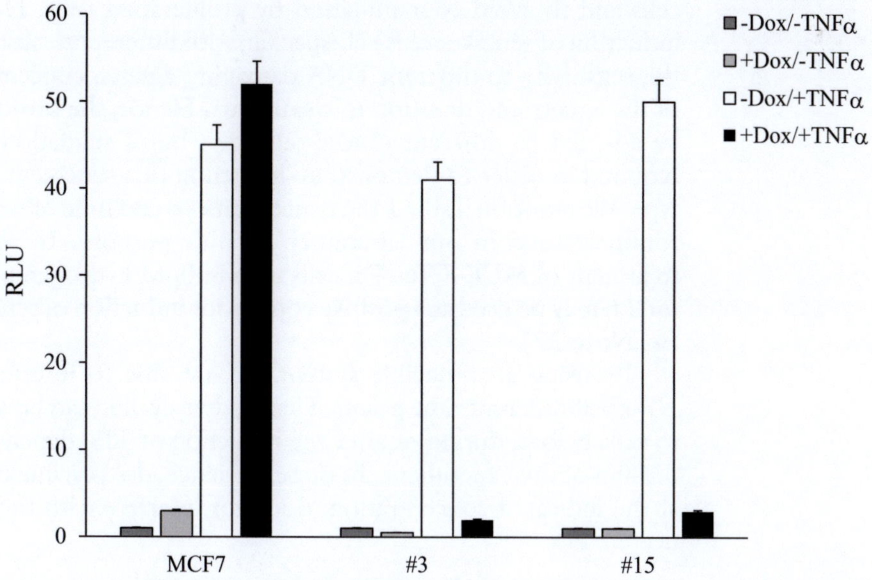

Fig. 6 Luciferase assay of NF-κB transcriptional activity in MCF7 IκBαM inducible clones: TNFα is able to promote NF-κB activation in the absence of doxycycline (Dox) in either MCF control cells or clones #3 and #15 (white bars). Instead, in the presence of Dox, TNFα induces NF-κB activation only in MCF7 control cells, because it drives FLAG-IκBαM expression in clones #3 and #15 (black bars). *RLU* relative luciferase units

Table 1
Drug concentrations and time of treatments for senescence induction

Cell line	Drug	Concentration (μM)	Treatment (h)
MCF-7 Tet-On	Doxorubicin	0.2	72
	Cisplatin	10	24
	5-FU	2.5	72
	H_2O_2	500	2

4. Analysis of NF-κB transcriptional activity in MCF7 IκBαM inducible clones: seed 4 × 10^5 cells/well in a 6-well plate. After 18 h, co-transfect cells with 1.5μg of the Ig-κB-firefly luciferase reporter gene plasmid, to analyze NF-κB transcriptional activity, and 0.5μg of pRL-renilla luciferase vector, to normalize for transfection efficiencies. Stimulate cells with Dox for 24 h to induce IκBαM and with TNFα to induce NF-κB activation (Fig. 6). Prepare cell extracts and analyze reporter gene activity.

3.2 Induction of Premature Senescence

The protocol for induction of senescence with DNA damaging chemotherapeutic agents that we describe in this chapter has been optimized for MCF-7 Tet-On cells. The duration of treatments and the concentrations of the drugs have been calibrated in order to produce culture dishes containing approximately 100% senescent cells and to avoid contamination by proliferating cells. However, induction of senescence is cell-specific, with differences observed in the sensitivity to different DNA damaging agents, concentrations of the agents and duration of treatments. Hence, the protocol can be adapted to different cancer cell lines, but a validation step is required in order to demonstrate induction of a senescent phenotype. We report in Table 1 the concentrations and time of treatment routinely used in our laboratory for this purpose. In addition, treatment of MCF-7 Tet-On cells with 500μM hydrogen peroxide for 2 h may be used as a positive control for induction of senescence (*see* **Note 22**).

By using an inducible system, we are able to interfere with NF-κB at different time points. Hence, doxycycline can be supplied to cells before, during or after the induction of TIS, depending on the aim of the experiment. In our experience, doxycycline by itself, at the indicated concentration, does not interfere with the induction of TIS.

1. Seed cells so that they are 40–60% confluent on the day of treatment (~2 × 10^6 cells/100 mm dish). It is not necessary to plate cells the day before treatment, just allow them to growth until they are 40–60% confluent. The density of cell

seeding have to be adjusted based upon duration of treatment. For instance, seed cells at 40% confluence for a 3-days treatment with doxorubicin, while using 60% confluent cells for a 24 h treatment with cisplatin (*see* **Note 23**).

2. Calculate the amount of media required for the assay and prepare culture media containing appropriate concentrations of the drugs.
3. Remove the growth medium, add medium containing senescence-inducing agents and incubate for indicated time.
4. At the end of the treatment period, gently wash the cells with pre-warmed PBS at 37 °C, for a total of three washes. Split cells as needed (*see* **Note 24**).
5. Starting from the day of release, characteristic morphological alterations can be observed at light microscopy (*see* Fig. 6a). To allow for the development of the senescent phenotype (*see* **Note 25**), cells should be analyzed starting from 3 to 4 days from the removal of the drug.
6. Keep senescent cells in culture, changing the medium every 72 h (*see* **Note 26**).

3.3 Senescence-Associated β-Galactosidase (SA-β-gal) Staining

1. Remove the growth medium. Add appropriate amount of formaldehyde fixation buffer (~2 ml per 60 mm dish) and incubate for 3–5 min at room temperature.
2. Gently wash the cells with pre-warmed PBS at 37 °C, for a total of three washes.
3. Add appropriate amount of pre-warmed SA-β-gal staining solution (~2 ml per 60 mm dish).
4. Incubate in a humidified 37 °C incubator without CO_2. Blue-greenish staining is visible after ~16 h (overnight). Prolonged incubation time increases the intensity of staining, but also increases background staining (*see* **Note 27**).
5. Count SA-β-gal-positive versus SA-β-gal-negative cells under a bright-field microscope using a 10× objective (*see* **Note 28**). We recommend counting at least three random fields for each sample (or minimum 200 cells). Calculate the percentage of cells positive for SA-β-gal.

3.4 BrdU Staining of Cells for Flow Cytometric Analysis

1. We recommend using at least a 60 mm dish of senescent cells for BrdU labeling and to use proliferating cells (1×10^6), as a positive control for the assay (*see* **Note 29**). Remove the growth medium and incubate cells with medium containing 30μM BrdU for 30–60 min (30 min for proliferating cells).
2. Remove medium and gently wash once with PBS. Detach cells with trypsin-EDTA solution, harvest cells with culture medium and transfer them to a 15 ml conical tube. Pellet at 1200 rpm

($338 \times g$) for 5 min. Resuspend cells in 300μl ice-cold PBS. Add 700μl cold ethanol dropwise while gently vortexing the cells. Store cells on ice for at least 30 min (*see* **Note 30**).

3. Wash the cells with cold PBS, for a total of three washes. Resuspend cells in 250μl cold PBS and transfer them to a 1.5 ml tube. Add 250μl freshly prepared 4 M HCl solution and tap gently. Incubate cells 30 min at room temperature.
4. Pellet cells at 1200 rpm ($100 \times g$) for 8 min at 4 °C in a microfuge. Wash the cells with 500μl cold TBST.
5. Repeat **step 4**.
6. Remove TBST, add 20μl FITC-conjugated anti-BrdU antibody, tap gently and incubate for 45–60 min at room temperature in dark. Protect samples from light for the remainder of the procedure.
7. Add 500μl cold TBST and pellet cells at 1200 rpm ($100 \times g$) for 8 min at 4 °C. Wash the cells with 500μl cold TBST twice.
8. Remove TBST, and resuspend cells in 200–500μl of Propidium Iodide staining solution. Incubate 20 min RT dark. Proceed to flow cytometric analysis.

3.5 Detection of DNA Damage Foci by Immunofluorescence Microscopy

The protocol that we describe below is designed for visualization of γ-H2AX foci (histone H2AX phosphorylated at Ser139) (Fig. 7). Detection of γ-H2AX nuclear foci is commonly used as marker of senescence, associated with a persistent DNA damage response (DDR) [19]. However, different proteins accumulate at repair foci in TIS cancer cells. Hence, we report minor modifications of

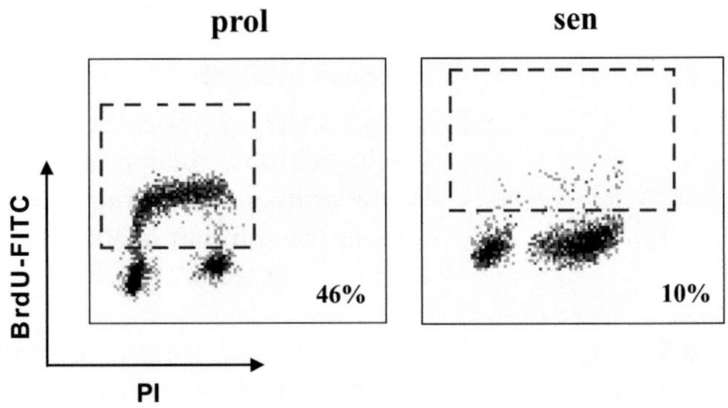

Fig. 7 Representative flow cytometric data. MCF-7 Tet-On cells were treated with 200 nM doxorubicin for 72 h. Cells were extensively washed and analyzed 5 days after drug removal. Proliferating (prol) and TIS cells (sen) were incubated with BrdU for 30 min and stained following the protocol described in Subheading 3.4. Samples were analyzed by flow cytometry and the percentage of BrdU-labeled cells is shown in the figure

the protocol to detect phosphorylated ATM (Ser1981) or phosphorylated Chk2 (Thr68) foci.

1. Seed TIS cells onto glass coverslips in 6-well multidishes and allow to adhere for 16–24 h (*see* **Note 31**).
2. Place coverslips on parafilm with cells facing up.
3. Fix cells with cold methanol (−20 °C) for 10 min (*see* **Note 32**). Remove excess methanol.
4. Permeabilize cells with cold acetone (4 °C) for 1 min. Do not allow cells to dry (*see* **Note 33**).
5. Wash twice for 5 min with TBS.
6. Wash three times for 10 min with TBST. Remove excess TBST.
7. Block samples for 15 min with TBST, 10% FBS.
8. Wash once with TBST. Remove excess TBST.
9. Incubate with anti-phospho H2AX (Ser139) antibody in a 1:200 dilution in TBST, 2% BSA for 2 h at room temperature. Occasionally tap gently to mix.
10. Wash three times for 5 min with TBST. Remove excess TBST.
11. Incubate with fluorophore-conjugated secondary antibody in a 1:500 dilution in TBST, 2% BSA for 1 h at room temperature in dark. Protect samples from light for the remainder of the procedure. Occasionally tap gently to mix.
12. Wash three times for 5 min with TBST. Remove excess TBST.
13. Stain with 1μg/ml DAPI in TBS for 10 min at room temperature (*see* **Note 34**).
14. Wash three times for 5 min with PBS.
15. Mount coverslips on a glass slide with 80% glycerol in PBS (~4μl/coverslip).
16. Count γ-H2AX foci visually in ≥100 cells by capturing images of randomly chosen fields. ImageJ, image processing software freely available at https://imagej.nih.gov/ij/, has a FindFoci tool that allows for automated identification and counting of foci per nucleus.

3.6 Conditioned Media Preparation

1. Seed TIS cells in 100 mm dishes and incubate for desired time period with normal culture medium (*see* **Note 35**).
2. Seed proliferating cells (~1–2 × 10^6 in 100 mm dishes) as negative controls.
3. Rinse proliferating and senescent tumor cells with serum-free medium.
4. Incubate cells in 7 ml medium supplemented with 1% FBS (or in serum-free Opti-MEM) per 100 mm dish and allow cells to condition the media for 48–72 h.

5. Gently collect conditioned media with attention not to disturb cells and transfer in 15 ml conical tubes on ice.
6. Wash the cells with PBS and detach cells with trypsin-EDTA solution. Count cells using a hemocytometer, to later normalize conditioned medium for cell number.
7. Centrifuge conditioned media from **step 29** at 1500 rpm (528 × *g*) for 5 min, in order to pellet any floating cells. Carefully, transfer the supernatant in a fresh 15 ml tube. Alternatively, filter-sterilize the conditioned media from **step 29**, using a membrane filter (0.22μm).
8. In order to normalize for cell number, count cells in TIS dishes and in negative control (proliferating) dishes. Dilute the conditioned medium appropriately with either medium supplemented with 1% FBS or Opti-MEM, according to **step 28**.
9. Aliquot and store at −80 °C (*see* **Note 36**).

4 Notes

1. Tet-On cloning vector pTRE2hyg is used to subclone FLAG-IκBαM cDNA (Fig. 8a) following PCR amplification from pcDNA3FLAG-IκBαM plasmid previously prepared in our lab.
2. BamHI and SalI are buffer compatible in digestion buffer D. Perform double digestion of both pTRE2hyg vector and PCR amplified FLAG-IκBαM cDNA.
3. MCF-7 Tet-On cells are commercially available. These cells are stably transfected with pTet-On Advanced vector expressing the tetracycline-regulated transactivator rtTA (Fig. 9) and are

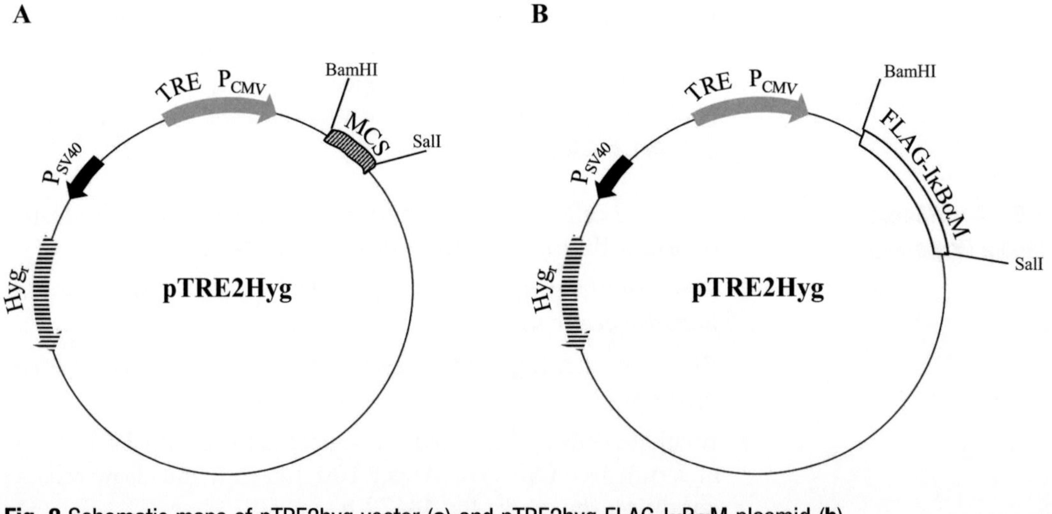

Fig. 8 Schematic maps of pTRE2hyg vector (**a**) and pTRE2hyg FLAG-IκBαM plasmid (**b**)

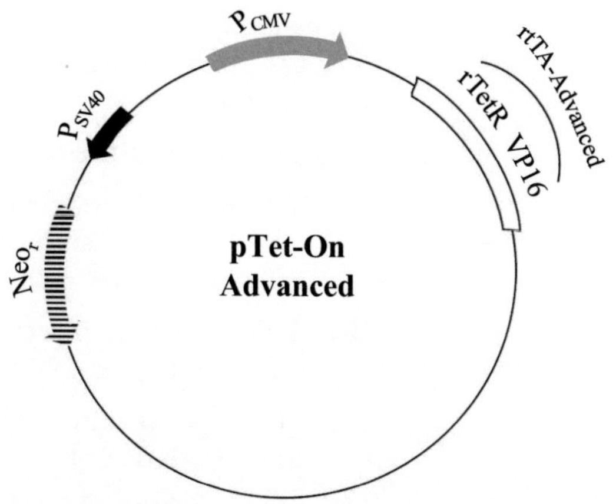

Fig. 9 Schematic map of pTet-On Advanced vector

used to transfect pTRE2hyg FLAG-IκBαM plasmid (Fig. 8b), in order to generate stable Dox-inducible clones. Hygromycin is used to select pTRE2hyg FLAG-IκBαM-transfected clones.

4. Anti-Flag antibodies are used to analyze by western blot the expression of FLAG-IκBαM fusion protein in hygromycin-selected MCF7 clones, in the absence and in the presence of 1μg/ml Dox.
5. Luciferase assay is used to analyze the ability of FLAG-IκBαM fusion protein to inhibit TNFα (2000 U/ml)-induced NF-κB activation in hygromycin-selected MCF7 clones, in the absence and in the presence of 1μg/ml Dox.
6. The tetracycline analogue doxycycline (Dox) is more stable than tetracycline and possesses a prolonged action. Dox is used at a final concentration of 1μg/ml.
7. Doxorubicin Hydrochloride is harmful if swallowed. May cause cancer. Use personal protective equipment. Wash hands thoroughly after handling.
8. Stability of cisplatin in aqueous solutions is enhanced by sodium chloride (NaCl) at concentration 0.9% .
9. Cisplatin is harmful if swallowed. Causes serious eye damage. May cause cancer. Use personal protective equipment. Wash hands thoroughly after handling.
10. 5-FU is harmful in contact with skin and if swallowed. Use personal protective equipment. Wash hands thoroughly after handling.
11. The 30% solution of hydrogen peroxide has a long shelf life, but once diluted should be used within a few hours. Prepare fresh dilute solution for each application. Dilute 30% H_2O_2 in sterile distilled water (30% hydrogen peroxide solution is 9.8 M). It is also possible to use hydrogen peroxide solutions containing a stabilizer.

12. Formaldehyde is harmful if swallowed. Causes serious eye damage. May cause cancer (inhalation). Use personal protective equipment.
13. Prepare separately the following concentrated stock solutions:
    (a) 10× citric acid-sodium phosphate buffer (400 mM citric acid, 120 mM sodium phosphate dibasic). Bring pH to 6.0 with NaOH pellets. Store at 4 °C.
    (b) 100× potassium ferrocyanide (500 mM $K_4[Fe(CN)_6]$). Store in dark at 4 °C. Handle with gloves.
    (c) 100× potassium ferricyanide (500 mM $K_3[Fe(CN)_6]$). Store in dark at 4 °C. Handle with gloves.
    (d) 10× sodium chloride (1.5 M NaCl). Store at room temperature.
    (e) 500× magnesium chloride (1 M $MgCl_2$). Store at 4 °C.
    (f) 50× X-Gal. Prepare a 50 mg/ml stock solution in *N,N*-dimethylformamide. Store at −20 °C. The solution is stable for 1 year. *N,N*-dimethylformamide causes eye irritation and is harmful in contact with skin. Use personal protective equipment.

Prepare the required amount of SA-β-gal staining solution from concentrated stock solutions (a) to (e). Check pH, and adjust to 6.0 if necessary. Add (f) to final concentration of 1 mg/ml. Bring to volume with water. Filter-sterilize using a membrane filter (0.22μm) e pre-warm at 37 °C.

14. Keep a tube of ethanol at −20 °C.
15. Once thawed, the solution can be stored at 4 °C for 1 month. Protect the tube from light (foil-wrapped).
16. We recommend using 4 M HCl solution freshly prepared. Calculate the amount of 4 M HCl solution required for the assay and dilute concentrated HCl 37% (12 M) 1:3 with distilled water. HCl causes severe skin burns and eye damage. May cause respiratory irritation. Use in a well-ventilated area. Use personal protective equipment.
17. PI is toxic if swallowed, in contact with skin or if inhaled. Use personal protective equipment. Wash hands thoroughly after handling.
18. HCl causes severe skin burns and eye damage. May cause respiratory irritation. Use in a well-ventilated area. Use personal protective equipment.
19. Methanol is toxic if swallowed and if inhaled. Use personal protective equipment.

20. If necessary sonicate to dissolve. DAPI stock solution may be stored at −20 °C for long periods. DAPI causes skin irritation. Use personal protective equipment.
21. In order to preserve viability and functions of TIS cells, we recommend using cell culture media supplemented with 1% FBS, or Opti-MEM, a commercial modification of Eagle's Minimum Essential Medium, supplemented with hypoxanthine, thymidine, sodium pyruvate, L-glutamine, trace elements, and growth factors. We observed that TIS cells were adversely affected when grown in serum-free media, even with the addition of bovine serum albumin (BSA).
22. Different senescence inducers result in different morphological changes (Fig. 10a).
23. Senescent cells develop a distinctive phenotype, characterized by a flat and enlarged morphology. Hence, plating of cells at high confluence may hamper the detection of senescent-associated morphological changes. In addition, many cancer cell lines stain positively for SA-beta-gal at confluence [28]. Finally, contact inhibition in high-density cultures leads

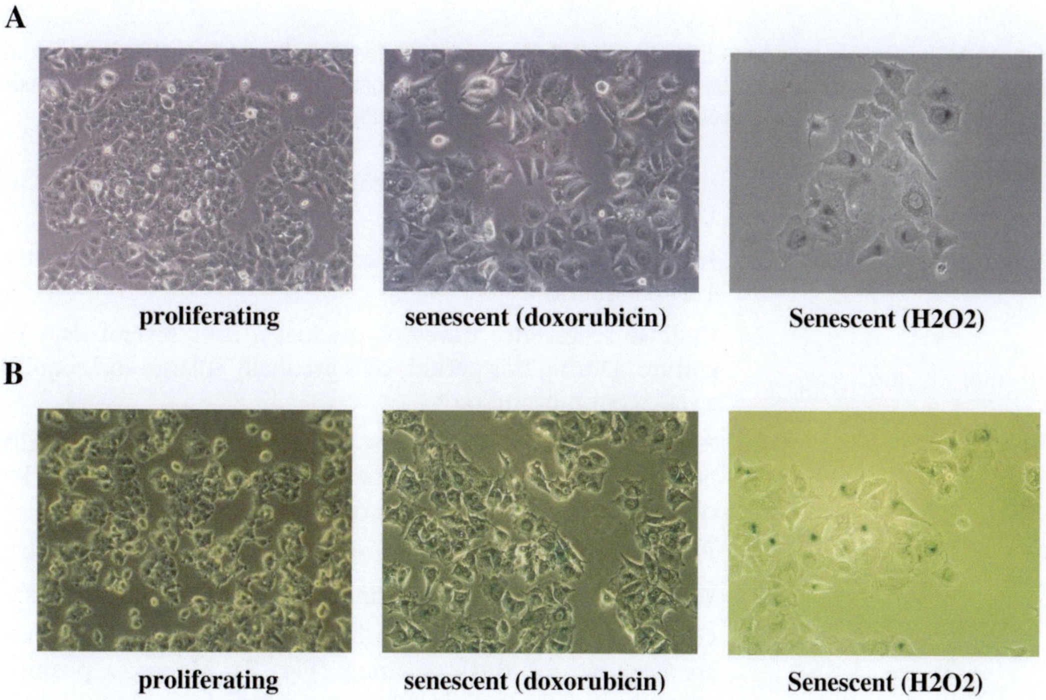

Fig. 10 (**a**) MCF-7 Tet-On cells were treated with 200 nM doxorubicin for 72 h or with 500μM hydrogen peroxide for 2 h. Cells were extensively washed and analyzed 6 days after release from the drug. Morphological alterations are observed at light microscopy. (**b**) MCF-7 Tet-On cells were treated with 200 nM doxorubicin for 72 h or with 500μM hydrogen peroxide for 2 h. Cells were extensively washed and stained for SA-beta-gal activity 6 days after release from the drug. Proliferating cells were analyzed as negative control

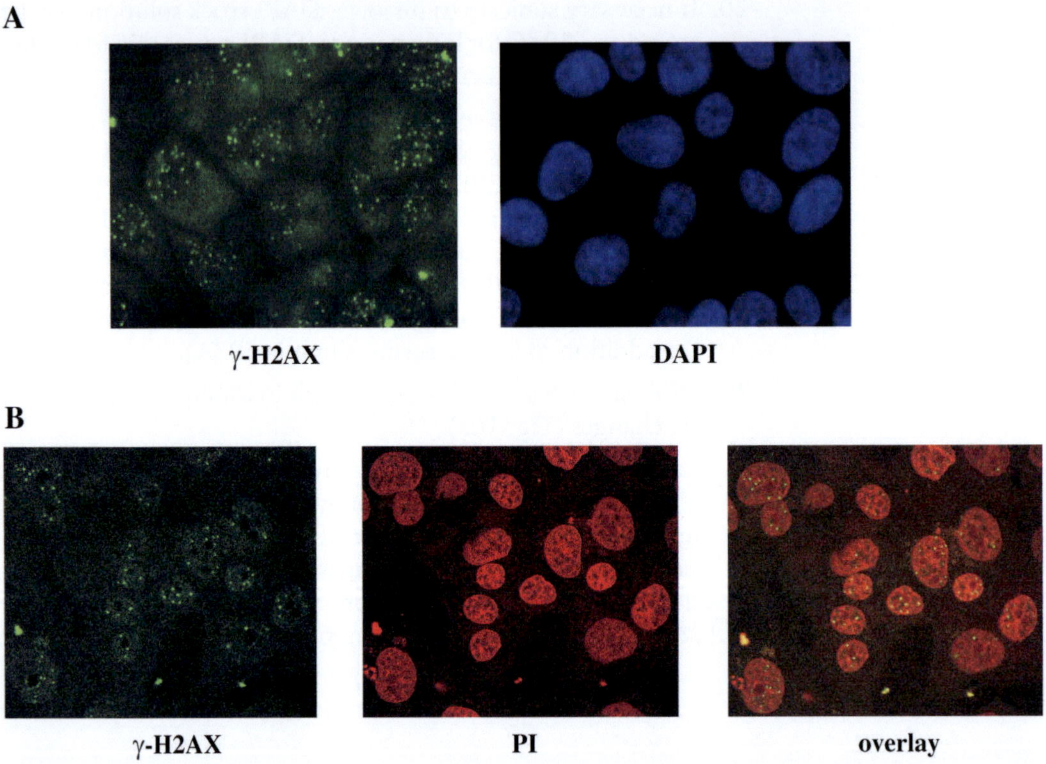

Fig. 11 MCF-7 Tet-On cells were treated with 200 nM doxorubicin for 72 h. Seven days after induction of senescence, cells were immunostained with an anti-gamma-H2AX monoclonal antibody followed by secondary fluorescein-conjugate antibodies. (**a**) Nuclei were stained with DAPI. (**b**) Nuclei were stained with PI

to reduced proliferation, which may affect cellular response to DNA damaging agents.

24. In our experience, MCF-7 Tet-On cells can be passaged at a 1:2–1:3 dilution.
25. Cellular senescence develops gradually, over several days in culture. During this period, cells gradually enlarge and acquire a senescent morphology.
26. After the removal of drug, the majority of the senescent cells persists. However, a progressive loss of viability may be observed, especially in long-term cultures (>15 days).
27. Proliferating cells can be used as a negative control (Fig. 10b).
28. We find that it is useful counting with a grid beneath the dish.
29. Growth-arrested, senescent cells do not incorporate BrdU and are negative for BrdU staining (Fig. 7). Hence, a positive control for the assay is highly recommended.
30. Cells may be stored at −20 °C for several weeks.
31. Prepare coverslips by washing with 1 M HCl and then rinse thoroughly with distilled water. Store coverslip in 70% ethanol.

Before cell seeding, set the coverslips in a 6-well plate and let them air-dry under a tissue culture hood for few minutes.

32. Use 50–80μl droplets of liquid for round 12 mm diameter coverslips.
33. For detection of phospho-ATM and phospho-Chk2 foci, exclude **steps 11** and **12**, and fix cells with 2% formaldehyde in TBS for 10 min at room temperature. Then, permeabilize cells with TBS, 0.5% NP-40 for 5 min at room temperature. Proceed to **step 14** of Subheading 3.
34. Choose the nuclear counterstaining dye according to fluorophores for secondary antibody, confocal laser lines and filters. DAPI has blue fluorescence. We successfully used propidium iodide (red fluorescence) as a nuclear counterstain (Fig. 11b). A propidium iodide stock solution (500μg/ml) is prepared in distilled water. Dilute 1:100 in TBS at final concentration of 5μg/ml.
35. The SASP develops gradually, over several days. Hence, the composition and, possibly, the effects of conditioned media change with time. We found that significant changes in cytokines can be detected starting from 3 days from the removal of the drug.
36. Long-term store may affect the stability of samples. However, the SASP comprises a plethora of cytokines, chemokines, growth factors and shed cell surface molecules, with different half-lives. Hence, the efficacy of conditioned media should be monitored for specific applications.

References

1. te Poele RH, Okorokov AL, Jardine L, Cummings J, Joel SP (2002) DNA damage is able to induce senescence in tumor cells in vitro and in vivo. Cancer Res 62:1876–1883
2. Mirzayans R, Scott A, Cameron M, Murray D (2005) Induction of accelerated senescence by gamma radiation in human solid tumor-derived cell lines expressing wild-type TP53. Radiat Res 163:53–62
3. Demaria M, O'Leary MN, Chang J, Shao L, Liu S, Alimirah F et al (2017) Cellular senescence promotes adverse effects of chemotherapy and cancer relapse. Cancer Discov 7:165–176
4. Coppé JP, Patil CK, Rodier F, Sun Y, Muñoz DP, Goldstein J et al (2008) Senescence-associated secretory phenotypes reveal cell-nonautonomous functions of oncogenic RAS and the p53 tumour suppressor. PLoS Biol 6:2853–2868
5. Kuilman T, Peeper DS (2009) Senescence-messaging secretome: SMS-ing cellular stress. Nat Rev Cancer 9:81–94
6. Chien Y, Scuoppo C, Wang X, Fang X, Balgley B, Bolden JE et al (2011) Control of the senescence-associated secretory phenotype by NF-κB promotes senescence and enhances chemosensitivity. Genes Dev 25:2125–2136
7. Crescenzi E, Pacifico F, Lavorgna A, De Palma R, D'Aiuto E, Palumbo G, Formisano S et al (2011) NF-κB-dependent cytokine secretion controls Fas expression on chemotherapy-induced premature senescent tumour cells. Oncogene 30:2707–2717
8. Wu PC, Wang Q, Grobman L, Chu E, Wu DY (2012) Accelerated cellular senescence in solid tumour therapy. Exp Oncol 34:298–305
9. Sidi R, Pasello G, Opitz I, Soltermann A, Tutic M et al (2011) Induction of senescence markers after neo-adjuvant chemotherapy of malignant pleural mesothelioma and association with

clinical outcome: an exploratory analysis. Eur J Cancer 47:326–332

10. Kim SB, Bozeman RG, Kaisani A, Kim W, Zhang L, Richardson JA et al (2016) Radiation promotes colorectal cancer initiation and progression by inducing senescence-associated inflammatory responses. Oncogene 35:3365–3375
11. Iannello A, Thompson TW, Ardolino M, Lowe SW, Raulet DH (2013) p53-dependent chemokine production by senescent tumor cells supports NKG2D-dependent tumor elimination by natural killer cells. J Exp Med 210:2057–2069
12. Eggert T, Wolter K, Ji J, Ma C, Yevsa T, Klotz S et al (2016) Distinct functions of senescence-associated immune responses in liver tumor surveillance and tumor progression. Cancer Cell 30:533–547
13. Gorgoulis V, Adams PD, Alimonti A, Bennett DC, Bischof O, Bishop C et al (2019) Cellular senescence: defining a path forward. Cell 179:813–827
14. Dimri GP, Lee X, Basile G, Acosta M, Scott G, Roskelley C et al (1995) A biomarker that identifies senescent human cells in culture and in aging skin in vivo: distinct roles for cyclin-dependent kinases in cell cycle control. Proc Natl Acad Sci U S A 92:9363–9367
15. Kurz DJ, Decary S, Hong Y, Erusalimsky JD (2000) Senescence-associated (beta)-galactosidase reflects an increase in lysosomal mass during replicative ageing of human endothelial cells. J Cell Sci 113:3613–3622
16. Muñoz-Espín D, Rovira M, Galiana I, Giménez C, Lozano-Torres B, Paez-Ribes M et al (2018) A versatile drug delivery system targeting senescent cells. EMBO Mol Med 10 (9):pii: e9355
17. Alimonti A, Nardella C, Chen Z, Clohessy JG, Carracedo A, Trotman LC et al (2010) A novel type of cellular senescence that can be enhanced in mouse models and human tumor xenografts to suppress prostate tumorigenesis. J Clin Invest 120:681–693
18. Muñoz-Espín D, Cañamero M, Maraver A, Gómez-López G, Contreras J, Murillo-Cuesta S et al (2013) Programmed cell senescence during mammalian embryonic development. Cell 155:1104–1118
19. Sedelnikova OA, Horikawa I, Zimonjic DB, Popescu NC, Bonner WM, Barrett JC (2004) Senescing human cells and ageing mice accumulate DNA lesions with unrepairable double-strand breaks. Nat Cell Biol 6:168–170
20. Rossiello F, Herbig U, Longhese MP, Fumagalli M, d'Adda di Fagagna F (2014) Irreparable telomeric DNA damage and persistent DDR signalling as a shared causative mechanism of cellular senescence and ageing. Curr Opin Genet Dev 26:89–95
21. Crescenzi E, Palumbo G, de Boer J, Brady HJ (2008) Ataxia telangiectasia mutated and p21CIP1 modulate cell survival of drug-induced senescent tumor cells: implications for chemotherapy. Clin Cancer Res 14:1877–1887
22. Gossen M, Bujard H (2001) Tetracyclines in the control of gene expression in eukaryotes. In: Nelson M, Hillen W, Greenwald RA (eds) Tetracyclines in biology, chemistry and medicine. Birkhäuser Verlag, Basel, pp 139–157
23. Baron U, Bujard H (2000) Tet repressor-based system for regulated gene expression in eukaryotic cells: principles and advances. Methods Enzymol 327:401–421
24. Berens C, Hillen W (2003) Gene regulation by tetracyclines. Constraints of resistance regulation in bacteria shape TetR for application in eukaryotes. Eur J Biochem 270:3109–3121
25. Baeuerle PA, Baltimore D (1988) I kappa B: a specific inhibitor of the NF-kappa B transcription factor. Science 242:540–546
26. Brown K, Gerstberger FS, Carlson L, Franzoso G, Siebenlist U (1995) Control of IκBα proteolysis by site-specific, signal-induced phosphorylation. Science 267:1485–1491
27. Traenckner EB, Pahl HL, Henkel T, Schmidt KN, Wilk S, Baeuerle PA (1995) Phosphorylation of human I kappa B-alpha on serines 32 and 36 controls I kappa B-alpha proteolysis and NF-kappa B activation in response to diverse stimuli. EMBO J 14:2876–2883
28. Severino J, Allen RG, Balin S, Balin A, Cristofalo VJ (2000) Is beta-galactosidase staining a marker of senescence in vitro and in vivo? Exp Cell Res 257:162–171

Chapter 13

Methods to Detect NF-κB Activity in Tumor-Associated Macrophage (TAM) Populations

Fan Sun, Zhaoxia Qu, and Gutian Xiao

Abstract

Macrophages are an abundant population in the tumor-infiltrating immune cells. The transcription factor NF-κB plays an important role in the response of tumor-associated macrophages (TAMs) to the tumor environmental cues. Detecting NF-κB activity in TAMs will help define the functional status of the TAMs. In this article, we describe several methods to detect NF-κB activity in TAM populations.

Key words Tumor-associated macrophages (TAMs), NF-κB; cancer, Immunohistochemistry (IHC), Immunofluorescence (IF), Chromatin immunoprecipitation sequencing (ChIP-Seq), RNA sequencing

1 Introduction

The normal role of immunity is to clear pathogens and damaged or transformed cells. However, during cancer pathogenesis, the immunity is transformed from tumor immunosurveillance to tumor-promoting inflammation. Macrophages, a major component in the tumor microenvironment, have emerged as main culprits for this transformation. Many studies have shown a positive correlation between the number of TAMs and poor prognosis in various types of human cancers, including lung, breast, brain, prostate, skin, colorectal, and bladder cancer [1–6]. On the other hand, macrophage depletion shows antitumor effects in various mouse cancer models [7–14].

The transcription factor NF-κB is a master regulator of immune responses [15]. It plays a pivotal role in the response of macrophages and other immune cells to diverse signals, including stress signals, inflammatory cytokines, and pathogens. The NF-κB family consists of five closely related DNA binding proteins: NF-κB1 (p105/p50), NF-κB2 (p100/p52), RelA (p65), RelB, and c-Rel, which function as various homodimers and heterodimers. Studies

Guido Franzoso and Francesca Zazzeroni (eds.), *NF-κB Transcription Factors: Methods and Protocols*, Methods in Molecular Biology, vol. 2366, https://doi.org/10.1007/978-1-0716-1669-7_13,

by our group and others have shown that myeloid cell-specific deficiency of RelA (p65), the prototypical member of NF-κB, leads to significant tumor reduction in mouse models of primary and metastatic lung cancer, and that RelA renders TAMs resistant to and capable of directly suppressing $CD8^+$ T cells [16, 17]. These studies unmasked a pro-tumoral role of NF-κB in myeloid cells, particularly TAMs. Given the important role of NF-κB activation in TAM function, we describe immunohistochemistry (IHC), immunofluorescence (IF) methods to detect NF-κB activity in TAM populations. We also describe methods of isolating TAMs from tumors, subsequent chromatin immunoprecipitation sequencing (ChIP-Seq) to identify NF-κB target genes in TAMs, and single cell multiome assay for transposase-accessible chromatin (ATAC) + gene expression to correlate the chromatin status of the NF-κB binding region of NF-κB target genes with their RNA expression in the same TAMs. TAMs will be identified by the macrophage marker F4/80 in mice or CD68 in human, and NF-κB activity in these cells will be determined by the nuclear localization of NF-κB members such as RelA, their specific binding to target DNA sequences, target DNA binding region chromatin status and the corresponding target gene expression. If sufficient amounts of TAMs can be obtained from large tumors, NF-κB activity in TAMs can also be examined using other assays, such as subcellular fraction immunoblotting (IB), electrophoretic mobility shift assay (EMSA), as we previously described [18].

2 Materials

2.1 Immunohistochemistry (IHC)

1. Neutral buffered formalin (NBF, 3.7% formaldehyde): dissolve 6.5 g of dibasic sodium phosphate, anhydrous (Na_2HPO_4), and 4 g of monobasic sodium phosphate, monohydrate ($NaH_2PO_4 \cdot H_2O$), in 900 ml of diH_2O. Adjust pH to 7.4, if necessary, with 1 M NaOH or 1 M HCl. Then add 100 ml of 37% formaldehyde solution (formalin) and mix.
2. Ethanol.
3. Xylene.
4. Paraffin.
5. Slides (positively charged).
6. Cover glass.
7. Hydrophobic barrier pen.
8. 10 mM sodium citrate buffer, pH 6.0: add 2.94 g of sodium citrate trisodium salt dihydrate to 1 l diH_2O. Adjust pH to 6.0 with 1 N HCl. Add 0.5 ml of Tween 20 and mix well. Store at room temperature for 3 months or at 4 °C for longer storage.

9. Tris Buffered Saline (TBS): 20 mM Tris, 150 mM NaCl. Prepare 1× TBS by diluting 10× TBS 1:10 with diH_2O and sterilize. To prepare 1 l of 10× TBS stock, dissolve 24.2 g of Tris, 87.6 g of NaCl in 800 ml of H_2O. Adjust pH to 7.6 with concentrated HCl and then add H_2O to 1 l. Autoclave and store at room temperature.
10. TBST: 20 mM Tris, 150 mM NaCl, 0.1% (v/v) Tween-20.
11. Blocking solution: TBST/5% normal serum from the species in which the secondary antibody was raised.
12. Peroxidase and alkaline phosphatase blocking solution (Vector Laboratories).
13. F4/80 antibody (Rat monoclonal, ThermoFisher).
14. Goat anti-rat IgG (mouse adsorbed) alkaline phosphatase (AP) polymer (Vector Laboratories).
15. Normal goat serum.
16. Alkaline phosphatase (AP) substrate ImmPACT® Vector® Red containing ImmPACT Vector Red Reagent 1, ImmPACT Vector Red Reagent 2 and ImmPACT Vector Red Diluent (Vector Laboratories).
17. NF-κB p65 antibody (Rabbit monoclonal, Cell Signaling Technology).
18. Horse anti-rabbit IgG horse radish peroxidase (HRP) polymer (Vector Laboratories).
19. Normal horse serum.
20. HRP substrate ImmPACT DAB EqV containing ImmPACT DAB EqV Reagent 1 (Chromogen) and ImmPACT DAB EqV Reagent 2 (Diluent) (Vector Laboratories).
21. Hematoxylin.
22. Nonaqueous mounting medium.

2.2 Immunofluorescence (IF)

1. O.C.T. (optimal cutting temperature).
2. Slides (gelatin or poly-L-lysine coated).
3. Cover glass.
4. Hydrophobic barrier pen.
5. 4% paraformaldehyde (PFA).
6. Phosphate buffered saline (PBS): 137 mM NaCl, 2.7 mM KCl, 10 mM Na_2HPO_4, 1.8 mM KH_2PO_4. Prepare 1× PBS by diluting 10× PBS 1:10 with diH_2O and sterilize. To prepare 1 l of 10× PBS stock, dissolve 80 g of NaCl, 2 g of KCl, 14.4 g of Na_2HPO_4, 2.4 g of KH_2PO_4, in 800 ml of diH_2O. Adjust the pH to 7.4 with HCl or NaOH, and then add diH_2O to 1 l. Dispense the solution into aliquots and sterilize by autoclaving for 20 min at 15 psi (1.05 kg/cm^2) on liquid cycle or by filter sterilization. Store at room temperature.

7. PBST: 0.1% (v/v) Tween-20 in PBS.
8. 5% normal serum from the species in which the secondary antibody was raised in PBST.
9. 1% BSA in PBST.
10. F4/80 antibody (rat monoclonal, ThermoFisher).
11. NF-κB p65 antibody (rabbit monoclonal, Cell Signaling Technology).
12. Anti-rat and anti-rabbit secondary antibodies conjugated with different fluorochromes.
13. Hoechst or DAPI DNA Dyes.

2.3 Chromatin Immunoprecipitation Sequencing (ChIP-Seq)

1. Tumor Dissociation Kit Mouse (Miltenyi), which contains Enzyme D, R, A and Buffer A.
2. RPMI 1640 or DMEM.
3. GentleMACS C Tube (Miltenyi).
4. PBS.
5. 1× PBS + 0.04% BSA.
6. GentleMACS Octo Dissociator with Heaters (Miltenyi).
7. 40 μm strainer and 70 μm strainer.
8. Red Blood Cell Removal Solution (10×, Miltenyi).
9. 0.4% trypan blue solution.
10. Automated cell counter or hemocytometer for cell concentration and viability determination.
11. Dead Cell Removal Kit (Miltenyi), which contains 10 ml Dead Cell Removal MicroBeads and 25 ml 20× Binding Buffer Stock Solution.
12. MACS Multistand (Miltenyi).
13. MACS LS columns (Miltenyi).
14. MidiMAC Separator (Miltenyi) or QuadroMACS Separator (Miltenyi) for use with MACS LS columns.
15. MACS MS columns (Miltenyi).
16. MiniMACS Separator (Miltenyi) or OctoMACS Separator (Miltenyi) for use with MACS MS columns.
17. Sterile, double-distilled water (ddH_2O).
18. Wash Buffer: 0.5% bovine serum albumin (BSA) and 2 mM EDTA in PBS, pH 7.2.
19. Anti-F4/80 MicroBeads UltraPure, mouse (Miltenyi).
20. NF-κB p65 antibody (rabbit monoclonal, Cell Signaling Technology).
21. RNA polymerase II antibody (Sigma-Aldrich).

22. Normal rabbit IgG (Cell Signaling Technology).
23. Magna ChIP A/G Chromatin Immunoprecipitation Kit (Sigma-Aldrich), which includes Magnetic Protein A/G Beads, ChIP Dilution Buffer, Low Salt Wash Buffer, High Salt Wash Buffer, LiCl Wash Buffer, TE Buffer, Cell Lysis Buffer, Nuclear Lysis Buffer, ChIP Elution Buffer (w/o Proteinase K), 10× Glycine, Spin Filters, Collection Tubes, Bind Reagent A, Wash Reagent B, Elution Reagent C, Protease Inhibitor Cocktail (PIC, 200×), Proteinase K (10 mg/ml), and RNAse A (10 mg/ml).
24. 37% Formaldehyde.
25. Sonicator.
26. Magnetic separator (e.g., Magna GrIP Rack (8 Well), Sigma-Aldrich).
27. DNase and RNase-free sterile H_2O.
28. Agarose gel electrophoresis apparatus.
29. 2% Agarose TAE gel with 0.5 μg/mL ethidium bromide and gel running buffer.
30. 100 bp DNA Ladder.
31. 6× gel Loading buffer.
32. qPCR reagents, e.g., SYBR Green Master Mix.
33. Real-time PCR Instrument and amplification plates.

2.4 TAM Single Cell ATAC Multiome ATAC + Gene Expression

1. Chromium Next GEM Single Cell Multiome ATAC Kit A (10× Genomics), which includes ATAC Enzyme B, ATAC Buffer B, and 20× Nuclei Buffer.
2. Chromium Next GEM Single Cell Multiome Reagent Kit A (10× Genomics), which includes Barcoding Reagent Mix, Barcoding Enzyme Mix, Template Switch Oligo, Reducing Agent B, Cleanup Buffer, Quenching Agent, Partitioning Oil, Single Cell Multiome Gel Beads, Dynabeads MyOne SILANE, Pre-Amp Mix, Pre-Amp Primers, Amp Mix, SI-PCR Primer B, cDNA Primers.
3. Chromium Next GEM Chip J Single Cell Kit (10× Genomics), which includes Next GEM Chip J and Gaskets.
4. Library Construction Kit (10× Genomics), which includes Fragmentation Enzyme, Fragmentation Buffer, Ligation Buffer, DNA Ligase, Adaptor Oligos, Amp Mix.
5. Single Index Plate N Set A, 96 rxns (10× Genomics).
6. Dual Index Plate TT Set A, 96 rxns (10× Genomics).
7. Chromium Controller (10× Genomics).
8. 10× Vortex Adapter (10× Genomics).

9. Chromium Next GEM Secondary Holder (10× Genomics).
10. 10× Magnetic Separator (10× Genomics).
11. 1× Lysis Buffer: 10 mM Tris–HCl, pH 7.4, 10 mM NaCl, 3 mM $MgCl_2$, 0.1% Tween-20, 0.1% Nonidet P40 Substitute, 1% BSA, 1 mM DTT, 1 U/μl RNase inhibitor in Nuclease-free Water.
12. Lysis Dilution Buffer: 10 mM Tris–HCl, pH 7.4, 10 mM NaCl, 3 mM $MgCl_2$, 1% BSA, 1 mM DTT, 1 U/μl RNase inhibitor in Nuclease-free Water.
13. 0.1× Lysis Buffer: 0.1× Lysis Buffer in Lysis Dilution Buffer.
14. Wash Buffer: 10 mM Tris–HCl (pH 7.4), 10 mM NaCl, 3 mM $MgCl_2$, 1% BSA, 0.1% Tween-20, 1 mM DTT, 1 U/μl RNase inhibitor in Nuclease-free Water.
15. Diluted Nuclei Buffer: 1× Nuclei Buffer (diluted from 20× Nuclei Buffer included in the 10× Genomics Single Cell Multiome ATAC Kit A), 1 mM DTT, 1 U/μl RNase inhibitor in Nuclease-free Water.
16. 40 μm strainer.
17. Automated cell counter.
18. Thermocycler.
19. 10× Vortex Adapter.
20. SPRIselect Reagent (Beckman Coulter).
21. 10% Tween 20 (Bio-Rad).
22. Glycerin (glycerol), 50% (v/v) Aqueous Solution (Ricca Chemical Company).
23. Buffer EB (Qiagen).

3 Methods

3.1 Immunohistochemistry (IHC)

3.1.1 Tissue Fixation, Dehydration, and Paraffin Embedding

1. Cut tissues into 3 mm thick slices and fix in 5–10 times of the tissue volume of neutral-buffered formalin for about 2 days at room temperature.
2. Wash one time with 70% ethanol. Tissues can then be kept in 70% ethanol for long-term storage.
3. Dehydrate the tissues through graded alcohols: 70% → 95% → 95% → 100% → 100% ethanol, 30 min each at room temperature.
4. Clear the ethanol from the tissues with xylene, 2 × 10 min.
5. Infiltrate the tissue with paraffin (58–60 °C), 3 × 1 h.
6. Embed the tissues in fresh paraffin.

3.1.2 Tissue Section Slide Preparation

1. Cut 4–6 μm thick tissue sections, spread the section ribbons in 45 °C water bath, and mount the sections on slides. Air-dry the slides overnight.
2. Incubate the slides at 56–58 °C for 2 h.
3. Deparaffinize the sections 2 × 10 min with xylene (*see* **Note 1**).
4. Hydrate the sections gradually through graded alcohols: 100% → 100% → 95% → 95% → 70% ethanol 10 min each.
5. Rinse the sections with distilled H_2O.

3.1.3 Heat-Induced Antigen Retrieval

Certain antigenic determinants are masked by formalin fixation and paraffin embedding and may be retrieved by this heat-induced method.

1. Preheat to 95–100 °C steamer or water bath with staining dish containing 10 mM sodium citrate buffer pH 6.0.
2. Immerse slides in the staining dish. Place the lid loosely on the staining dish and incubate for 15 min.
3. Turn off steamer or water bath and remove the staining dish to room temperature and allow the slides to cool for more than 30 min to room temperature.
4. Wash 3 × 5 min with distilled H_2O.

3.1.4 Staining for Macrophage Marker F4/80 and NF-κB p65

1. Incubate the sections for 10 min in peroxidase and alkaline phosphatase blocking solution.
2. Draw a circle on the slide around the tissue with a hydrophobic barrier pen.
3. Wash for 2 × 5 min in TBST.
4. Incubate for 1 h with 2.5% normal goat serum in humidified chamber (*see* **Note 2**).
5. Remove excess serum from sections.
6. Incubate for 1 h at room temperature with rat F4/80 monoclonal antibody diluted in 2.5% normal goat serum or 1% BSA in TBS in humidified chamber (*see* **Note 3**).
7. Wash for 3 × 5 min with TBST.
8. Incubate for 30 min with goat anti-rat IgG (mouse adsorbed) alkaline phosphatase (AP) polymer (*see* **Note 4**).
9. Wash 3 × 5 min with TBST.
10. Prepare AP substrate solution: add 2 drops (~ 80 μl) of ImmPACT Vector Red Reagent 1 and 2 drops (~ 60 μl) of ImmPACT Vector Red Reagent 2 to 5 ml ImmPACT Vector Red Diluent (*see* **Note 4**).
11. Incubate in alkaline phosphatase substrate solution. Examine under microscope. Stop the incubation when the desired red stain intensity develops.

12. Wash 3 × 5 min with TBST.
13. Wash 3 × 5 min with TBST.
14. Incubate for 1 h with 2.5% normal horse serum (*see* **Note 2**).
15. Remove excess serum from sections.
16. Incubate overnight at 4 °C with NF-κB p65 rabbit antibody diluted in 2.5% normal horse serum or 1% BSA in TBS in humidified chamber (*see* **Note 4**).
17. Wash for 3 × 5 min with TBST.
18. Incubate for 30 min with horse anti-rabbit IgG HRP polymer.
19. Wash for 3 × 5 min with TBST.
20. Combine equal volumes of ImmPACT DAB EqV Reagent 1 with Reagent 2 and mix well to make the HRP substrate working solution (*see* **Note 4**).
21. Incubate in HRP substrate working solution. Examine under microscope. Stop the incubation when the desired brown stain intensity develops.
22. Wash for 3 × 5 min in TBST.
23. Rinse sections in tap water.
24. Counterstain with diluted hematoxylin for 5 s for nuclear colocalization. DO NOT over stain with hematoxylin.
25. Rinse thoroughly with tap water for 1 min.
26. Dehydrate the sections through graded alcohols: 70% → 95% → 95% → 100% → 100% ethanol, 1 min each.
27. Clear the ethanol from the tissue sections with xylene, 2 × 1 min.
28. Apply 1–2 drops (50–100 μl) or enough volume of nonaqueous mounting medium to cover tissue and apply glass coverslip.
29. Examine the slides under light microscope after drying. Colocalization of p65 brown stain with hematoxylin blue nuclear stain in F4/80 red stained cells indicate NF-κB activation in $F4/80^+$ macrophages.

3.2 Double Immunofluorescence (IF)

3.2.1 Slide Preparation

1. Freeze the tissue in cryo-embedding medium (e.g., O.C.T.) on top of a block mostly immersed in liquid nitrogen.
2. Cut 4–8 μm thick cryostat sections. Mount the cold tissue sections onto gelatin or poly-L-lysine coated warm slides. Slides may be stored at −70 °C.
3. Thaw slides at room temperature prior to fixing and staining.
4. Draw a circle on the slide around the tissue with a hydrophobic barrier pen.
5. Fix slides for 15 min with 4% PFA at room temperature.
6. Wash 3 × 5 min with PBS.

3.2.2 Double Immunofluorescence (IF) Staining

1. Incubate for 10 min at room temperature with permeabilization buffer.
2. Wash 3 × 5 min with PBST.
3. Block for 1 h in a humidified chamber at room temperature with 5% normal serum from the species in which the secondary antibodies to be used was raised (*see* **Note 2**).
4. Remove excess serum from sections.
5. Incubate for 1 h at room temperature or overnight at 4 °C with rat F4/80 monoclonal antibody and rabbit NF-κB p65 antibody diluted in blocking solution or 1% BSA in PBST (*see* **Note 3**).
6. Wash 3 × 5 min with PBST.
7. Incubate in dark for 1 h in a humidified chamber at room temperature with fluorochrome-conjugated anti-rat secondary antibody and anti-rabbit secondary antibody conjugated with a different fluorochrome diluted in 5% normal serum from the species in which the secondary antibody was raised or 1% BSA in PBST (*see* **Note 5**).
8. Wash 3 × 5 min with PBST.
9. Counter stain with 0.1–1 μg/ml Hoechst or DAPI (nuclear DNA stain) for 5 min. Rinse with PBS.
10. Mount the coverslip with aqueous mounting medium. Allow mountant to cure overnight.
11. Seal the coverslip with nail polish to prevent drying and movement under microscope.
12. Examine the slides under a fluorescence microscope with appropriate filters. Nuclear localization of p65 staining in F4/80 stained cells indicate NF-κB activation in TAMs. Store the slides in dark at 4 °C.

3.3 Chromatin Immunoprecipitation Sequencing (ChIP-Seq)

Tumor tissue in a range of 0.04–1 g is dissociated in a volume of approximately 2.5 ml enzyme mix.

3.3.1 Tumor Tissue Cell Dissociation

1. Wash the tissue in a 50-ml centrifuge tube by adding 10 ml chilled 1× PBS.
2. Place in a petri dish and cut the tissue to small pieces of ~2–4 mm^3.
3. Prepare enzyme mix in a gentleMACS C Tube by adding:
   - 2.35 ml RPMI 1640 or DMEM.
   - 100 μl Enzyme D.
   - 50 μl Enzyme R.
   - 12.5 μl Enzyme A.
4. Transfer the tissue pieces to the C Tube containing the enzyme mix.

5. Tightly close the C Tube and attach upside down to a sleeve of a gentleMACS Octo Dissociator with Heaters. The tissue in the C Tube should be close to the stator of the dissociator (*see* **Note 6**).
6. Select the dissociator program based on tumor tissue texture (soft, medium or hard tissue).
7. Run the program 37C_m_TDK_1 for soft and medium tissue. Run the program 37C_m_TDK_2 for hard tissue.
8. At the end of the run, detach the C Tube from the dissociator (*see* **Note 7**).
9. Centrifuge at 300 × *g* for 30 s at room temperature.
10. Remove the supernatant without disturbing the cell pellet.
11. Add 10 ml RPMI 1640 or DMEM and gently pipette mix to resuspend the cell pellet.
12. Filter the cell suspension through a prewetted 70 μm strainer placed on a 50 ml centrifuge tube.
13. Wash the strainer with 10 ml RPMI1640 or DMEM and collect the wash in the tube with the cell suspension.
14. Centrifuge the cell suspension at 300 × *g* for 7 min at room temperature.
15. Remove supernatant without disturbing the cell pellet. Proceed immediately to Red Blood Cell Lysis.

3.3.2 Red Blood Cell Lysis

1. Prepare 1× Red Blood Cell Lysis Solution by diluting Red Blood Cell Lysis Solution (10×) 1:10 with double-distilled water (ddH_2O) (*see* **Note 8**).
2. Add 1 ml chilled 1× Red Blood Cell Removal Solution to the cell pellet and gently pipette mix to resuspend the cells.
3. Incubate for 10 min at 4 °C.
4. Add 10 ml chilled 1× PBS + 0.04% BSA.
5. Remove supernatant without disturbing the cell pellet.
6. Add 5 ml chilled 1× PBS + 0.04% BSA and gently pipette mix to resuspend the cell pellet.
7. Pass cells through 30 μm nylon mesh to remove cell clumps. Moisten filter with 1× PBS + 0.04% BSA before use.
8. Determine the cell number and viability using an automated cell counter or hemocytometer with trypan blue. If the percentage of viable cells <70%, proceed to dead cell removal.

3.3.3 Dead Cell Removal

1. Prepare 1× Binding Buffer by diluting 20× Binding Buffer Stock Solution 1:20 with sterile, double-distilled water. Store at 2–8 °C (*see* **Note 9**).
2. Centrifuge the cell sample at 300 × *g* for 10 min.

3. Remove the supernatant without disturbing pellet.
4. Add 100 μl Dead Cell Removal MicroBeads per 10^7 total cells and resuspend cell pellet using a wide-bore pipette tip (*see* **Note 10**).
5. Incubate for 15 min at room temperature.
6. During incubation, place column in the magnetic field of a suitable MACS Separator attached to a MACS Multistand, prepare column by rinsing with the appropriate amount of 1× Binding Buffer (*see* **Note 11**). MS column: 500 μl; LS column: 3 ml.
7. After incubation is complete, add 1× Binding Buffer to the cell suspension to reach a minimum volume of 500 μl for separation if necessary.
8. Apply cell suspension to the prepared column. The positively selected dead cells will be retained on the column while the negatively selected live cells pass through the column (*see* **Note 12**).
9. Collect the effluent (effluent 1) containing the live cell fraction in a 15 ml centrifuge tube.
10. Wash column with the appropriate amount of 1× Binding Buffer and collect the effluent (effluent 2). MS column: 4 × 500 μl; LS column: 4 × 3 ml.
11. Combine effluents 1 and 2.
12. Centrifuge cells at 300 × *g* for 10 min.
13. Remove the supernatant without disturbing the pellet.
14. Using a wide-bore pipette tip, add 1 ml Wash Buffer to each tube and gently pipette mix 5× to resuspend cell pellet.
15. Determine the cell number and viability using an automated cell counter or hemocytometer with trypan blue. Place the cells on ice.

3.3.4 Macrophage Isolation

1. Centrifuge cell suspension at 300 × *g* for 10 min. Aspirate supernatant completely.
2. Resuspend cell pellet in 90 μl of Wash Buffer per 10^7 total cells.
3. Add 10 μl of Anti-F4/80 MicroBeads UltraPure per 10^7 total cells.
4. Mix well and incubate for 15 min in the dark in the refrigerator (2–8 °C).
5. Wash cells by adding 1–2 ml of Wash Buffer per 10^7 cells and centrifuge at 300 × *g* for 10 min. Aspirate supernatant completely.
6. Resuspend up to 10^8 cells in 500 μl of Wash Buffer (*see* **Note 13**).

7. Place column in the magnetic field of a suitable MACS Separator attached to a MACS Multistand (*see* **Note 11**).
8. Prepare column by rinsing with Wash Buffer: MS column: 500 μl; LS column: 3 ml.
9. Apply cell suspension onto the column. Collect flow-through containing unlabeled cells.
10. Wash column with the appropriate amount of Wash Buffer. Collect unlabeled cells that pass through and combine with the flow-through from **step 3** (*see* **Note 14**). MS column: 3 × 500 μl; LS column: 3 × 3 ml.
11. Remove column from the separator and place it on a suitable collection tube.
12. Pipette 1 ml (for MS column) or 5 ml (for LS column) of Wash Buffer onto the column. Immediately flush out the magnetically labeled cells by firmly pushing the plunger into the column.
13. To increase the purity of F4/80$^+$ cells, the eluted fraction must be enriched over a second MS or LS Column. Repeat the magnetic separation procedure as described in **steps 7–12** by using a new column.
14. Transfer the cell suspension to a new tube.
15. Determine the cell number and viability using an automated cell counter or hemocytometer with trypan blue. Place the cells on ice.

3.3.5 Chromatin Preparation and Immunoprecipitation

Cross Linking and Cell Lysis

1. Centrifuge cell suspension at 300 × *g* for 10 min. Aspirate supernatant completely.
2. Resuspend up to 2 × 10^7 cells in 1 ml of Wash Buffer + PIC.
3. To cross-link proteins to DNA, add 37% formaldehyde to a final concentration of 1% (28 μl per 1 ml of cells) and rock at room temperature for 10 min.
4. Stop cross-linking by adding 100 μl of 10× Glycine per 1 ml of cells and rock for 5 min at room temperature.
5. Centrifuge cells at 800 × *g* for 5 min. Remove the supernatant.
6. Wash pellet two times with 1 ml ice-cold PBS + 1× PIC.
7. Resuspend the cell pellets in Cell Lysis Buffer + 1× PIC (0.5 ml per 10^7 cells). Incubate on ice for 15 min, vortex the cell suspension briefly every 5 min, to break the cell membrane.
8. Centrifuge at 800 × *g* for 5 min at 4 °C to separate nuclei from the cytoplasmic component. Remove the supernatant.
9. Resuspend the nuclear pellets in 0.5 ml of Nuclear Lysis Buffer +1× PIC.

Sonication of Isolated Chromatin to Shear DNA

1. Sonicate the nuclear lysates on ice to shear the DNA to lengths of between 200 and 600 bp (*see* **Note 15**).
2. Centrifuge at least 10,000 × *g* for 10 min at 4 °C to remove insoluble material. However, do not exceed 15,000 × *g* to prevent loss of chromatin.
3. Carefully transfer the supernatant to a new tube. Store in 50 μl aliquots at −80 °C if needed (*see* **Note 16**).

Immunoprecipitation (IP) of Cross-Linked Protein/DNA

1. Add 450 μl Dilution Buffer to each tube containing 50 μl sheared cross linked chromatin.
2. Remove 25 μl (5%) of the diluted chromatin as "Input" and save at 4 °C until Subheading "Elution of Protein/DNA Complexes and Protein/DNA Complex Crosslink Reversal", **step 1**.
3. Add 2 μl of NF-κB p65 antibody to one tube of 500 μl diluted sheared cross linked chromatin for one ChIP reaction. Prepare multiple ChIP reactions for pooling to obtain sufficient immunoprecipitated DNA for library construction (*see* **Note 17**). Similar to NF-κB p65 ChIP, prepare one ChIP reaction for RNA Polymerase II using 1.0 μg RNA Polymerase II antibody and one ChIP reaction for mock IgG using 1 μ g normal rabbit IgG.
4. Resuspend protein A/G magnetic beads by gently vortexing. Immediately add 20 μl of fully resuspended Protein A/G magnetic beads to each IP reaction and incubate overnight at 4 °C with rotation (*see* **Note 18**).
5. Separate Protein A/G magnetic beads with a magnetic separation device compatible with 1.5 ml tubes (e.g., Millipore Magna GrIP Rack (8 Well)). Allow the beads to completely separate then carefully remove the supernatant without disturbing the beads.
6. Wash the Protein A/G bead-antibody/chromatin complex by resuspending the beads in 0.5 ml each of the cold buffers in the order listed below and incubating for 3–5 min on a rotating platform between each wash followed by magnetic separation and careful removal of the supernatant fraction:
   (a) Low Salt Immune Complex Wash Buffer, one wash.
   (b) High Salt Immune Complex Wash Buffer, one wash.
   (c) LiCl Immune Complex Wash Buffer, one wash.
   (d) TE Buffer, one wash.

Elution of Protein/DNA Complexes and Protein/DNA Complex Crosslink Reversal

1. Prepare final elution buffer for all IP tubes as well as the Input tube (prepared in Subheading "Immunoprecipitation (IP) of Cross-Linked Protein/DNA", **step 2**). For each tube, prepare elution buffer with 100 μl ChIP Elution Buffer and 1 μl RNase A.

2. Add 100 μl ChIP Elution Buffer/RNase A mixture to each tube and incubate at 37 °C for 30 min with shaking or agitation (approximately 950 rpm on orbital mixer). During this incubation period thaw the Proteinase K for use in **step 3**. Once thawed store Proteinase K on ice).
3. Add 1 μl proteinase K to each sample and Incubate at 62 °C for 2 h with shaking or agitation (approximately 950 rpm on orbital mixer).
4. Incubate at 95 °C for 10 min.
5. Allow sample to cool to room temperature over an approximately 5-min period.
6. Separate beads using a magnet separation device. Carefully remove and transfer the supernatant to a new 1.5 ml microcentrifuge tube.

DNA Purification

1. Remove one Spin Filter with Collection Tube and a second Collection Tube for each sample tube from Subheading "Elution of Protein/DNA Complexes and Protein/DNA Complex Crosslink Reversal".
2. Add 0.5 mL of Bind Reagent "A" to each 100 μl DNA sample tube (Immunoprecipitations and Inputs) and mix well (*see* **Note 19**).
3. Transfer the sample/Bind Reagent "A" mixture to the Spin Filter in Collection Tube.
4. Centrifuge for 30 s at a minimum of 10,000 × *g*. Do not exceed 15,000 × *g*.
5. Remove the Spin Filter from the Collection Tube, save the Collection Tube and discard the liquid (*see* **Note 20**).
6. Put the Spin Filter back into the same Collection Tube.
7. Add 500 μl of the Wash Reagent "B" to the Spin Filter in Collection Tube.
8. Centrifuge for 30 s at a minimum of 10,000 × *g*. Do not exceed 15,000 × *g*.
9. Remove the Spin Filter from the Collection Tube, save the Collection Tube and discard the liquid.
10. Place the Spin Filter back into the same Collection Tube.
11. Centrifuge for 30 s at a minimum of 10,000 × *g*. Do not exceed 15,000 × *g*.
12. Discard the Collection Tube and liquid.
13. Put the Spin Filter into a clean Collection Tube.
14. Add 50 μl of Elution Buffer "C" directly onto the center of the white Spin Filter membrane.
15. Centrifuge for 30 s at a minimum of 10,000 × *g*. Do not exceed 15,000 × *g*.

16. Remove and discard Spin Filter. Eluate is now purified DNA. It can be analyzed immediately or stored frozen at −20 °C. Determine DNA quality and quantity using Qubit system (Life Technologies) or Agilent 2100 Bioanalyzer. Sequence the NF-κB p65 antibody ChIP DNA and control input DNA after enrichment verification.

Verify ChIP DNA Enrichment by Real-Time Quantitative PCR (RT-qPCR)

It is extremely important to verify enrichment of the ChIP DNA before sending out for sequencing. Verification of enrichment can be performed by RT-qPCR comparing PCR amplicons for NF-κB p65 binding region vs nonbinding region with p65 antibody ChIP DNA or comparing p65 antibody ChIP DNA vs IgG mock ChIP DNA for p65 binding region PCR amplicons. Input DNA will be used as a control.

1. Add 2 μl of the ChIP or input sample to a PCR plate suitable for your real time instrument (Performing triplicate of qPCR reactions per ChIP sample is recommended).
2. Prepare a qPCR master mix as shown in Table 1 using example primers shown in Table 2. Prepare a volume of master mix at least 10% greater than required for all samples to account for pipette carryover.
3. Add 18 μl of qPCR master mix to 2 μl of the sample.
4. Use caps or an optical tape to seal the plate and start the qPCR reactions as the following: 94 °C 10 min → (94 °C 20 s → 60 °C 1 min) × 50 cycles.

3.4 Single Cell Multiome ATAC + Gene Expression

Single cell Multiome ATAC + Gene Expression can simultaneously profile open chromatin landscape and gene expression in the same single TAM nuclei. The method described here uses the Chromium Next GEM Single Cell Multiome ATAC + Gene Expression kits and majorly follows the Chromium Next GEM Single Cell Multiome ATAC + Gene Expression Reagent Kits User Guide (10× Genomics). TAMs are isolated from tumor tissue(s) as described in Subheadings 3.3.1–3.3.4. Nuclei are then isolated from these TAMs. Nuclei suspensions are incubated in a Transposition Mix that includes an adapter sequence loaded transposase, which preferentially fragments the chromatin DNA in the open regions, and

Table 1
qPCR reagents for one reaction

ddH_2O	6μl
SYBR Green Master Mix	10μl
Primer mix	2μl
Total	18μl

Table 2
Example primers for NF-κB p65 binding regions

Gene name	Forward primer (5′–3′)	Reverse primer (5′–3′)
IL-10 (−95/+37)	TAGAAGAGGGAGGAGGAGCC	TGTGGCTTTGGTAGTGCAAG
B7H4 (−11,939/−11,752)	TTCCCAACCCCAAGGATTGTT	CAGATGGGTGACTGAGGACAC

adds adapter sequences to the ends of the DNA fragments. The nuclei are then partitioned into Gel Beads-in-emulsion (GEMs) with a single multiome ATAC + gene expression (GEX) gel bead, which contains a unique 10× Barcode to index the transposed DNA and cDNA of the nucleus in the same GEM. ATAC and GEX libraries are generated from the same pool of preamplified transposed DNA/cDNA and sequenced. The 10× Barcodes in each library type are used to associate individual reads back to each individual nucleus, enabling simultaneous profiling of open chromatin landscape and gene expression in the same single nuclei. NF-κB activity in TAM populations are revealed by open chromatin status in NF-κB binding regions and the corresponding NF-κB target gene expression in the TAMs. Alternatively, single cell multiome ATAC + GEX are performed on fresh or frozen tumor tissue without TAM isolation. When analyzing Single cell Multiome ATAC + Gene Expression results, TAMs are identified by TAM specific gene expression.

3.4.1 Nuclei Isolation

1. Add cells to a 2-ml microcentrifuge tube and centrifuge at 500 × *g* for 5 min at 4 °C.
2. Remove the supernatant without disrupting the cell pellet.
3. Add 100 μl chilled 0.1× Lysis Buffer. Pipette mix 5×. Incubate for 5 min on ice.
4. Add 1 ml chilled Wash Buffer to the lysed cells. Pipette mix 5×. Centrifuge cells at 500 × *g* for 5 min at 4 °C. Remove the supernatant without disrupting the nuclei pellet.
5. Repeat **step 4** two more times for a total of three washes.
6. Based on the original cell concentration and assuming ~50% nuclei loss during cell lysis, resuspend in chilled Diluted Nuclei Buffer to achieve nuclei stock concentration of 0.2–0.5 × Targeted Nuclei Recovery × 1.61 (nuclei/μl, *see* **Note 21**). Maintain on ice.
7. OPTIONAL If cell debris and large clumps are observed, pass through a 40 μm cell strainer.
8. Determine the nuclei concentration using an automated cell counter.
9. Proceed immediately to Chromium Next GEM Single Cell Multiome ATAC + Gene Expression.

3.4.2 Nuclei Transposition

1. Prepare on ice a master mix of Transposition Mix for the nuclei samples, with10 μl Transposition Mix (7 μl ATAC Buffer B, 3 μl ATAC Enzyme B, add reagents in the order listed) for each sample and extra 10% of 1× to offset pipetting loss. Pipette mix 10× and centrifuge briefly.
2. Add 10 μl Transposition Mix to a PCR tube for each sample. Centrifuge briefly and maintain on ice.
3. Calculate the volume of Nuclei Stock and Diluted Nuclei Buffer to be added (*see* **Note 21**):
   (a) Volume of Nuclei Stock (μl) = Targeted Nuclei Recovery × 1.61 (Recovery efficiency factor)/Nuclei Stock Concentration (nuclei/μl).
   (b) Volume of Diluted Nuclei Buffer (20× Nuclei Buffer diluted 1:20 in nuclease-free water, μl) = 5 μl − volume of Nuclei Stock (μl).
4. Add the calculated volume of Diluted Nuclei Buffer to the Transposition Mix. Pipette mix. Centrifuge briefly.
5. Gently pipette mix the Nuclei Stock. Add the calculated volume of the Nuclei Stock to the tube containing the Transposition Mix. Gently pipette mix 6× (pipette set to 10 μl). DO NOT centrifuge.
6. Incubate in a thermal cycler using the following program: lid temperature: 50 °C; incubate at 37 °C for 60 min and hold at 4 °C. Prepare **steps 1** and **2** in next Section during the incubation.
7. Proceed immediately to the next step.

3.4.3 GEM Generation and Barcoding

1. Equilibrate Single Cell Multiome Gel Beads to room temperature 30 min before loading the Chromium Next GEM Chip J.
2. Prepare on ice a Master Mix for the number of reactions needed with 60 μl (49.5 μl Barcoding Reagent Mix, 1.9 μl Reducing Agent B, 1.1 μl Template Switch Oligo, 7.5 μl Barcoding Enzyme Mix, add reagents in the order listed) for each reaction and extra 10% of 1× to offset pipetting loss. Mix well and centrifuge briefly.
3. Assemble Chromium Next GEM Chip J on Chromium Next GEM Secondary Holder (*see* **Note 22**).
4. Load Chromium Next GEM Chip J.
   (a) Dispense 50% Glycerol into Unused Chip Wells (if <8 samples per chip): 70 μl to unused wells in row labeled 1, 50 μl to unused wells in row labeled 2, 40 μl to unused wells in row labeled 3.
   (b) Add 60 μl Master Mix to each tube containing Transposed Nuclei for a total of 75 μl in each tube. Gently pipette mix

5×. Using the same pipette tip, dispense 70 μl Master Mix + Transposed Nuclei into the bottom center of each well in row labeled 1 without introducing bubbles.

(c) Snap the tube strip holder with the Gel Bead strip into a 10× Vortex Adapter. Vortex 30 s. Centrifuge the Gel Bead strip for ~5 s. Confirm there are no bubbles at the bottom of the tubes and the liquid levels are even. Place the Gel Bead strip back in the holder. Secure the holder lid.

(d) Puncture the foil seal of the Gel Bead tubes. Slowly aspirate 50 μl Gel Beads. Dispense into the wells in row labeled 2 without introducing bubbles. Wait 30 s.

(e) Dispense 45 μl Partitioning Oil into the wells in row labeled 3 (*see* **Note 23**).

(f) Attach the 10× gasket.

5. Run the chip in the Chromium Controller immediately after loading the Partitioning Oil (~18 min). Immediately proceed to the next step on completion of the run.
6. Slowly aspirate 100 μl GEMs from the lowest points of the recovery wells in the top row labeled 3 without creating a seal between the pipette tips and the wells, then slowly dispense into to a PCR 8-tube strip on ice with the pipette tips against the sidewalls of the wells (*see* **Notes 24** and **25**).
7. Incubate the GEMs in a thermal cycler with the following program:

   Lid temperature: 50 °C; incubate at 37 °C for 45 min, then 25 °C for 30 min, and hold at 4 °C.

   Retrieve Quenching Agent from −20 °C and equilibrate to room temperature during the PCR.
8. Add 5 μl Quenching Agent to each sample to stop the reaction. Slowly pipette mix 10× (pipette set to 90 μl). The solution will be viscous. Ensure that no liquid remains along the tube sidewalls and pipette tips. If necessary, aspirate the entire volume and dispense back slowly into the tube. Store at −80 °C for 4 weeks or proceed to the next step.

3.4.4 Post-GEM Cleanup

Post-GEM Incubation Cleanup: Dynabeads

1. Add 125 μl Recovery Agent to each sample at room temperature. DO NOT pipette mix or vortex the biphasic mixture. Gently invert tube 10× to mix. Centrifuge briefly.
2. Slowly remove and discard 125 μl Recovery Agent/Partitioning Oil (pink) from the bottom of the tube. DO NOT aspirate any aqueous sample.
3. Prepare a master mix of Dynabeads Cleanup Mix for the number of cleanups needed with 200 μl (182 μl Cleanup Buffer,

13 μl Dynabeads MyOne SILANE (*see* **Note 26**), 5 μl Reducing Agent B, add reagents in the order listed) for each sample and extra 10% of 1× to offset pipetting loss.

4. Vortex and add 200 μl Dynabeads Cleanup Mix to each sample. Pipette mix 10× (pipette set to 200 μl).
5. Incubate 10 min at room temperature (keep caps open).
6. Prepare a master mix of Elution Solution I with 50 μl (49.0 μl Buffer EB, 0.5 μl 10% Tween 20, 0.5 μl Reducing Agent B, add reagents in the order listed) for each sample and extra 10% of 1× to offset pipetting loss. Vortex and centrifuge briefly.
7. At the end of the 10-min incubation, place on the 10× Magnetic Separator, high position (magnet · High) until the solution clears. Remove the supernatant.
8. Add 300 μl freshly prepared 80% ethanol to the pellet. Wait 30 s. Remove the ethanol.
9. Add 200 μl 80% ethanol to the pellet. Wait 30 s. Remove the ethanol.
10. Centrifuge briefly. Place on the magnet•Low. Remove remaining ethanol.
11. Remove from the magnet. Immediately add 50.5 μl Elution Solution I to avoid clumping. Pipette mix (pipette set to 50 μl) without introducing bubbles. Incubate 1 min at room temperature.
12. Centrifuge briefly. Place on the magnet · Low until the solution clears. Transfer 50 μl sample to a new tube strip.

Post-GEM Incubation Cleanup: SPRIselect

1. Vortex the SPRIselect reagent until fully resuspended. Add 90 μl SPRIselect reagent to each sample. Pipette mix thoroughly. Incubate 5 min at room temperature.
2. Centrifuge briefly. Place on the magnet · High until the solution clears. Remove the supernatant.
3. Add 200 μl 80% ethanol to the pellet. Wait 30 s. Remove the ethanol.
4. Repeat **step 2** for a total of two washes.
5. Centrifuge briefly. Place on the magnet · Low. Remove any remaining ethanol. Residual ethanol can inhibit Pre-Amplification PCR and impact assay performance.
6. Remove the tube strip from the magnet. Immediately add 46.5 μl Buffer EB. Pipette mix (pipette set to 45 μl) without introducing bubbles. Incubate 2 min at room temperature.
7. Centrifuge briefly. Place on the magnet · Low until the solution clears. Transfer 46 μl sample to a new tube strip (*see* **Note 27**).

3.4.5 Pre-amplification PCR

Pre-amplification PCR

1. Prepare on ice a master mix of Pre-Amplification Mix, with 54 μl (50 μl Pre-Amp Mix, 4 μl Pre-Amp Primers, add reagents in the order listed) for each PCR reaction and extra 10% of 1× to offset pipetting loss. Pipette mix 10× and centrifuge briefly.
2. Add 54 μl Pre-Amplification Mix to each sample. Pipette mix and centrifuge briefly.
3. Incubate in a thermal cycler with the following program.

   Lid Temperature: 105 °C; Reaction Volume: 100 μl; Run Time: 30 min.

   72 °C 5 min → 98 °C 3 min → (98 °C 20 s → 63 °C 30 s → 72 °C 1 min) × 7 cycles → 72 °C 1 min → 4 °C hold.
4. Store at 4 °C for up to 18 h or proceed to the next step.

Post Pre-amplification PCR Cleanup

1. Vortex the SPRIselect reagent until fully resuspended. Add 160 μl SPRIselect reagent to each sample. Pipette mix thoroughly. Incubate 5 min at room temperature.
2. Centrifuge briefly. Place on the magnet · High until the solution clears. Remove the supernatant.
3. Add 300 μl 80% ethanol to the pellet. Wait 30 s. Remove the ethanol.
4. Add 200 μl 80% ethanol to the pellet. Wait 30 s. Remove the ethanol.
5. Centrifuge briefly. Place on the magnet · Low. Remove any remaining ethanol.
6. Remove the tube strip from the magnet. Immediately add 160.5 μl Buffer EB. Pipette mix (pipette set to 150 μl) without introducing bubbles. Incubate 2 min at room temperature.
7. Centrifuge briefly. Place on the magnet · High until the solution clears. Transfer 160 μl sample to a new tube strip. Store at 4 °C for up to 72 h or at −20 °C for long-term storage, or proceed to the next step.

3.4.6 ATAC Library Construction

Sample Index PCR

1. Choose the appropriate sample index sets to ensure that no sample indices overlap in a multiplexed sequencing run.
2. Prepare on ice a master mix of Sample Index PCR Mix for the number of PCR reactions needed, with 57.5 μl (50 μl Amp Mix + 7.5 μl SI-PCR Primer B) for each PCR reaction and extra 10% of 1× to offset pipetting loss. Pipette mix 10× and centrifuge briefly.
3. Transfer 40 μl pre-amplified sample (**step** 7 of Subheading "Post Pre-amplification PCR Cleanup") to a new tube strip. Add 57.5 μl Sample Index PCR Mix to the sample. Pipette mix and centrifuge briefly.
4. Add 2.5 μl of an individual Sample Index N, Set A to each sample. Record assignment. Pipette mix and centrifuge briefly.

Table 3
Cycle number optimization

Targeted nuclei recovery	Total cycles
≤2000	9
2001–6000	8
6001–10,000	7

5. Incubate in a thermal cycler with the following program:

   Lid Temperature: 105 °C; Reaction Volume: 100 μl; Run Time: ~30 min.

   98 °C 45 s → (98 °C 20 s → 67 °C 30 s → 72 °C 20 s) × n cycles → 72 °C 1 min → 4 °C hold.

   Use the Table 3 for cycle number optimization starting point based on Targeted Nuclei Recovery.

6. Store at 4 °C for up to 72 h or proceed to the next step.

Post Sample Index PCR Double-Sided Size Selection

1. Vortex to resuspend SPRIselect reagent. Add 60 μl SPRIselect reagent to each sample. Pipette mix. Incubate 5 min at room temperature.
2. Place on the magnet · High until the solution clears. Transfer 150 μl supernatant to a new strip tube.
3. Vortex to resuspend SPRIselect reagent. Add 95 μl SPRIselect reagent to each sample (supernatant). Pipette mix. Incubate 5 min at room temperature.
4. Place on the magnet · High until the solution clears. Remove the supernatant.
5. Add 300 μl 80% ethanol to the pellet. Wait 30 s. Remove the ethanol.
6. Add 200 μl 80% ethanol to the pellet. Wait 30 s. Remove the ethanol.
7. Centrifuge briefly. Place on the magnet · Low. Remove remaining ethanol.
8. Remove from the magnet. Immediately add 20.5 μl Buffer EB. Pipette mix. Incubate 2 min at room temperature.
9. Centrifuge briefly. Place on the magnet · Low until the solution clears. Transfer 20 μl sample to a new tube strip. Store at 4 °C for up to 72 h or at −20 °C for long-term storage.
10. Run 1 μl sample on the Agilent Bioanalyzer High Sensitivity DNA chip to determine fragment size. Lower molecular weight product (≤150 bp) and/or a high molecular weight product (~2000 bp) may be present. This does not affect sequencing.

3.4.7 cDNA Amplification

cDNA Amplification

1. Prepare on ice a master mix of cDNA Amplification Mix for the number of PCR reactions needed, with 65 μl (50 μl Amp Mix + 15 μl cDNA Primers) for each PCR reaction and extra 10% of 1× to offset pipetting loss. Mix well and centrifuge briefly.
2. Transfer 35 μl pre-amplified sample (**step** 7 of Subheading "Post Pre-amplification PCR Cleanup") to a new PCR tube. Add 65 μl cDNA Amplification Reaction Mix to the sample. Mix well and centrifuge briefly.
3. Incubate in a thermal cycler with the following protocol.

   Lid Temperature: 105 °C; Reaction Volume: 100 μl; Run Time: ~30–45 min.

   98 °C 3 min → (98 °C 15 s → 63 °C 20 s → 72 °C 1 min) × n cycles → 72 °C 1 min → 4 °C hold.

   Recommended starting point for cycle number optimization is shown in Table 4 (*see* **Note 28**):
4. Store at 4 °C for up to 72 h or −20 °C for ≤1 week, or proceed to the next step.

cDNA Cleanup

1. Vortex to resuspend the SPRIselect reagent. Add 60 μl SPRIselect reagent (0.6×) to each sample and pipette mix 15× (pipette set to 150 μl). Incubate 5 min at room temperature.
2. Place on the magnet · High until the solution clears. Remove the supernatant.
3. Add 200 μl 80% ethanol to the pellet. Wait 30 s. Remove the ethanol.
4. Repeat **step** 7 for a total of 2 washes.
5. Centrifuge briefly and place on the magnet · Low. Remove any remaining ethanol. Air-dry for 2 min. DO NOT exceed 2 min as this will decrease elution efficiency.
6. Remove from the magnet. Add 40.5 μl Buffer EB. Pipette mix 15×. Incubate 2 min at room temperature.
7. Place the tube strip on the magnet · High until the solution clears. Transfer 40 μl sample to a new tube strip. Store at 4 °C for up to 72 h or at −20 °C for up to 4 weeks, or proceed to the next step.

Table 4
Cycle number optimization

Targeted nuclei recovery	Total cycles
≤2000	9
2001–6000	7
≥6001	6

8. Run 1 μl undiluted sample on an Agilent Bioanalyzer High Sensitivity chip.
9. Multiply the cDNA concentration [pg/μl] reported via the Agilent 2100 Expert Software by the elution volume (40 μl) of the Post cDNA Amplification Reaction Clean Up sample (taking any dilution factors into account) and then divide by 1000 to obtain the total cDNA yield in ng. Carry forward ONLY 25% of total cDNA yield into 3′ Gene Expression Library Construction.

3.4.8 Gene Expression Library Construction

Fragmentation, End Repair, and A-Tailing

1. Prepare a thermal cycler with the following incubation protocol.

   Lid Temperature: 65 °C; Reaction Volume: 50 μl; Run Time: ~35 min.

   Pre-cool block to 4 °C hold (Pre-cool block prior to preparing the Fragmentation Mix) → Fragmentation 32 °C 5 min → End Repair and A-tailing 65 °C 30 min → 4 °C hold.
2. Vortex Fragmentation Buffer. Verify there is no precipitate.
3. Prepare on ice a master mix of Fragmentation Mix for the number of Fragmentation reactions needed, with 15 μl (5 μl Fragmentation Buffer, 10 μl Fragmentation Enzyme, add reagents in the order listed) for each reaction and extra 10% of 1× to offset pipetting loss. Pipette mix and centrifuge briefly.
4. Transfer ONLY 10 μl purified cDNA sample from cDNA Cleanup to a tube strip (*see* **Note 29**).
5. Add 25 μl Buffer EB to each sample. Add 15 μl Fragmentation Mix to each sample. Pipette mix on ice. Centrifuge briefly.
6. Transfer into the pre-cooled thermal cycler (4 °C) and press "SKIP" to initiate the protocol.

Post Fragmentation, End Repair, and A-Tailing Double-Sided Size Selection

1. Vortex to resuspend SPRIselect reagent. Add 30 μl SPRIselect (0.6×) reagent to each sample. Pipette mix. Incubate 5 min at room temperature.
2. Place on the magnet · High until the solution clears. Transfer 75 μl supernatant to a new tube strip.
3. Vortex to resuspend SPRIselect reagent. Add 10 μl SPRIselect reagent (0.8×) to each sample. Pipette mix. Incubate 5 min at room temperature.
4. Place on the magnet · High until the solution clears. Remove 80 μl supernatant. DO NOT discard any beads.
5. Add 125 μl 80% ethanol to the pellet. Wait 30 s. Remove the ethanol.
6. Repeat **step 5** for a total of two washes.

7. Centrifuge briefly. Place on the magnet · Low until the solution clears. Remove remaining ethanol. Air-dry for 2 min. DO NOT exceed 2 min as this will decrease elution efficiency.
8. Remove from the magnet. Add 50.5 μl Buffer EB to each sample. Pipette mix. Incubate 2 min at room temperature.
9. Place on the magnet · High until the solution clears. Transfer 50 μl sample to a new tube strip.

Adaptor Ligation

1. Prepare on ice a master mix of Adaptor Ligation Mix for the number of reactions needed, with 50 μl (20 μl Ligation Buffer, 10 μl DNA Ligase, 20 μl Adaptor Oligos, add reagents in the order listed) for each reaction and extra 10% of 1× to offset pipetting loss. Mix well and centrifuge briefly.
2. Add 50 μl Adaptor Ligation Mix to 50 μl sample. Pipette mix. Centrifuge briefly.
3. Incubate in a thermal cycler with the following program:
   Lid Temperature: 30 °C; Reaction Volume: 100 μl; Run Time: 15 min.
   20 °C 15 min → 4 °C hold.

Post Ligation Cleanup

1. Vortex to resuspend SPRIselect Reagent. Add 80 μl SPRIselect Reagent (0.8×) to each sample. Pipette mix. Incubate 5 min at room temperature.
2. Place on the magnet · High until the solution clears. Remove the supernatant.
3. Add 200 μl 80% ethanol to the pellet. Wait 30 s. Remove the ethanol.
4. Repeat **step 3** for a total of two washes.
5. Centrifuge briefly. Place on the magnet · Low. Remove any remaining ethanol. Air-dry for 2 min. DO NOT exceed 2 min as this will decrease elution efficiency.
6. Remove from the magnet. Add 30.5 μl Buffer EB. Pipette mix. Incubate 2 min at room temperature.
7. Place on the magnet · Low until the solution clears. Transfer 30 μl sample to a new tube strip.

Sample Index PCR

1. Choose the appropriate sample index sets to ensure that no sample indices overlap in a multiplexed sequencing run.
2. Add 50 μl Amp Mix to 30 μl sample.
3. Add 20 μl of an individual Dual Index TT Set A to each well and record sample index name (PN-3000431 Dual Index Plate T Set A well ID) used. Pipette mix 5× (pipette set to 90 μl). Centrifuge briefly.

Table 5
Cycle number optimization

cDNA input, ng	Recommended cycle numbers
0.25–25	14–16
25–150	12–14
150–500	10–12
500–1000	8–10
1000–1500	6–8
>1500	5

4. Incubate in a thermal cycler with the following protocol.

   Lid Temperature: 105 °C; Reaction Volume: 100 μl; Run Time: ~25–40 min.

   98 °C 45 s → (98 °C 20 s → 54 °C 30 s → 72 °C 20 s) × *n* cycles → 72 °C 1 min → 4 °C hold.

   The total cycles should be optimized based on 25% carry forward cDNA yield/input calculated during Post cDNA Amplification QC and Quantification as shown in Table 5.

5. Store at 4 °C for up to 72 h or proceed to the next step.

Post Sample Index PCR Double-Sided Size Selection

1. Vortex to resuspend the SPRIselect reagent. Add 60 μl SPRIselect Reagent (0.6×) to each sample. Pipette mix. Incubate 5 min at room temperature.
2. Place on the magnet · High until the solution clears. Transfer 150 μl supernatant to a new tube strip.
3. Vortex to resuspend the SPRIselect reagent. Add 20 μl SPRIselect Reagent (0.8×) to each sample (supernatant). Pipette mix. Incubate 5 min at room temperature.
4. Place on the magnet · High until the solution clears. Remove 165 μl supernatant. DO NOT discard any beads.
5. Add 200 μl 80% ethanol to the pellet. Wait 30 s. Remove the ethanol.
6. Repeat **step 5** for a total of two washes.
7. Centrifuge briefly. Place on the magnet · Low. Remove remaining ethanol. Air-dry for 2 min. DO NOT exceed 2 min as this will decrease elution efficiency.
8. Remove from the magnet. Add 35.5 μl Buffer EB. Pipette mix. Incubate 2 min at room temperature.
9. Place on the magnet · Low until the solution clears. Transfer 35 μl to a new tube strip. Store at 4 °C for up to 72 h or at −20 °C for long-term storage.

10. Run 1 μl sample at 1:3 dilution on an Agilent Bioanalyzer High Sensitivity chip. Determine the average fragment size from the Bioanalyzer trace. This will be used as the insert size for library quantification.

3.4.9 Sequencing

It is recommended to sequence the ATAC Library at 25,000 read pairs per nucleus, and the Gene Expression Library at 20,000 read pairs per nucleus.

4 Notes

1. Do not allow slides to dry at any time hereafter. Drying out will cause nonspecific antibody binding and therefore high background staining.
2. Use normal serum from the species in which the secondary antibody was raised prior to primary antibody application to minimize cross reaction of the secondary antibody with endogenous immunoglobulins in the tissue and to eliminate Fc receptor binding of both the primary and secondary antibody.
3. Optimize antibody concentration as a single staining procedure first before initiation of the double staining. High antibody concentration tends to increase nonspecific staining, whereas low antibody concentration tends to lower staining intensity.
4. The combination of AP polymer/ImmPACT Vector Red, HRP polymer/ImmPACT DAB EqV and hematoxylin gives a good color contrast of red, brown and blue. Other HRP/AP substrate combination that gives good color contrast may also be used.
5. Steps afterwards all need to be performed in dark.
6. It has to be ensured that the sample material is located in the area of the rotor/stator.
7. When working with tough tumors, some larger pieces of tissue may remain. To further increase the cell yield, allow the remaining tissue to settle and transfer the supernatant to a new tube. Add 4 ml RPMI 1640 or DMEM to the C Tube with the remaining tissue pieces. Insert tube onto the sleeve of the gentleMACS Dissociator. Run program m_impTumor_01. Combine the resulting cell suspension with the previously removed supernatant.
8. Do not dilute with deionized water. Discard unused solution at the end of the day.
9. Use only sterile double-distilled water but not deionized water for 1× Binding Buffer preparation. Binding of Dead Cell Removal MicroBeads requires Ca^{2+}. The presence of the ion chelator EDTA will abolish binding.

10. Dead Cell Removal MicroBeads are susceptible to bacterial contamination. Handle under sterile conditions. When working with fewer than 10^7 cells, use the same volumes as indicated.
11. MS column can be used for a maximum number of 2×10^8 total cells and a maximum number of 10^7 labeled cells. LS column can be used for a maximum number of 2×10^9 total cells and a maximum number of 10^8 labeled cells. Use higher capacity column if needed.
12. DO NOT apply the plunger supplied with the column, otherwise positively selected dead cells will be collected in the effluent.
13. For higher cell numbers, scale up buffer volume accordingly.
14. Perform washing steps by adding buffer aliquots as soon as the column reservoir is empty.
15. When performing ChIP-seq it is critical to carefully monitor the fractionation of your chromatin to ensure that the optimal size range of 200–600 bp is produced. Sonication conditions must be determined empirically for each cell or tissue type, and sonicator model. Avoid foaming by keeping the tip end of the sonicator near the bottom of the sample tubes, but not touching. Placing the tip end near the surface induces foaming, which can result in unequal shearing of DNA samples.
16. It is strongly recommended that chromatin quality be evaluated before proceeding. To evaluate chromatin quality, remove one 5 μl aliquot, purify as directed in Subheading "Elution of Protein/DNA Complexes and Protein/DNA Complex Cross-link Reversal", and determine the purified DNA fragment size by electrophoresis on a 2% agarose gel.
17. The amount of antibody providing the highest level of enrichment should be titrated by performing ChIP in advance of the ChIP-Seq experiment. Although good sequencing libraries can be generated from as little as 1 ng of starting material, whenever possible use 10 ng of purified ChIP DNA, which may be obtained from pooling of multiple ChIP reactions.
18. Make sure the magnetic bead slurry is well mixed before removing volume for IP, as magnetic beads will settle on the bottom of the tube over a short period of time.
19. 5 volumes of Bind Reagent "A" should be used for every 1 volume of sample. A white precipitate may be observed. This will not interfere with this procedure.
20. If a precipitate formed in **step 2**, it may be observed in the bottom of the Collection Tube and this will not interfere with this procedure.

21. Multiplet rate increase as the targeted nuclei recovery increase. With the targeted nuclei recovery ranging from 500 to 10,000, the multiplet rate increase from 0.4% to 7.7%.
22. After removing the chip from the sealed Chromium Next GEM Chip J bag, use the chip in $\leq$24 h.
23. Failure to add Partitioning Oil to the top row labeled 3 will prevent GEM generation and can damage the Chromium Controller.
24. Since some plastics can destabilize GEMs, when handling GEMs, use 10× Genomics validated emulsion-safe plastic consumables, such as USA Scientific# 1402-4700, ThermoFisher# N8010580, or Eppendorf# 951010022 PCR 8-tube strips, and Rainin# 30389240 pipette tips.
25. GEMs should appear opaque and uniform across all channels. Excess Partitioning Oil (clear) in the pipette tips indicates a potential clog.
26. Vortex Dynabeads MyOne SILANE thoroughly ($\geq$30 s) immediately before adding to the mix. Aspirate the full liquid volume with a pipette tip to verify that the beads have not settled in the bottom of the tube. If clumps are present, pipette mix to resuspend completely. DO NOT centrifuge before use.
27. Residual SPRI beads can inhibit Pre-Amplification PCR and impact assay performance.
28. The optimal number of cycles is a trade-off between generating sufficient final mass for library construction and minimizing PCR amplification artifacts. The number of cDNA cycles should also be reduced if large numbers of nuclei are sampled.
29. Only 10 μl (25%) cDNA sample is sufficient for generating 3′ Gene Expression library. The remaining 30 μl (75%) cDNA sample can be stored at 4 °C for up to 72 h or at −20 °C for up to 4 weeks for generating additional Gene Expression libraries.

References

1. Sorensen MD, Dahlrot RH, Boldt HB et al (2018) Tumour-associated microglia/macrophages predict poor prognosis in high-grade gliomas and correlate with an aggressive tumour subtype. Neuropathol Appl Neurobiol 44:185–206
2. Jackute J, Zemaitis M, Pranys D et al (2018) Distribution of M1 and M2 macrophages in tumor islets and stroma in relation to prognosis of non-small cell lung cancer. BMC Immunol 19:3
3. Miyasato Y, Shiota T, Ohnishi K et al (2017) High density of CD204-positive macrophages predicts worse clinical prognosis in patients with breast cancer. Cancer Sci 108:1693–1700
4. Nonomura N, Takayama H, Nakayama M et al (2011) Infiltration of tumour-associated macrophages in prostate biopsy specimens is predictive of disease progression after hormonal therapy for prostate cancer. BJU Int 107:1918–1922
5. Bailey C, Negus R, Morris A et al (2007) Chemokine expression is associated with the accumulation of tumour associated macrophages (TAMs) and progression in human colorectal cancer. Clin Exp Metastasis 24:121–130
6. Hansen BD, Schmidt H, von der Maase H et al (2006) Tumour-associated macrophages are related to progression in patients with

metastatic melanoma following interleukin-2 based immunotherapy. Acta Oncol 45:400–405

7. Duong L, Radley-Crabb HG, Gardner JK et al (2018) Macrophage depletion in elderly mice improves response to tumor immunotherapy, increases anti-tumor T cell activity and reduces treatment-induced cachexia. Front Genet 9:526
8. Bader JE, Enos RT, Velázquez KT et al (2018) Macrophage depletion using clodronate liposomes decreases tumorigenesis and alters gut microbiota in the AOM/DSS mouse model of colon cancer. Am J Physiol Gastrointest Liver Physiol 314:G22–G31
9. Tham M, Khoo K, Yeo KP et al (2015) Macrophage depletion reduces postsurgical tumor recurrence and metastatic growth in a spontaneous murine model of melanoma. Oncotarget 6:22857–22868
10. Gabrusiewicz K, Hossain MB, Cortes-Santiago N et al (2015) Macrophage ablation reduces M2-like populations and jeopardizes tumor growth in a MAFIA-based glioma model. Neoplasia 17:374–384
11. Droller MJ, Gomolka D (1984) Macrophage depletion and manipulation of enhanced immune response in an animal model of bladder cancer. Cell Immunol 83:433–441
12. Griesmann H, Drexel C, Milosevic N et al (2017) Pharmacological macrophage inhibition decreases metastasis formation in a genetic model of pancreatic cancer. Gut 66:1278–1285
13. Niu M, Valdes S, Naguib YW et al (2016) Tumor-associated macrophage-mediated targeted therapy of triple-negative breast cancer. Mol Pharm 13:1833–1842
14. Zaynagetdinov R, Sherrill TP, Polosukhin VV et al (2011) A critical role for macrophages in promotion of urethane-induced lung carcinogenesis. J Immunol 187:5703–5711
15. Xiao G, Fu J (2011) NF-kappaB and cancer: a paradigm of Yin-Yang. Am J Cancer Res 1:192–221
16. Li L, Han L, Sun F et al (2018) NF-kappaB RelA renders tumor-associated macrophages resistant to and capable of directly suppressing CD8(+) T cells for tumor promotion. Onco Targets Ther 7:e1435250
17. Li D, Beisswenger C, Herr C et al (2014) Myeloid cell RelA/p65 promotes lung cancer proliferation through Wnt/beta-catenin signaling in murine and human tumor cells. Oncogene 33:1239–1248
18. Qu Z, Xiao G (2015) Systematic detection of noncanonical NF-kappaB activation. Methods Mol Biol 1280:121–154

Chapter 14

Methods to Study the Effect of IKK Inhibition on TNF-Inducing Apoptosis and Necroptosis in Cultured Cells

Alessandra Pescatore, Carmela Casale, Francesca Fusco, and Matilde Valeria Ursini

Abstract

The engagement of TNF on TNFR can result in cell survival or cell death depending on the different complex formation downstream this interaction. Here we describe reagents and assay procedures that can be used to study caspase-independent cell death (necroptosis) in cultured cells, in response to pharmacological interventions with NF-kappaB and death inhibitors. We provide protocol to detect death-specific proteins using immunoblot and to dissect necrosome complex by sequential co-immunoprecipitation of death-specific components during necroptosis.

Key words NEMO, NF-kappaB, Necroptosis, Cell death, Caspase-8, Effector caspases, MLKL

1 Introduction

The molecular mechanism leading IkappaB kinase (IKK) complex activation has been well characterized in the context of tumor necrosis factor (TNF) receptor-1 (TNFR1) signaling. Depending on the cell type, cell activation state and microenvironment factors, soluble TNF administration can result in cell survival or death, reflecting an intricate network of signals that operate downstream of TNFR1. Following ligand binding, TNFR1 trimers undergo a conformational change that triggers recruitment of multiple adaptors, ubiquitin ligases and kinases. In particular, the TNFR1 cytosolic death domain (DD) represents the apical component through which a pro-survival or pro-death signal pathways are transduced [1]. The TNFR1 associated complex (also referred as complex I) includes the adaptor proteins tumor necrosis factor receptor type

Guido Franzoso and Francesca Zazzeroni (eds.), *NF-κB Transcription Factors: Methods and Protocols*, Methods in Molecular Biology, vol. 2366, https://doi.org/10.1007/978-1-0716-1669-7_14,

1-associated DD (TRADD), the Receptor interacting protein kinase-1 (RIPK1), TRAF2, and the E3 ubiquitin ligases cellular inhibitor of apoptosis protein (cIAP) 1, cIAP2, and the linear ubiquitin chain assembly complex (LUBAC). The ubiquitination of multiple substrates within the complex subsequently attracts the TAK and IKK complexes, through the ability of TAB2/3 and NEMO respectively, to specifically recognize non-degradative ubiquitin chains [2, 3]. The pro-survival IKK complex activity relies on its ability to activate NF-kappaB signaling. Following internalization of TNF-R1, a cytosolic complex II consisting of RIPK1, Fas-Associated protein with Death Domain (FADD) and Caspase-8 forms [4]. The pro-survival cellular FLICE-like inhibitor protein (cFLIP) expression prevents extrinsic apoptosis by inducing the formation of inhibitory cFLIP/caspase-8 heterodimers [5, 6]. However, recent evidence suggests that IKK complex phosphorylates RIPK1 at multiple sites in the complex I reducing its ability to bind FADD/caspase-8 and to induce apoptosis [7, 8]. When IKK activity is inhibited, e.g., in the absence of NEMO, in the IKK-1 and -2 double knock out, or by using IKK inhibitors, active dimers of Caspase-8 are formed leading to apoptotic cell death [7, 9, 10]. Active Caspase-8 not only drives the apoptotic program but also inactivates essential necroptosis mediators, RIPK1 and RIPK3 [11, 12]. TNF promotes necroptosis when Caspase-8 activity is absent or pharmacologically inhibited, such as by using Z-VAD-FMK. The integrity of RIPK1 and RIPK3 is preserved and the consequent RIPK1-RIPK3 oligomerization triggers RIPK3 autoactivation [13, 14]. Activated RIPK3 in turn phosphorylates MLKL (S345 of mouse MLKL) in a complex called necrosome. The MLKL phosphorylation is required for its translocation to the membrane where it leads to the execution of necroptosis [15, 16]. RIPK1 inhibition by using necrostatin-1s (Nec-1s) rescues TNF/zVAD-induced necrosome formation [10, 17]. Recently, small-molecule acting as RIPK3 kinase activity inhibitors, such as GW440139B (GW′39B) and GSK′872, have been widely used to block necroptosis [18, 19]. Here we provide detailed advices for in vitro induction and detection of TNF-induced apoptosis and necroptosis in murine cultured cells by using combination of IKK inhibitors and caspase inhibitor zVAD. In particular, we describe methods for detecting MLKL phosphorylation (pMLKL) by western blotting and the concomitant detection of RIPK1 and pMLKL to MLKL-containing complexes in wild type and Nemo knock out (NEMO-KO) Mouse Embryonic Fibroblasts (MEFs). This protocol was developed in MEFs cells and can be applied adequately on other murine cell types expressing RIPK3.

2 Materials

Prepare all solutions using ultrapure water (prepared by purifying deionized water, to attain a sensitivity of 18 MΩ cm at 25 °C) and analytical grade reagents. Prepare and store all reagents at room temperature (unless indicated otherwise).

2.1 Inducing Cell Death in Cultured Cell Lines

1. Spontaneously immortalized wild type and Nemo knock out Mouse Embryonic Fibroblasts (MEF).
2. Dulbecco's Modified Eagle Medium (DMEM): Supplement DMEM with 2 mM glutamine, 100 units/ml penicillin, 100 μg/ml streptomycin, 0.25 μg/ml amphotericin B, 10% heat inactivated Fetal Bovine Serum (FBS).
3. Sterile Dulbecco's phosphate-buffered saline (DPBS) and trypsin–EDTA.
4. 5 μg/ml recombinant murine TNF-α (mTNF) solution in PBS and 1 mg/ml BSA, filtered through a 0.2 μm pore size membrane.
5. Dimethyl sulfoxide (DMSO).
6. 5 mM TPCA-1 in DMSO (IKK inhibitor).
7. 1 mM IKK-16 in DMSO (IKK inhibitor).
8. 10 mM Z-Val-Ala-DL-Asp-fluoromethylketone (Z-VAD-FMK, zVAD) in DMSO (Caspase inhibitor).
9. 50 mM GSK′872 in DMSO (RIPK3 inhibitor).
10. Necrostatin-1 (Nec1).

2.2 Evaluation of ; Cell Death by ATP Level and Caspase Assay

1. 96-well white assay plates, clear bottom with lid.
2. CellTiter-Glo® Luminescent Cell Viability Assay and Caspase Glo 3/7 assay (Promega).
3. Orbital shaker for mixing multiwell plates.
4. Microplate reader, such as Victor PerkinElmer.

2.3 Co-immunoprecipitation and Immunoblotting

1. Lysis buffer: 30 mM Tris–HCl, pH 7.4, 120 mM NaCl, 2 mM EDTA, 2 mM KCl, 10% Glycerol. The solution is filtered through a 0.2 μm pore size membrane and stored at 4 °C. 1% Triton X-100, EDTA free protease-inhibitor cocktail (Roche) and 1× phosphatase inhibitor (Sigma) should be add fresh before use.
2. Anti-RIPK1 monoclonal antibody (BD).
3. Anti-FADD antibody (Santa Cruz Biotechnology).
4. Anti-RIPK3 rabbit polyclonal antibody (Enzo LifeSciences).
5. Anti-MLKL rabbit polyclonal antibody (Abgent).

6. Anti-phospho-MLKL (S345) rabbit monoclonal antibody (Abcam).
7. Anti-Cleaved Caspase-8 antibody (Cell Signaling).
8. Protein A and Protein G-beads.
9. Microfuge tube rotator placed in cold room.
10. Bolt Bis-Tris 4–12% gradient polyacrylamide gels.
11. Running buffer: Bolt™ MOPS SDS Running Buffer.
12. iBlot2 Dry Blotting System transfer apparatus.
13. Nitrocellulose membranes or iBlot 2 transfer stacks.
14. 4× Sample loading Buffer (LDS): 2.5 ml of 1 M Tris–HCl, pH 6.8, 0.5 ml of ddH_2O, 1.0 g of SDS, 0.8 ml of 0.1% Bromophenol Blue, 4 ml of 100% glycerol, 2 ml of 14.3 M 100% β-mercaptoethanol stock solution.
15. Reducing agent: 500 mM dithiothreitol (DTT) for a 10× DTT.
16. PBS-T: 137 mM NaCl, 2.7 mM KCl, 10 mM Na_2HPO_4, 1.8 mM KH_2PO_4, dH_2O (or PBS 1×), 0.05% Tween 20.
17. Blocking buffer: PBS 1×, 0.05% Tween 20, 0.05% NaN_3, 5% (w/v) of Non-fat dry milk.
18. HRP-conjugated secondary antibodies.

3 Methods

3.1 Cell Viability and Caspase Activity Detection

1. Seed MEF cells in 96-well plates (white with clear bottoms) at 8–10,000 cells per well, using 100 μl of complete DMEM. Incubate the plates at 37 °C with 5% CO_2 for at least 6 h. Consider a duplicate experiment to examine separately but in parallel the cell viability and caspases 3/7 activity (*see* **Note 1**).
2. Pre-treatment with the various inhibitor combinations at 37 °C for 0.5–1 h. Remove media and replace with 50 μl of media and inhibitors for each treatment conditions and untreated control (DMSO). Prepare 1.5 ml Eppendorf tube for each desired treatment. The inhibitors will be added at the final concentration wanted in cell culture medium (*see* **Note 2**). Consider a triplicate for control (cell-free) and treated conditions (*see* **Note 3**). The final DMSO concentration in the media should not exceed 0.5–1%.
3. Add Mouse TNF to the wells to achieve a final concentration of 25 ng/ml and incubate for 6–24 h at 37 °C with 5% CO_2. The final volume in each well of the 96-well plates should be 100 μl (*see* **Note 4**).

4. Allow a frozen Caspases-Glo 3/7 and CellTiter-Glo reagents to equilibrate to ambient temperature (22 °C).
5. Remove the multiwell assay plates from the cell incubator and equilibrate the plate and its content to ambient temperature (approximately for 30 min).
6. Add a volume of Caspases-Glo 3/7 reagent equal to the plate. The volume of cell culture medium present into each well (Add 100 μl reagent for the 96-well plate format containing 100 μl of cell in culture medium). After 20 min, add a volume of CellTiter-Glo reagent equal to the volume of cell culture medium present into each well.
7. Allow the assay plate to ambient temperature for additional 10 min to stabilize the luminescent signal.
8. Record luminescence signals by using appropriate multiwell plate reader. For adherent cells set a step of 2-min shaking time before measuring the luminescence to ensure uniform distribution and complete cell lysis (typical integration time 0.5–1 s, Victor PerkinElmer).
9. Average the triplicate values for each treatment condition. Cell viability of each individual treatment group is express as a percentage relative to unstimulated control cells (100%). Caspase 3/7 activity is express as fold increase relative to unstimulated cells (*see* **Note 5**). Results from a typical experiment are shown in Fig. 1.

3.2 Isolation of FADD- and MLKL-Containing Complexes

After cell viability and caspase assay have indicated the presence of apoptosis versus necroptosis, ulterior validation is mandatory.

3.2.1 Cell Stimulation and Detection of Phosphorylated MLKL (pSer345-MLKL)

1. Plates 1.5×10^6 MEF cells on to 100 mm tissue culture plate per experimental condition in complete DMEM. Incubate the plates at 37 °C with 5% CO_2 for at least 12 h.
2. Pre-treat with the various inhibitor combinations (note that the inhibitors concentrations are taken from Subheading 3.1, **step 1**). Discard the media of overnight incubation and replace with 5–8 ml of fresh 1% FBS DMEM media and inhibitors for each treatment conditions and untreated control (DMSO). Return the plate at 37 °C with 5% CO_2 for 0.5–1 h.
3. Add Mouse TNF to achieve a final concentration of 10 ng/ml and incubate for 6 h at 37 °C (*see* **Note 6**). Leave one untreated plate as control.
4. Place the plates on ice to block activation, collect the media from each plate to a chilled falcon tube and spin at $1000 \times g$ for 3 min. Remove the supernatant and keep the pellets on ice for later (*see* **Note 7**).

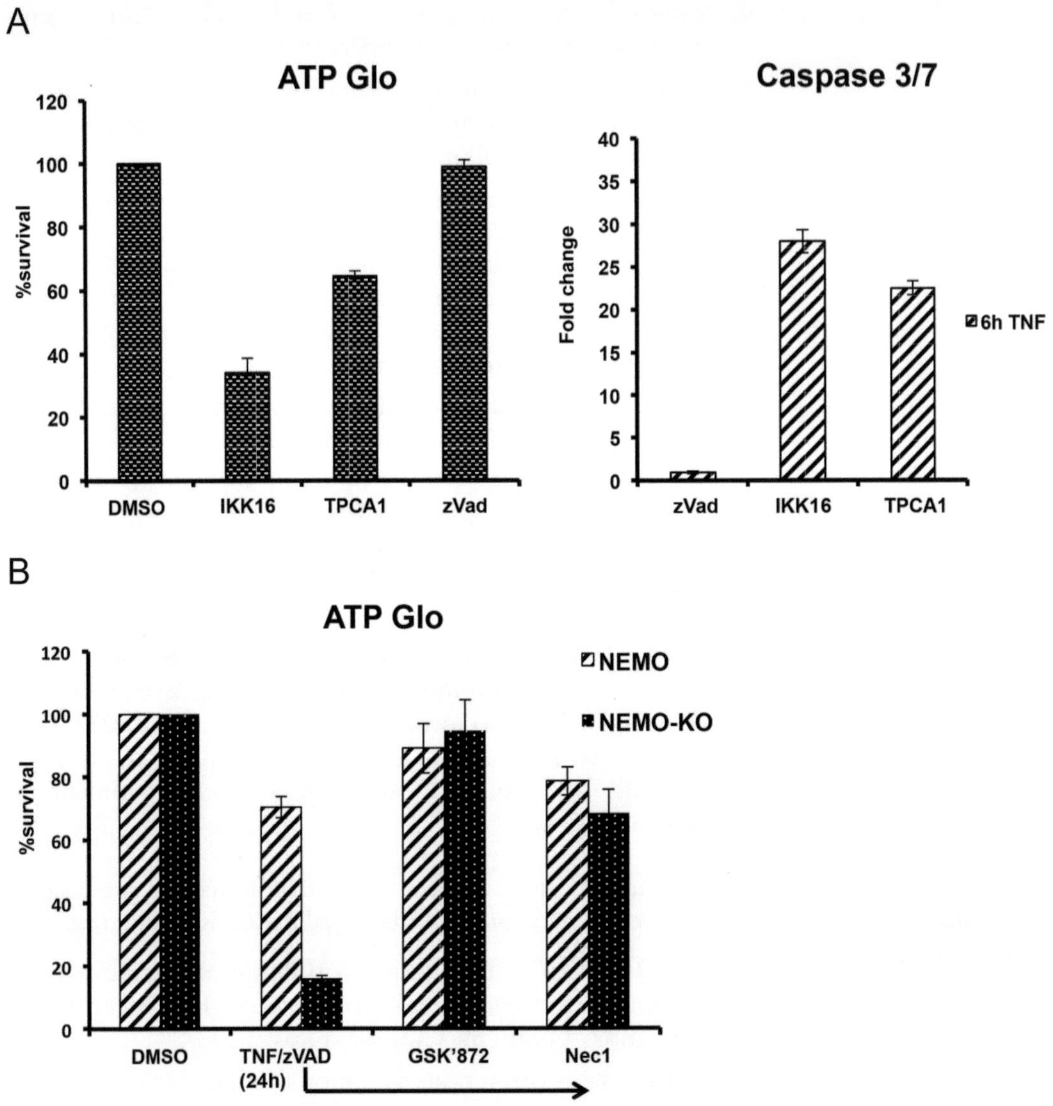

Fig. 1 (**a**) MEF cells were pretreated with 20μM z-VAD, 5μM TPCA-1, 1μM IKK-16 or DMSO before stimulation with TNF 25 ng/ml for 6 h. Cell viability and caspase 3/7 activity were determined by using CellTiter-Glo® Assays. (**b**) Relative viability of wild type and NEMO-KO MEFs 24 h after treatment with IKK-16, TNF and z-VAD (concentration as in **a**), in presence of Nec-1 (30 μM) or GSK′872 (5 μM) as indicated. Data are presented as the mean of an experiment performed in triplicate, ±standard deviation. The experiment show is representative of at least three independent experiments

5. Keep the plates on ice and wash the cells once gently with PBS. After thorough aspiration of PBS from the plate, lyse and scrape cells by adding 400 μl ice-cold LB with Inhibitors directly on the dish. Collect the lysate and combine it with the corresponding pellet of **step 4** and transfer into a chilled microcentrifuge tube (*see* **Note 8**). Leave the tube on a wheel at 4 °C for 15 min.

6. Clarify lysates by centrifuging at 13,000 × *g* for 5 min at 4 °C. Collect the supernatant into a new microcentrifuge tube, kept on ice (*see* **Note 9**).
7. Separate input samples of whole cell lysates by SDS-PAGE: Denaturate and load the samples on 4–12% SDS-PAGE gel, use constant 100 V for 30 min. Increase the voltage up to 160 V once the samples have run into the resolving gel. Stop the run when bromophenol blue start run outside the gel.
8. Transfer to nitrocellulose and incubate the membrane with 5% milk in PBS-T for more than 30 min.
9. Rinse the membrane with PBS-T and proceed to immunoblotting. Prepare dilutions of 0.5–1.0 μg/ml of the primary antibodies in PBS-T, 2% BSA (*see* **Note 10**). Incubate membrane with primary antibody dilution overnight with rotation at 4 °C (*see* **Note 11**).

3.2.2 Sequential Immunoprecipitation of FADD- and MLKL-Containing Complexes

Co-immunoprecipitation addresses a direct association of apoptotic and necroptotic components. Carry out all procedure on ice, unless otherwise specified.

1. Prepare protein A-coupled sepharose beads (A-beads) or G-coupled sepharose beads (G-beads) by washing the beads two times with IP-Lysis buffer. Add primary antibody to the beads (0.5–1 μg per sample). Use A-beads for anti-MLKL polyclonal antibody (IP-1), G-beads for the anti-FADD (IP-2). Incubate for 2–4 h at 4 °C with rotation.
2. Wash the beads with IP-Lysis buffer three times to remove non-specific binding. For each wash, gently mix the beads and centrifuge 2500 × *g* at 4 °C and discard the supernatant.
3. Add the MLKL-beads mix to each sample from **step 6** of Subheading 3.2.1 and incubate overnight at 4 °C with rotation.
4. Centrifuge the tubes 2500 × *g* at 4 °C. Recover the supernatants on a new chilled tubes (*see* **Note 12**).
5. Wash the beads with IP-Lysis buffer three times as in **step 2**. Remove as much wash buffer as possible from the beads. Add 30–50 μl 2× LDS buffer with 100 mM DTT to MLKL-beads, 5 min at 95 °C and centrifuge at 10,000 × *g* for 5 min. Collect the supernatant and discard the beads. At this stage, samples may be stored at −80 °C for future analysis.
6. Add the FADD-beads mix to the samples keep in **step 4** and incubate 2 h at 4 °C with rotation.
7. Centrifuge the tubes 2500 × *g* at 4 °C, discard the supernatant and wash the beads with IP-Lysis buffer three times as in **step 2**. Remove as much wash buffer as possible from the beads.

Add 30–50 μl 2× LDS buffer with 100 mM DTT to FADD-beads, boil the eluted samples for 5 min and centrifuge at 10,000 × *g* for 5 min. Collect the supernatant and discard the beads. At this stage samples may be stored at −80 °C for future analysis.

8. Input samples and IP-samples from **steps 5** and 7 are now ready for immunoblot analysis and detection of FADD- and MLKL-associated components (*see* **Note 13**).
9. Electrophorese samples as indicated in Subheading 3.2.1.
10. Transfer to nitrocellulose and incubate the membrane with 5% milk in PBS-T for more than 30 min.
11. Rinse the membrane with PBS-T and proceed to immunoblotting. Prepare dilutions of 0.5–1.0 μg/ml of the primary antibodies in PBS-T, 2% BSA. Incubate membrane with primary antibody dilution overnight with rotation at 4 °C (*see* **Note 14**).
12. Incubate the blot with species-specific secondary antibody–horseradish peroxidase (HRP) conjugates in 5% milk PBS-T for 1 h.
13. Allow chemiluminescent detection (Fig. 2, near here).

4 Notes

1. We prepare a duplicate of each experiment in the same 96-well plate to ensure the same number of cells, activation condition and time of stimulation for both assays.
2. In case of adherent cells, we prepare Eppendorf tube for each desired treatment (mix). The remaining mix will be stored at 4 °C and used for the subsequent TNF stimulation to guarantee the final concentration of inhibitors wanted in the cell culture medium. Calibration of cell toxicity in respond to zVAD, Nec-1s and GSK′872 is necessary in establishing this assay. DMEM containing one of the following: DMSO (control); 20 μM zVAD; 5 μM TPCA-1; 1 μM IKK-16; zVAD/TPCA-1; zVAD/IKK-16; Necroptosis prevention: zVAD/IKK-16 in combination with 5 μM GSK′872; zVAD/IKK-16 in combination with 30 μM Nec1.
3. Control wells containing medium without cells are used to determine the background signal from serum-supplemented medium. This step is not necessary when serum-free medium is used.
4. We add TNF at 2× concentration, e.g., if the final concentration in the well is 25 ng/ml, we add to the mix of inhibitors (*see* **Note 2**) the amount of TNF to achieve a concentration of

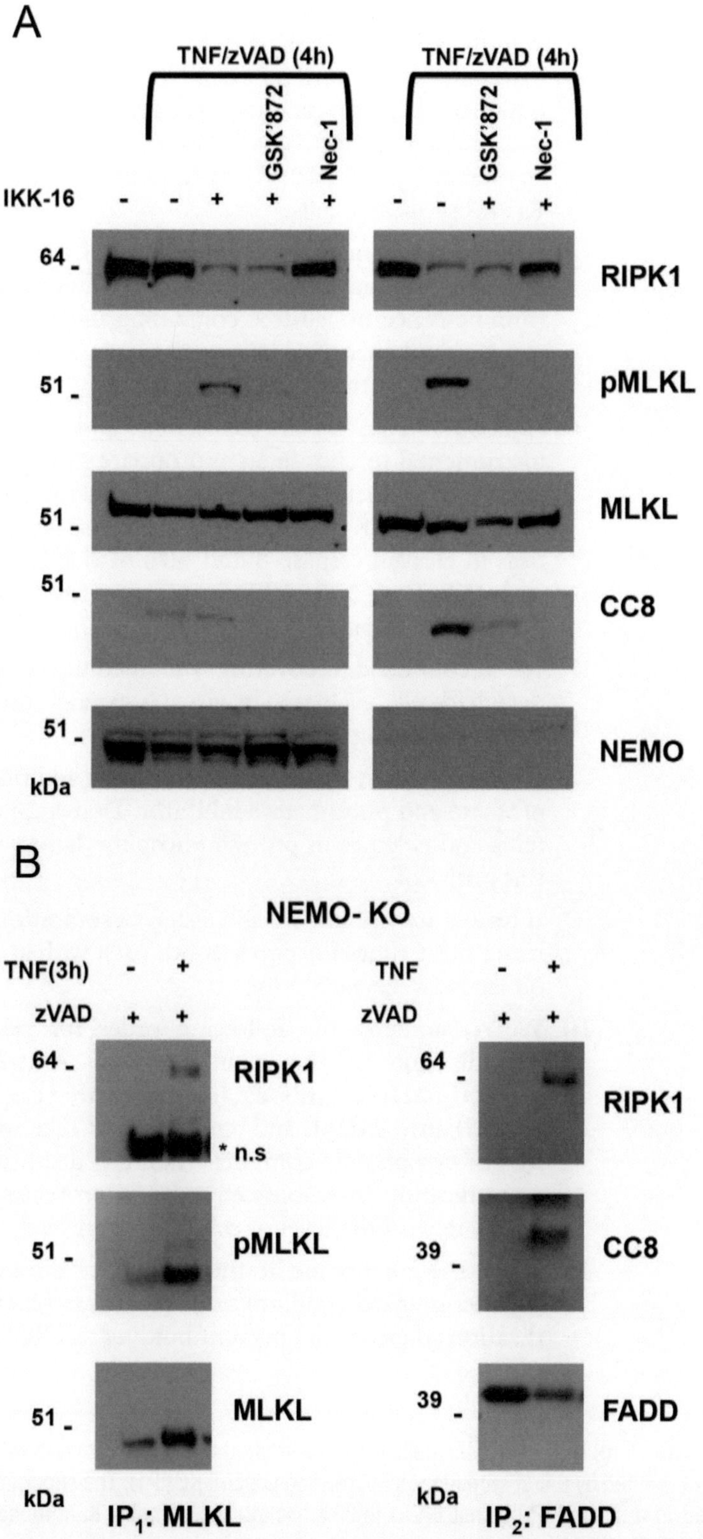

Fig. 2 (**a**) Wild type and NEMO-KO MEF cells were treated as indicated. Whole cell lysates were separated by reducing SDS-PAGE to detect RIPK1, phosphorylated Ser345-MLKL, total MLKL, cleaved Caspase 8 and

50 ng/ml. Add 50 μl of 2× TNF mix to each well. The final volume is 100 μl. Sensitivity to TNF-induced cell death varies markedly among different cell types so a titration of TNF and inhibitors is recommended. Because RIPK3 represents a key determinant to transduce necroptosis signal through the canonical RIPK1-RIPK3-MLKL axis, it is mandatory to use RIPK3 proficient cells.

5. Before determine the relative cell viability in response to each treatment condition, we subtract the average background value (luminescence of cell-free containing media) from each treatment absorbance evaluation, then we average the triplicate values for each treatment condition.

6. Data obtained using the cell viability and caspase 3/7 assay are instrumental to choose an appropriate time point for samples collection. Generally, 10–20 folds induction of Caspases 3/7 activity after IKK-16/TNF treatment is sufficient for the detection of cleaved caspase-8 and 60% of viability after treatment with IKK-16/TNF/zVAD is sufficient for pMLKL detection. Necroptosis is prevented by using Nec-1 or GSK′872.

7. We recommend recovering the medium when it contains detached cells, evaluated by visual inspection under microscope before collecting samples.

8. We complement the IP-lysis buffer with addition of both fresh protease and phosphatase inhibitor. Detection of cleaved proteins and changes of protein phosphorylation are key goals of this protocol.

9. If lysates are viscous, we add a step of sonication: sonicate each lysate three times for two seconds each with at least 1 min rest on ice between each pulse.

10. We recommend the following order for primary antibody hybridization of the membrane: (a) Anti-RIP1 antibody (b) Anti-pMLKL antibody; (c) Anti-cleaved caspase 8 antibody; and (d) anti- MLKL and anti-NEMO. Take into account the loss of cytoplasmic contents when cell death is induced. The quantification of whole cell extracts protein should negative correlate with the level of cell death observed.

11. Wash the membrane if the secondary antibodies used are HRP-conjugated, sodium azide present in blocking and hybridization solutions can prevent binding of HRP to its substrate.

Fig. 2 (continued) NEMO. (**b**) NEMO-KO MEF cells were pre-treated with zVAD (1h) and with TNF for 3 h. The MLKL-complex was isolated by a first immunoprecipitation against MLKL (IP1). The supernatants derived from IP1 were subjected to a second IP against FADD (IP2) to isolate the complex II. Immunoblot analysis was performed to reveal the indicated proteins

12. Keep 10% of each sample (input control) to verify the quality of protein extracts after IP.
13. We load the half amount of the immunoprecipitation and 15–30 μg of input.
14. We recommend the following order for primary antibody hybridization of the membrane: (a) Anti-RIP1 antibody (b) Anti-pMLKL antibody; (c) Anti-cleaved caspase 8 antibody; and (d) anti-total MLKL and anti-FADD. The use of zVAD stabilizes the FADD-containing complex and cleaved Caspase 8 (CC8) and RIPK1 are easily detectable. As mentioned in the introduction section, detecting RIPK1 and pMLKL into MLKL-IP is currently the most specific and sensitive indicator of necroptosis. Normally, 60% of viability that decrease after treatment with zVAD is recommended for pMLKL detection and RIK1 recruitment to the MLKL.

Acknowledgments

We thank Dr. G. Courtois for MEF cells. The work was supported by the MIUR PRIN 2017WLKYAM_3 to A.P. and by France Incontinentia Pigmenti Foundation to M.V.U.

References

1. Brenner D, Blaser H, Mak TW (2015) Regulation of tumour necrosis factor signalling: live or let die. Nat Rev Immunol 15(6):362–374. https://doi.org/10.1038/nri3834
2. Kanayama A, Seth RB, Sun L, Ea CK, Hong M, Shaito A, Chiu YH, Deng L, Chen ZJ (2004) TAB2 and TAB3 activate the NF-kappaB pathway through binding to poly- ubiquitin chains. Mol Cell 15(4):535–548. https://doi.org/10.1016/j.molcel.2004.08.008
3. Ea CK, Deng L, Xia ZP, Pineda G, Chen ZJ (2006) Activation of IKK by TNFalpha requires site-specific ubiquitination of RIP1 and polyu- biquitin binding by NEMO. Mol Cell 22(2):245–257. https://doi.org/10.1016/j.molcel.2006.03.026
4. Micheau O, Tschopp J (2003) Induction of TNF receptor I-mediated apoptosis via two sequential signaling complexes. Cell 114(2):181–190. https://doi.org/10.1016/S0092-8674(03)00521-X
5. Micheau O, Lens S, Gaide O, Alevizopoulos K, Tschopp J (2001) NF-kappaB signals induce the expression of c-FLIP. Mol Cell Biol 21(6):5299–5305. https://doi.org/10.1128/MCB.21.16.5299-5305.2001
6. Kreuz S, Siegmund D, Scheurich P, Wajant H (2001) NF-kappaB inducers upregulate cFLIP, a cycloheximide-sensitive inhibitor of death receptor signaling. Mol Cell Biol 21(12):3964–3973. https://doi.org/10.1128/MCB.21.12.3964-3973.2001
7. Dondelinger Y, Jouan-Lanhouet S, Divert T, Theatre E, Bertin J, Gough PJ, Giansanti P, Heck AJ, Dejardin E, Vandenabeele P, Bertrand MJ (2015) NF-kappaB-independent role of IKKalpha/IKKbeta in preventing RIPK1 kinase-dependent apoptotic and necroptotic cell death during TNF signaling. Mol Cell 60(1):63–76. https://doi.org/10.1016/j.molcel.2015.07.032
8. Annibaldi A, Meier P (2018) Checkpoints in TNF-Induced Cell Death: Implications in Inflammation and Cancer. Trends Mol Med 24(1):49–65. https://doi.org/10.1016/j.molmed.2017.11.002
9. Legarda-Addison D, Hase H, O'Donnell MA, Ting AT (2009) NEMO/IKKgamma regulates an early NF-kappaB-independent cell-death checkpoint during TNF signaling. Cell Death Differ 16(9):1279–1288. https://doi.org/10.1038/cdd.2009.41

10. Pescatore A, Esposito E, Draber P, Walczak H, Ursini MV (2016) NEMO regulates a cell death switch in TNF signaling by inhibiting recruitment of RIPK3 to the cell death-inducing complex II. Cell Death Dis 7(8): e2346. https://doi.org/10.1038/cddis.2016.245
11. Lin Y, Devin A, Rodriguez Y, Liu ZG (1999) Cleavage of the death domain kinase RIP by caspase-8 prompts TNF-induced apoptosis. Genes Dev 13(19):2514–2526. https://doi.org/10.1101/gad.13.19.2514
12. Feng S, Yang Y, Mei Y, Ma L, Zhu DE, Hoti N, Castanares M, Wu M (2007) Cleavage of RIP3 inactivates its caspase-independent apoptosis pathway by removal of kinase domain. Cell Signal 19(10):2056–2067. https://doi.org/10.1016/j.cellsig.2007.05.016
13. Zhang DW, Shao J, Lin J, Zhang N, Lu BJ, Lin SC et al (2009) RIP3, an energy metabolism regulator that switches TNF-induced cell death from apoptosis to necrosis. Science 325 (5938):332–336. https://doi.org/10.1126/science.1172308
14. He S, Wang L, Miao L, Wang T, Du F, Zhao L et al (2009) Receptor interacting protein kinase-3 determines cellular necrotic response to TNF-alpha. Cell 137(6):1100–1111. https://doi.org/10.1016/j.cell.2009.05.021
15. Sun L, Wang H, Wang Z, He S, Chen S, Liao D, Wang L, Yan J, Liu W, Lei X, Wang X (2012) Mixed lineage kinase domain-like protein mediates necrosis signaling downstream of RIP3 kinase. Cell 148(1–2):213–227. https://doi.org/10.1016/j.cell.2011.11.031
16. Murphy JM, Czabotar PE, Hildebrand JM, Lucet IS, Zhang JG, Alvarez-Diaz S, Lewis R, Lalaoui N, Metcalf D, Webb AI, Young SN, Varghese LN, Tannahill GM, Hatchell EC, Majewski IJ, Okamoto T, Dobson RC, Hilton DJ, Babon JJ, Nicola NA, Strasser A, Silke J, Alexander WS (2013) The pseudokinase MLKL mediates necroptosis via a molecular switch mechanism. Immunity 39(3):443–453. https://doi.org/10.1016/j.immuni.2013.06.018
17. Dillon CP, Weinlich R, Rodriguez DA, Cripps JG, Quarato G, Gurung P, Verbist KC, Brewer TL, Llambi F, Gong YN, Janke LJ, Kelliher MA, Kanneganti TD, Green DR (2014) RIPK1 blocks early postnatal lethality mediated by caspase-8 and RIPK3. Cell 157 (5):1189–1202. https://doi.org/10.1016/j.cell.2014.04.018
18. Mandal P, Berger SB, Pillay S, Moriwaki K, Huang C, Guo H, Lich JD, Finger J, Kasparcova V, Votta B, Ouellette M, King BW, Wisnoski D, Lakdawala AS, DeMartino MP, Casillas LN, Haile PA, Sehon CA, Marquis RW, Upton J, Daley-Bauer LP, Roback L, Ramia N, Dovey CM, Carette JE, Chan FK, Bertin J, Gough PJ, Mocarski ES, Kaiser WJ (2014) RIP3 induces apoptosis independent of pronecrotic kinase activity. Mol Cell 56 (4):481–495. https://doi.org/10.1016/j.molcel.2014.10.021
19. Rodriguez DA, Weinlich R, Brown S, Guy C, Fitzgerald P, Dillon CP, Oberst A, Quarato G, Low J, Cripps JG, Chen T, Green DR (2016) Characterization of RIPK3-mediated phosphorylation of the activation loop of MLKL during necroptosis. Cell Death Differ 23 (1):76–88. https://doi.org/10.1038/cdd.2015.70

Chapter 15

Use of ChIP-qPCR to Study the Crosstalk Between HIF and NF-κB Signaling in Hypoxia and Normoxia

Dilem Shakir, Michael Batie, and Sonia Rocha

Abstract

Hypoxia and inflammation are intensely connected in a functional crosstalk. Within this crosstalk, two major transcription factors take center stage: HIF and NF-κB. To investigate transcription factor function, an important aspect is its ability to bind DNA. The most appropriate method to study this property in cells is the use of chromatin immunoprecipitation followed by qPCR and/or next generation sequencing. This allows identification of potentially directly regulated genes as well as enhancer regions. Here we describe the ChIP-qPCR method in detail, including key aspects important for the success of the technique.

Key words Hypoxia, Inflammation, HIF, NF-kappaB, ChIP, qPCR

1 Introduction

Hypoxia is the dysregulation of homeostatic balance between oxygen supply and consumption in cells, leading to insufficient oxygen for regular metabolic activity and survival of multicellular organisms. Hypoxia inducible factors (HIFs) are a transcription factor family that control a specific transcriptional program in response to decrease in oxygen levels [1]. HIFs typically form heterodimeric complexes composed of oxygen sensitive HIF-α subunits (HIF-1α, HIF-2α, or HIF-3α), and a constitutively expressed HIF-1β subunit (gene name, aryl hydrocarbon nuclear translocator (ARNT)) [2]. The stability and transcriptional activity of HIF-α subunits is controlled in part by post-translational modifications. Under sufficiently oxygenated conditions (normoxia), HIF-α proteins are degraded by the proteasome following prolyl hydroxylation by 2-oxoglutarate (2-OG) dependent dioxygenases, called prolyl hydroxylases (PHDs) [3]. In normoxia, HIF-α proteins also undergo asparaginyl hydroxylation by another 2-OG dependent dioxygenase, factor inhibiting HIF (FIH), which prevents the recruitment of the HIF coactivators (CBP/p300) in the nucleus

Guido Franzoso and Francesca Zazzeroni (eds.), *NF-κB Transcription Factors: Methods and Protocols*, Methods in Molecular Biology, vol. 2366, https://doi.org/10.1007/978-1-0716-1669-7_15,

[4]. Whereas, under hypoxic conditions, PHDs and FIH have reduced activity due to their dependence on molecular oxygen, thus HIF-α is stabilized and forms a heterodimer with HIF-1β in the nucleus [5]. The HIF complex then binds to the hypoxia response element (HRE) nucleotide sequences on DNA and initiates the transcription of HIF target genes [5, 6]. The HIF heterodimer can regulate the expression of wide ranges of genes involved in various physiological functions such as angiogenesis, erythropoiesis, cell proliferation and survival [7–9]. Dysregulation of this transcription mechanism has been associated with pathological conditions including different cancer types [10–12].

Hypoxia is also associated with another important transcription factor family, Nuclear Factor kappa-light-chain-enhancer of B cells (NF-κB), which is known to have importance in maintaining homeostasis in immune and inflammatory responses [13, 14]. Activation of the NF-κB pathway can occur through the stimulation of canonical, non-canonical, or atypical pathways [15]. The NF-κB family is composed of five distinct members, including RelA (p65), RelB, c-Rel, NF-κB1 (p105/p50), and NF-κB2 (p100/p52) [15]. All NF-κB proteins have a Rel homology domain (RHD), which enables DNA-binding ability and homo- or heterodimer formation [16]. In an unstimulated canonical pathway, inhibitor of kappa B (IκB) holds NF-κB proteins inactive in the cytoplasm. Upon stimulation of the pathway, the upstream kinase complex, IκB Kinase (IKK) phosphorylates the IκBα, which leads to its proteasomal degradation and releases NF-κB to translocate into the nucleus [17].

Besides pro-inflammatory cytokines and pathogen associated molecular stimuli, hypoxia can also lead to activation of the NF-κB pathway. The mechanism of hypoxia-induced inflammation has been extensively described in many studies [18–20]. Also, Tumor Necrosis Factor alpha (TNF-α) is able to stimulate the expression of HIF subunits [21, 22]. Apart from sharing common stimuli, both HIF and NF-κB transcription factors are known to have common target genes and regulators [23]. HIF subunits can have different roles in regulating the transcriptional response of NF-κB. HIF-1α is able to restrict the NF-κB transcriptional activity, whereas HIF-1β is found to be required for full NF-κB activation [24, 25]. Various methods have been used to discover the coregulation of target genes by HIF and NF-κB subunits; however, a comprehensive analysis on a genome-wide scale is still lacking. Genome-wide sequencing approaches such as Chromatin Immunoprecipitation (ChIP)-sequencing can be employed to investigate coregulation of NF-κB and HIF target genes in response to hypoxia and inflammation. ChIP assay coupled with quantitative polymerase chain reaction (qPCR) can verify the direct protein-DNA interactions and functional connections between HIF, NF-κB and their target sites in response to different stimuli.

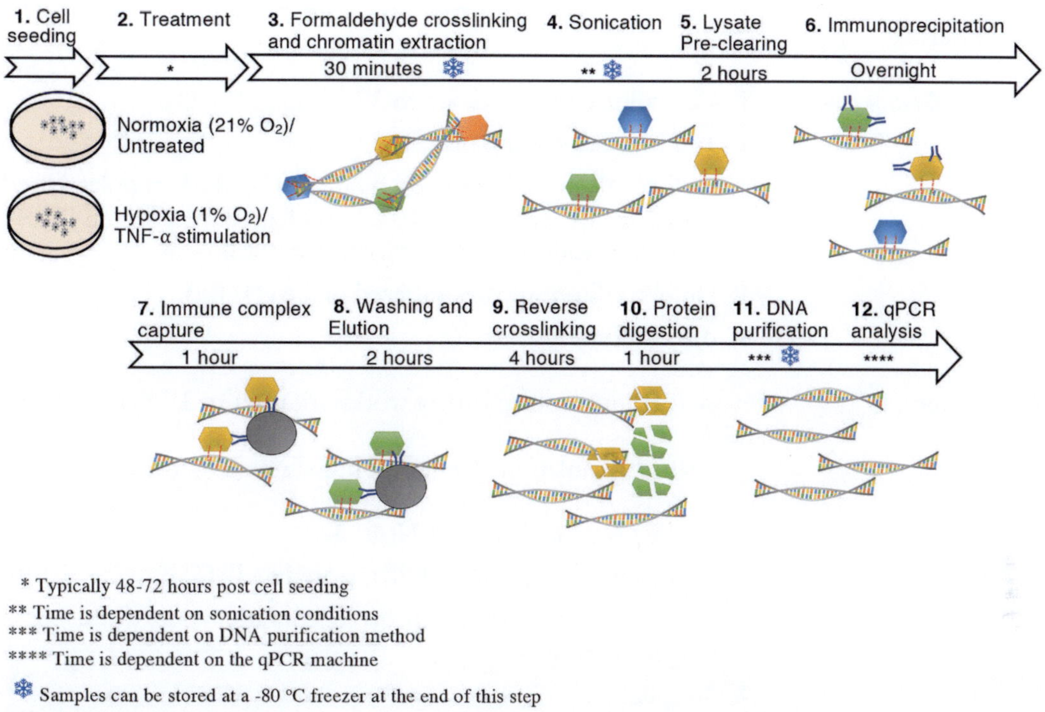

Fig. 1 Workflow of ChIP-qPCR

Here we present the ChIP-qPCR method that we have been successfully applying to analyze the crosstalk between HIF and NF-κB signaling pathways in human cell lines. ChIP-qPCR includes multiple steps, which are summarized in Fig. 1. Main steps include cell culture and stimulation of endogenous HIFs and NF-κB with hypoxia or TNF-α. After the completion of the treatment, proteins are fixed to the DNA with formaldehyde cross-linking. Then chromatin is extracted with a lysis buffer and fragmented via sonication. After the chromatin fragments have been prepared, appropriate antibodies and protein G-sepharose beads are used to select DNA-bound proteins of interests via immunoprecipitation (IP) and immune complex capture. DNA and protein complexes are detached by reversing the cross-links and proteins are eliminated by a protein degradation step. After purifying the DNA, qPCR is used to analyze the relative amount of specific genomic DNA sequences enriched for binding by the protein of interest. Although ChIP assays have been a routine molecular technique for identifying DNA-binding proteins, all of its steps, including pre-clearing, washing, and elution need to be followed carefully to achieve a successful result (Fig. 1).

2 Materials

2.1 Cell Culture (See Note 1)

1. Cell culture incubator set to 37 °C and 5% CO_2.
2. 150 mm cell culture plates.
3. Dulbecco's modified eagle's medium (DMEM) supplemented with 2 mM L-glutamine, 10% (v/v) fetal bovine serum (FBS), 50 U/ml penicillin, and 50 μg/ml streptomycin.
4. Dulbecco's phosphate-buffered saline (DPBS).
5. 0.25% Trypsin.

2.2 Treatments

Hypoxia stimulation: Hypoxia workstation set to 1% O_2, 5% CO_2, and 37 °C.

TNF-α stimulation: 10 ng/ml human recombinant TNF-α.

2.3 ChIP Assay (See Note 2)

1. 36.5% Formaldehyde (*see* **Note 3**).
2. Phosphate buffer saline (PBS), stored at 4 °C.
3. 1.5 M Glycine, filtered to avoid contamination during prolonged storage.
4. Scraper.
5. Lysis buffer: 1% sodium dodecyl sulfate (SDS), 10 mM ethylenediaminetetraacetic acid (EDTA), 50 mM Tris–HCl (pH 8.1), 1 protease inhibitor tablet (1 mM Phenylmethylsulfonyl fluoride (PMSF), 1 mg/ml leupeptin, and 1 mg/ml aprotinin).
6. Dilution buffer: 1% Triton X-100, 2 mM EDTA, 150 mM sodium chloride (NaCl), 20 mM Tris–HCl (pH 8.1).
7. Wash buffer 1: 0.1% SDS, 1% Triton X-100, 2 mM EDTA, 20 mM Tris–HCl (pH 8.1), 150 mM NaCl.
8. Wash buffer 2: 0.1% SDS, 1% Triton X-100, 2 mM EDTA, 20 mM Tris–HCl (pH 8.1), 500 mM NaCl.
9. Wash buffer 3: 0.25 M (lithium chloride) LiCl, 1% Nonidet P-40, 1% sodium deoxycholate, 1 mM EDTA, 10 mM Tris–HCl (pH 8.1).
10. TE Buffer: 10 mM Tris–HCl (pH 8.1), 1 mM EDTA.
11. Elution buffer: 1% SDS, 0.1 M $NaHCO_3$.
12. Sheared salmon sperm DNA (ssDNA).
13. Protein G-Sepharose slurry (2 mg recombinant Protein G/ml resin (50% (v/v) resin suspension in 20% ethanol).
14. Antibodies (*see* **Note 4**).
15. 5 M NaCl.
16. Proteinase K: 20 μg/μl stock.
17. 1 M Tris–HCl (pH 6.5).

18. 0.5 M EDTA (PH 8).
19. 10% Polyoxyethylene laural ether (BRIJ35) solution.
20. Benchtop centrifuge.
21. Rotating wheel set to 40 rpm.
22. Thermomixer.
23. Sonicator.
24. Vacuum aspirator.
25. Nuclease-free water.
26. QIAquick PCR Purification kit (Qiagen).

2.4 qPCR Analysis

1. 100 μM forward and reverse primers.
2. Brilliant II Sybr green kit.
3. Mx3005P qPCR platform (Stratagene/Agilent).
4. Mx3005P 96-well plates.
5. Mx3005P 96-well plate lids.
6. Plate centrifuge.
7. Nuclease-free water.

3 Methods

3.1 Cell Culture Growth Conditions

1. Cells are maintained in 150 mm tissue culture plates containing DMEM and kept in 5% CO_2 and 37 °C.
2. For preparing the cells for the experiment, remove culture medium and wash cells with 5 ml PBS. Add 5 ml of pre-warmed trypsin and incubate the cells at 37 °C until the cells have detached (*see* **Note 5**).
3. Resuspend cells by adding 5 ml DMEM and move the 10 ml cell solution to a 15 ml canonical tube. Pipette the cells up and down to have a single cell suspension.
4. Plate cells to a final density of ~4.5 × 10^6 cells into two 150 mm culture plates in a final volume of 15 ml of medium (*see* **Note 6**).
5. Apply the appropriate treatment for desired length of time prior to chromatin extraction.

3.2 Treatments

1. For TNF-α stimulation, add 10 ng/ml of human recombinant TNF-α to the cells.
2. Mix plates by moving up and down and side to side.
3. For hypoxia stimulation, put the cells into a hypoxia workstation set to 1% O_2, 5% CO_2 and 37 °C.

3.3 ChIP-qPCR

All ChIP steps are carried out in 4 °C temperature, unless otherwise specified. Samples are kept on ice at all times. Rotating wheel and sonicator are processed in a cold room set to 4 °C.

3.3.1 Formaldehyde Cross-Linking and Chromatin Extraction

1. Cross-link proteins to DNA by adding formaldehyde drop-wise directly to the cell medium to a final concentration of 1% (i.e., 410 μl of 36.5% formaldehyde into 15 ml of culture medium in a 150 mm plate). Mix by moving the plate up and down and side by side and incubate at room temperature for 10 min (*see* **Note 7**).
2. Stop the cross-linking reaction by adding 1.5 M glycine drop-wise to the cell medium to a final concentration of 0.125 M (i.e., 1.284 ml of 1.5 M glycine into 15.410 ml). Mix by moving the plate up and down and side by side and incubate it at room temperature for 5 min.
3. Discard the media and wash twice with cold DPBS by disposing the waste to an appropriate formaldehyde waste container.
4. Add 1 ml of cold DPBS and scrape the cells into a 15 ml canonical tube. Centrifuge at 100 × *g* for 5 min at room temperature.
5. Aspirate out the supernatant carefully and resuspend the pellet in 450 μl lysis buffer. Allow the cells to lyse for 10 min on ice.
6. After **step 5**, lysed cells can be carried on to the sonication procedure or they can be stored in a −80 °C freezer for later use.

3.3.2 Sonication

1. Sonicate the HeLa cell samples at 4 °C, 20 times for 15 s with a 30 s gap between each sonication at 50% amplitude (*see* **Note 8**).
2. After sonication, spin down the samples at 13,800 × *g* for 10 min at 4 °C and discard the pellet.
3. Reserve 10 μl of the sheared chromatin from each condition, as the input samples in a separate 1.5 ml eppendorf tube. The input samples can be kept in a −20 °C freezer until processing with rest of the samples in Subheading 3.3.7 (*see* **Note 9**).
4. Split the samples into 100 μl aliquots, one for each IP.
5. Dilute each of the 100 μl samples tenfold by adding 900 μl dilution buffer to each sample.
6. After **step 5**, samples can be carried on to the pre-clearing procedure or they can be stored in a −80 °C freezer for later use.
7. Before continuing to the pre-clearing section, prepare the protein G-sepharose slurry by following the steps in Subheading 3.3.3.

3.3.3 Preparing the Protein G-Sepharose Slurry

1. Calculate the total starting volume of the slurry required for all of the samples (i.e., 20 μl per sample), put the required volume into an Eppendorf tube (*see* **Note 10**).
2. Spin down the beads by centrifugation at 100 × *g* for 1 min at room temperature and discard the supernatant without disturbing the beads.
3. Wash the beads by resuspending them in 500 μl cold PBS.
4. Centrifuge the solution at 100 × *g* for 1 min at room temperature and discard the PBS without disturbing the beads.
5. Resuspend the beads in cold PBS by adding 50% of the starting volume.
6. Prepare a master mix by adding 2 μg ssDNA per 20 μl protein G-sepharose slurry and proceed to the next section.

3.3.4 Lysate Pre-clearing

1. Add 22 μl of the prepared master mix into each sample (*see* **Note 11**).
2. Incubate the samples for 2 h on the rotating platform at 4 °C.
3. Centrifuge at 100 × *g* for 1 min at 4 °C. Then, transfer the supernatant to a new 1.5 ml eppendorf tube and discard the pelleted beads.

3.3.5 Immuno-precipitation (IP)

1. Add BRIJ35 at a final concentration of 0.1% (10 μl of 10% BRIJ35 solution) to each sample and appropriate amount of antibody (*see* **Note 12**).
2. Incubate the samples overnight on the rotating wheel at 4 °C.

3.3.6 Immune Complex Capture, Washing, and Elution

1. Prepare a master mix containing 30 μl protein G-sepharose slurry and 2 μg ssDNA per sample (follow the **steps 1–5** in Subheading 3.3.3 for preparing the protein G-sepharose slurry).
2. Add 32 μl of the prepared master mix to each sample and incubate for 1 h on the rotating wheel at 4 °C.
3. Centrifuge the samples at 100 × *g* for 1 min at 4 °C and store the supernatant in a −80 °C freezer (*see* **Note 13**).
4. Resuspend pelleted beads in 1 ml wash buffer 1 and incubate the samples for 5 min on the rotating wheel at 4 °C.
5. Centrifuge the samples at 100 × *g* for 1 min at 4 °C and discard the supernatant without disturbing the beads.
6. Resuspend pelleted beads in 1 ml wash buffer 2 and incubate them for 5 min on the rotating wheel at 4 °C.
7. Repeat **step 5** and resuspend the pelleted beads in 1 ml wash buffer 3.

8. Incubate the samples for 5 min on the rotating wheel at 4 °C and repeat **step 5**.
9. Wash the beads twice by resuspending them in 1 ml TE buffer.
10. Repeat **step 5** and resuspend the pelleted beads with 100 μl elution buffer and incubate them for 1 h at room temperature on a rocker.
11. Spin down the beads at 100 × *g* for 1 min at room temperature.
12. Transfer the supernatant to a new eppendorf tube and discard the pelleted beads.

3.3.7 Reverse Cross-Linking and DNA Purification

1. Thaw 10 μl of the input samples reserved after the sonication (Subheading 3.3.2, **step 3**) and add 90 μl nuclease-free water to each. Input samples are included for the rest of the protocol.
2. For reverse cross-linking, add 0.2 M NaCl into each sample (i.e., 4 μl of 5 M NaCl stock in 100 μl of each sample). Incubate them for 4 h or overnight on a thermomixer pre-set to 65 °C and 500 rpm.
3. For digesting the proteins, prepare a master mix to add the following into each 100 μl IP and input samples: 20 μg Proteinase K, 40 mM Tris–HCl (pH 6.5), 10 mM EDTA (pH 8) (i.e., 1 μl Proteinase K, 4 μl 1 M Tris–HCl, 2 μl 0.5 M EDTA). After adding 7 μl of the master mix into the tubes, incubate them for 1 h on a thermomixer pre-set at 45 °C and 500 rpm.
4. Purify DNA using a QIAquick PCR Purification kit (Qiagen) following manufacturer's instructions, eluting the samples in 50 μl nuclease-free water or elution buffer (*see* **Note 14**).
5. Eluted DNA can be stored in a −80 °C freezer for later use.

3.3.8 qPCR Analysis

1. Prepare a master mix to add the following into each qPCR reaction well: 3.9 μl nuclease-free water, 0.6 μl of 10 μM prepared primer mix (5 μM forward and 5 μM reverse), and 7.5 μl Sybr green.
2. Put 3 μl of the purified DNA samples to the 96-well plate reaction wells and add 12 μl of the prepared master mix (*see* **Note 15**).
3. Close the plate and centrifuge briefly using a plate centrifuge to collect the reaction at the bottom of the wells.
4. Program the qPCR instrument as follows: 10 min 95 °C (1 cycle); 30 s 95 °C, 1 min 60 °C, 1 min 72 °C (50 cycles); 1 min 95 °C, 30 s 55 °C, 30 s 95 °C (1 cycle).
5. Data obtained for the ChIP should be normalized to the input sample DNA (*see* **Note 16**).

4 Notes

1. All cell culture procedure should be performed in a tissue culture hood with sterile practice. All material should be sprayed with 70% ethanol before placing into the tissue culture hood. DMEM and trypsin are stored at 4 °C fridge and should be warmed to 37 °C in a water bath before use. Although DMEM is used here for the human cervical cancer (HeLa) cells, different cell lines may need to be maintained in a different medium.
2. Throughout the ChIP-qPCR assay, always use the low protein binding and DNase/RNase-free eppendorf tubes. Use filter tips and nuclease-free water for preparing solutions to prevent cross contamination.
3. Formaldehyde is a highly toxic and flammable chemical. Thus, it should be used with appropriate safety measures, such as protective gloves, and a laboratory coat. All procedure that includes the formaldehyde should be carried on in a tissue culture hood. Waste should be disposed according to regulations for hazardous waste.
4. Selection and the concentration of the ChIP antibodies are one of the most critical factors that will contribute to the success of the ChIP assay. In order to check for the protein binding specificity of the antibodies, western blot method can be applied after the last step of Subheading 3.3.6.
5. Detachment of the HeLa cells from the tissue culture plate takes 3–5 min incubation in trypsin. This should be visible by eye, which can also be confirmed microscopically.
6. The number of cells stated is for lysis 48 h post cell seeding. This number should be optimized for experiments with different timescales and/or cells lines, with aim of reaching 90% confluency at the time of lysis.
7. If doing a hypoxia treatment, add formaldehyde in the hypoxia chamber, then incubate for 10 min at room temperature inside the tissue culture hood.
8. Sonication step needs to be optimized to successfully fragment the DNA to 200–500 bp. These conditions are dependent on cell line, concentration of chromatin in the lysate, and the sonicator used. The indicated settings are specifically based on the sonication of 90% confluent HeLa cells in a 150 mm plate in 450μl ChIP lysis buffer via Sonics Vibra Cell #VCX130 sonicator.
9. Cut the end of the pipette tip before using the protein G-sepharose slurry as a narrow ending pipette tip can harm the beads. Protein G-sepharose slurry comes in 50% ethanol, thus it is important to remove the ethanol before using it.

10. The input sample represents 10% of the amount of chromatin used in the ChIP without any specific selection of DNA fragments through IP. This sample will be used to do the final analysis in the last section of the experiment.
11. Mix the master mix gently by pipetting up and down before adding to each sample to ensure beads are evenly distributed. Remember to cut the pipette tip before while using the master mix as otherwise it can harm the beads.
12. In addition to using antibodies to detect the proteins of interests (specific antibodies), a non-specific antibody (IgG) is used as a negative control. IgG needs to be from the same species as the specific antibody is used to account for non-specific binding. The amount and species of the IgG must be the same as the specific antibody.
13. The supernatant can be used to determine IP efficiency (e.g., how much of the protein of interest is captured and how much of it is lost following the IP and immune capture steps). These can be investigated by western blotting.
14. Alternative methods can be used to purify the DNA such as magnetic bead or ethanol precipitation-based DNA purification procedures.
15. Typically input DNA is diluted to 2% with nuclease-free water prior to using it in qPCR.
16. When analyzing the data, normalize the IgG (negative control) and IP samples to the input sample. In order to determine if there is significant enrichment of protein of interest binding to a specific DNA sequence, input normalised IgG is compared to the IP (specific antibody IP) within a treatment. Also, input normalized IP signal can be compared across different treatments to determine if the binding is affected by the treatments.

References

1. Smith TG, Robbins PA, Ratcliffe PJ (2008) The human side of hypoxia-inducible factor. Br J Haematol 141(3):325–334. https://doi.org/10.1111/j.1365-2141.2008.07029.x
2. Rocha S (2007) Gene regulation under low oxygen: holding your breath for transcription. Trends Biochem Sci 32(8):389–397. https://doi.org/10.1016/j.tibs.2007.06.005
3. Salceda S, Caro J (1997) Hypoxia-inducible factor 1alpha (HIF-1alpha) protein is rapidly degraded by the ubiquitin-proteasome system under normoxic conditions. Its stabilization by hypoxia depends on redox-induced changes. J Biol Chem 272(36):22642–22647. https://doi.org/10.1074/jbc.272.36.22642
4. Lando D, Peet DJ, Gorman JJ, Whelan DA, Whitelaw ML, Bruick RK (2002) FIH-1 is an asparaginyl hydroxylase enzyme that regulates the transcriptional activity of hypoxia-inducible factor. Genes Dev 16(12):1466–1471. https://doi.org/10.1101/gad.991402
5. Jiang BH, Rue E, Wang GL, Roe R, Semenza GL (1996) Dimerization, DNA binding, and transactivation properties of hypoxia-inducible factor 1. J Biol Chem 271(30):17771–17778. https://doi.org/10.1074/jbc.271.30.17771
6. Semenza GL (2000) HIF-1: mediator of physiological and pathophysiological responses to hypoxia. J Appl Physiol (1985) 88

(4):1474–1480. https://doi.org/10.1152/jappl.2000.88.4.1474

7. Manalo DJ, Rowan A, Lavoie T, Natarajan L, Kelly BD, Ye SQ, Garcia JG, Semenza GL (2005) Transcriptional regulation of vascular endothelial cell responses to hypoxia by HIF-1. Blood 105(2):659–669. https://doi.org/10.1182/blood-2004-07-2958
8. Semenza G (2002) Signal transduction to hypoxia-inducible factor 1. Biochem Pharmacol 64(5–6):993–998. https://doi.org/10.1016/s0006-2952(02)01168-1
9. Jiang BH, Zheng JZ, Leung SW, Roe R, Semenza GL (1997) Transactivation and inhibitory domains of hypoxia-inducible factor 1alpha. Modulation of transcriptional activity by oxygen tension. J Biol Chem 272 (31):19253–19260. https://doi.org/10.1074/jbc.272.31.19253
10. Talks KL, Turley H, Gatter KC, Maxwell PH, Pugh CW, Ratcliffe PJ, Harris AL (2000) The expression and distribution of the hypoxia-inducible factors HIF-1alpha and HIF-2alpha in normal human tissues, cancers, and tumor-associated macrophages. Am J Pathol 157 (2):411–421
11. Zagzag D, Zhong H, Scalzitti JM, Laughner E, Simons JW, Semenza GL (2000) Expression of hypoxia-inducible factor 1alpha in brain tumors: association with angiogenesis, invasion, and progression. Cancer 88 (11):2606–2618
12. Volm M, Koomagi R (2000) Hypoxia-inducible factor (HIF-1) and its relationship to apoptosis and proliferation in lung cancer. Anticancer Res 20(3A):1527–1533
13. Zhang Q, Lenardo MJ, Baltimore D (2017) 30 Years of NF-kappaB: a blossoming of relevance to human pathobiology. Cell 168 (1–2):37–57. https://doi.org/10.1016/j.cell.2016.12.012
14. Cummins EP, Taylor CT (2005) Hypoxia-responsive transcription factors. Pflugers Arch 450(6):363–371. https://doi.org/10.1007/s00424-005-1413-7
15. Perkins ND (2007) Integrating cell-signalling pathways with NF-kappaB and IKK function. Nat Rev Mol Cell Biol 8(1):49–62. https://doi.org/10.1038/nrm2083
16. Wan F, Lenardo MJ (2009) Specification of DNA binding activity of NF-kappaB proteins. Cold Spring Harb Perspect Biol 1(4):a000067. https://doi.org/10.1101/cshperspect.a000067
17. Adhikari A, Xu M, Chen ZJ (2007) Ubiquitin-mediated activation of TAK1 and IKK. Oncogene 26(22):3214–3226. https://doi.org/10.1038/sj.onc.1210413
18. Culver C, Sundqvist A, Mudie S, Melvin A, Xirodimas D, Rocha S (2010) Mechanism of hypoxia-induced NF-kappaB. Mol Cell Biol 30 (20):4901–4921. https://doi.org/10.1128/MCB.00409-10
19. Fitzpatrick SF, Tambuwala MM, Bruning U, Schaible B, Scholz CC, Byrne A, O'Connor A, Gallagher WM, Lenihan CR, Garvey JF, Howell K, Fallon PG, Cummins EP, Taylor CT (2011) An intact canonical NF-kappaB pathway is required for inflammatory gene expression in response to hypoxia. J Immunol 186(2):1091–1096. https://doi.org/10.4049/jimmunol.1002256
20. Melvin A, Mudie S, Rocha S (2011) Further insights into the mechanism of hypoxia-induced NFkappaB. [corrected]. Cell Cycle 10(6):879–882. https://doi.org/10.4161/cc.10.6.15157
21. van Uden P, Kenneth NS, Webster R, Muller HA, Mudie S, Rocha S (2011) Evolutionary conserved regulation of HIF-1beta by NF-kappaB. PLoS Genet 7(1):e1001285. https://doi.org/10.1371/journal.pgen.1001285
22. van Uden P, Kenneth NS, Rocha S (2008) Regulation of hypoxia-inducible factor-1alpha by NF-kappaB. Biochem J 412(3):477–484. https://doi.org/10.1042/BJ20080476
23. Bandarra D, Rocha S (2013) Tale of two transcription factors: NF-κB and HIF crosstalk. OA Mol Cell Biol 1(1):1–7
24. Bandarra D, Biddlestone J, Mudie S, Muller HA, Rocha S (2015) HIF-1alpha restricts NF-kappaB-dependent gene expression to control innate immunity signals. Dis Model Mech 8(2):169–181. https://doi.org/10.1242/dmm.017285
25. D'Ignazio L, Shakir D, Batie M, Muller HA, Rocha S (2020) HIF-1beta positively regulates NF-kappaB activity via direct control of TRAF6. Int J Mol Sci 21(8). https://doi.org/10.3390/ijms21083000

Chapter 16

Methods to Analyze the Roles of TAK1, TRAF6, and NEMO in the Regulation of NF-κB Signaling by RANK Stimulation During Osteoclastogenesis

Gaurav Swarnkar, Manoj Arra, Suresh Adapala, and Yousef Abu-Amer

Abstract

The skeletal system is constantly undergoing turnover in order to create strong, organized structures, requiring the bone breakdown and building properties by osteoclasts and osteoblasts, respectively. However, in pathological disease states, excessive osteoclast activity can cause bone loss leading to increase in morbidity and mortality. Osteoclasts differentiate from macrophages in the presence of various factors. M-CSF is a cytokine that is required to maintain the survival of macrophages. However, RANKL is the critical factor required for differentiation of osteoclasts. RANKL is produced from a variety of different cell types such as osteoblasts and osteocytes. RANKL binds to RANK, its receptor, on the surface of osteoclast precursors, which activates various signaling pathways to drive the transcription and production of genes important for osteoclast formation. The major signaling pathway activated by RANKL–RANK interaction is the NF-κB pathway. The NF-κB pathway is the principle inflammatory response pathway activated by a variety of stimuli such as inflammatory cytokines, genotoxic stress, and other factors. This likely explains the finding that inflammatory diseases often present with some component of increased osteoclast formation and activity, driving bone loss. Determining the signaling mechanisms downstream of RANKL can provide valuable therapeutic targets for the treatment of bone loss in various disease states.

Key words NF-κB, NEMO, TAK1, ISG15, Osteoclast, RANKL

1 Introduction

Various studies have shown that NF-κB signaling mediates RANKL induced osteoclast differentiation. NF-κB signaling plays a crucial role in skeletal development and function under physiological and pathological condition [1–6]. As per current knowledge, binding of RANKL to its receptor RANK recruits various adaptor and effector proteins and forms a signaling complex called IKK-signaling complex. This signaling complex includes multiple proteins like TRAF6, TAK1, TAB2, IKK1, IKK2, and IKKγ/NEMO. This complex in turn phosphorylates IκBα leading to its

Guido Franzoso and Francesca Zazzeroni (eds.), *NF-κB Transcription Factors: Methods and Protocols*, Methods in Molecular Biology, vol. 2366, https://doi.org/10.1007/978-1-0716-1669-7_16,

proteasomal degradation and release of NF-κB subunits (p65/RelA and p50) [7–12]. The active or phosphorylated p65 translocates to nucleus and regulates transcription of target genes [1]. In this chapter, we have discussed the methodology to investigate RANKL induced NF-κB signaling during osteoclast differentiation.

2 Materials

2.1 Isolation of Bone Marrow Macrophage Cells (BMMs) and Osteoclast Culture

1. 4- to 8-week-old mouse.
2. Sterile Dulbecco's PBS (DPBS).
3. Minimal Essential Medium, alpha (α-MEM).
4. Fetal bovine serum (FBS).
5. 70 μM cell strainers (BD).
6. M-CSF (R&D Systems).
7. RANKL (R&D Systems).
8. 96-well tissue culture plate (MidSci).
9. Cell counter.
10. Leukocyte acid phosphatase kit/TRAP staining kit (Sigma-Aldrich).
11. Inverted microscope for imaging.

2.2 Real-Time PCR Analyzing Osteoclast Differentiation

1. 4- to 8-week-old mouse.
2. Sterile Dulbecco's PBS (DPBS).
3. Minimal Essential Medium, alpha (α-MEM).
4. Fetal bovine serum (FBS).
5. 70 μM cell strainers (BD).
6. M-CSF (R&D Systems).
7. RANKL (R&D Systems).
8. 12-well tissue culture plate (MidSci).
9. Cell counter.
10. mRNA isolation kit (Purelink mRNA mini, Invitrogen).
11. cDNA synthesis kit (High capacity reverse transcription kit, ABI, ThermoFisher).

12. iTaq universal SYBR green kit (Biorad) and real-time PCR machine.
13. Real-time PCR primers for osteoclast differentiation associated genes. *(Acp5)* TRAP-F: *CGACCATTGTTAGCCACATACG*, TRAP-R: *CACATAGCCCACACCGTTCTC*, *Ctsk (CathepsinK)*-F: *ATGTGGGTGTTCAAGTTTCTGC*, *Ctsk*-R: *CCACAAGATTCTGGGGACTC*, *Nfatc1*-F: *CCGGGACGCCCATGCAATCTGTTAGT*, *Nfatc1*-R: *GCGGGTGCCCTGAGAAAGCTACTCTC*, *DC-Stamp F: GCACCTACTTCTGTCAGATTGATG*, *DC-Stamp R: GCGGAGTGGCAACAGACT*.

2.3 Western Blot for NF-κB Activation and Osteoclast Differentiation

1. 4- to 8-week-old mouse.
2. Sterile Dulbecco's PBS (DPBS).
3. Minimal Essential Medium, alpha (α-MEM).
4. Fetal bovine serum (FBS).
5. 70 μM cell strainers (BD).
6. Cell scraper (MidSci).
7. M-CSF (R&D Systems).
8. RANKL (R&D Systems).
9. 95 mm tissue culture plate/p100 plate (MidSci).
10. Cell lysis buffer: 20 mM Tris–HCl, pH 7.5, 150 mM NaCl, 1 mM Na2EDTA, 1 mM EGTA, 1% Triton, 2.5 mM sodium pyrophosphate, 1 mM beta-glycerophosphate, 1 mM Na3VO4, 1 μg/ml leupeptin (Cell signaling technology).
11. Halt Protease and Phosphatase Inhibitor Cocktail (ThermoFisher).
12. BCA assay kit (Pierce, Rapid Gold BCA protein assay kit, ThermoFisher).
13. Primary antibodies: pIKK2, IKK2, p-P65, p65, p-IκBα, IκBα, Ub-K63, TAK1 (Cell signaling Technology), Actin (Sigma).

2.4 NF-κB Luciferase Reporter Assay

1. 4- to 8-week-old NF-κB (RelA)-luciferase reporter mice (Taconic) or NF-κB-GFP-luciferase (The Jackson Laboratory).
2. Sterile Dulbecco's PBS (DPBS).
3. Minimal Essential Medium, alpha (α-MEM).
4. Fetal bovine serum (FBS).
5. 70 μM cell strainers (BD).
6. M-CSF and RANKL (R&D Systems).
7. Illumination™ Firefly & Renilla Luciferase Enhanced Assay Kit (Gold bio).

8. BCA assay kit (Pierce, Rapid Gold BCA protein assay kit, ThermoFisher).
9. Colorimetric plate reader.

2.5 Co-immuno-precipitation to Analyze the Interaction of Different Proteins Involved in NF-κB Signaling

1. BMMs from 4- to 8-week-old mice.
2. Sterile Dulbecco's PBS (DPBS).
3. Minimal Essential Medium, alpha (α-MEM).
4. Fetal bovine serum (FBS).
5. 70 μM cell strainers (BD).
6. Cell scraper (MidSci).
7. M-CSF and RANKL (R&D Systems).
8. Lysis buffer: 10 mM Tris–HCL, pH 8.0, 1 mM EDTA, 125 mM NaCl, and 1% NP-40.
9. Washing buffer: 10 mM Tris–HCL, pH 8.0, 1 mM EDTA, 20 mM NaCl, and 1% NP-40.
10. Protein-G beads (Genescript).

2.6 Micro-CT Analysis

1. Long bones isolated from 4- to 8-week-old mice. Animals with different ages can be used based on the experimental design.
2. Sterile Dulbecco's PBS (DPBS).
3. 10% neutral Formalin.
4. 70% Ethanol.
5. Micro-CT instrument such as μCT 40 (Scanco Medical).

2.7 Histo-morphometric Analysis

1. Long bones isolated from 4- to 8-week-old mice. Animals with different ages can be used based on the experimental design.
2. Sterile Dulbecco's PBS (DPBS).
3. 10% neutral Formalin.
4. 70% Ethanol.
5. Graded alcohol solutions and xylene to dehydrate decalcified bones.
6. Paraffin for embedding tissue.
7. Hematoxylin and eosin stain.
8. Tartrate-resistant acid phosphatase (TRAP).

2.8 Generation of Retroviral Particles to Study Role for TAK1 and NEMO Modulation

1. BMMs isolated from 4- to 8-week-old mice.
2. pMX-retroviral vectors (cell biolabs).
3. Plat-E cells (Cell biolabs, CA, USA).
4. Transfection reagents such as x-tremegene 9 (Roche).

5. 8 μg/ml Polybrene (Santacruz, USA).
6. 0.45 μm syringe filters with syringe.

2.9 Immunofluorescence for ISGylated NEMO

1. 1× Dulbecco Phosphate Buffered Saline (DPBS).
2. 4% Formaldehyde (Polysciences Inc).
3. Blocking Buffer: 1× DPBS/5% normal serum/0.3% Triton X-100.
4. Antibody Dilution Buffer: 1× DPBS/1% BSA/0.3% Triton™ X-100.
5. Primary antibodies: NEMO and ISG15, 1:200 (santacruz).
6. Fluorochrome-conjugated secondary antibodies (Alex fluor Rabbit-488, Mouse-568, Invitrogen).
7. Biotinylated-anti-ISG15 (using biotinylation kit Cat# 90407, ThermoFisher).
8. Prolong Gold Anti-fade Reagent with DAPI (Cell signaling technology).

3 Methods

3.1 RANKL Induced NF-κB Activation During Osteoclast Differentiation

1. Day 0: Isolate long bones (femur and tibia) from 4- to 8-week-old mouse and place them in petri-plate with sterile DPBS on ice. Carefully, clean the bone and remove muscles and other tissues.

3.1.1 Osteoclast Culture

Isolation of Bone Marrow Macrophage Cells (BMMs)

1. Poke 3–4 holes in a 0.6 ml tube using a 20-G needle and place it inside a 1.5 ml tube and add 500 μl of α-MEM without FBS. Cut the ends of the femur and tibia carefully and place them vertically inside the 0.5 ml tube.
2. Centrifuge at 9000 × *g* at 4 °C for 3 min. While the cells are being centrifuged, place a cell strainer on 50 ml tube.
3. The bone marrow will pass from the 0.6 ml tubes and pellet down in the 1.5 ml tubes after centrifugation. Discard the 0.6 ml tubes with bones. Suspend the BMMs pellet in α-MEM and filter through a 70 μm cell strainer to get rid of any debris.

Differentiation of BMMs to Osteoclast (TRAP Staining)

1. Plate the cells in 150 mm petri-plate in α-MEM with 10% FBS and 50 ng/ml M-CSF (*see* **Note 1**).
2. Day 1: After 24 h, collect the non-adherent cells in sterile 50 ml tube. Rinse the plate twice with sterile DPBS to collect remaining non-adherent cells. Centrifuge at 300 × *g* for 5 min. Suspend the cells in 5 ml of α-MEM with 10% FBS. These cells can be differentiated to osteoclasts (OC) in the presence of 50 ng/ml M-CSF and 50 ng/ml RANKL.

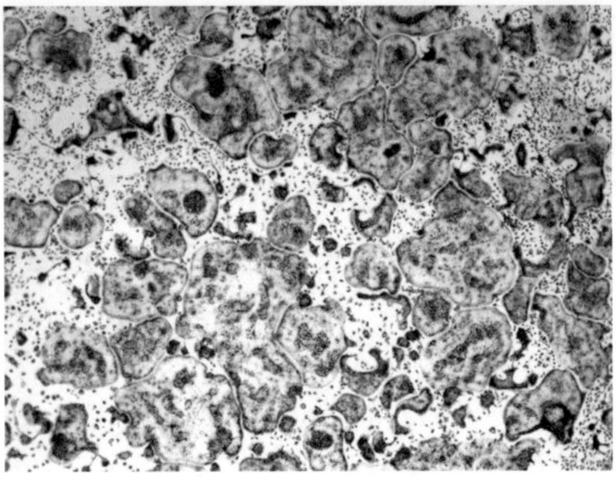

Fig. 1 Mature multinucleatedTRAP positive osteoclast from bone marrow cells

3. Count and plate the cells in a 96-well plate at different densities (10,000, 20,000, 40,000 cells/well, etc.). Add 50 ng/ml M-CSF and 50 ng/ml RANKL to wells containing the cells. Replenish M-CSF and RANKL every alternate day.
4. On day 4, mature multinucleated OC will be readily visible in the plate. Fix and stain the cells for Tartrate-Resistant Acid Phosphatase (TRAP) using leukocytes acid phosphatase kit according to the manufacturer's protocol (Fig. 1).

Real-Time PCR for Analyzing Osteoclast Differentiation Markers

Gene expression of specific osteoclast markers such as TRAP, DC-STAMP, CTSK, and NFATC1 (primers sequence in Subheading 2.2) can also be useful for measuring RANKL induced osteoclast formation. Gene expression is effective for measuring osteoclast formation and the changes are often seen earlier than changes visible by TRAP staining [9, 13–16]. One of the drawbacks of gene expression changes is that they sometime do not directly translate to protein levels. In addition, qPCR does not provide information about the upstream signaling mechanism.

1. For real-time PCR, on day 1, collect the cells in a 50 ml tube and centrifuge at 300 × *g* for 5 min. Suspend the cells in 5 ml of α-MEM with 10% FBS.
2. Count and plate the cells in a 12-well plate at different densities (50,000, 100,000, 200,000 cells/well, etc.). Add 50 ng/ ml M-CSF to all the wells every alternate day. Treat the cells with 50 ng/ml RANKL for 0, 2, and 4 days.
3. At the end of experiment, isolate the RNA from different treatment conditions using RNA isolation kit.
4. Using equal amount of RNA (500–1000 ng), synthesize cDNA as per the kit recommendation.

5. Dilute the cDNA prepared using 500–1000 ng RNA to 10 times for real-time PCR reaction.
6. Perform real-time PCR to analyze the expression of osteoclast marker genes such as TRAP, Cathepsin-K (CTSK), DC-STAMP, and NFATc1, using ACTIN as housekeeping gene.

Western Blotting for NF-κB Activation and Osteoclastogenesis

Western blotting is useful for measuring changes in protein levels and determining NF-κB signaling during RANKL induced osteoclast differentiation. NF-κB pathway activation can be measured at different steps. The principle methods for measuring NF-κB activation include measuring phosphorylation of IKK2, phosphorylation and degradation of IkB-a, and phosphorylation of p65. In addition, expression of osteoclast-specific proteins NFATc1, CTSK, and TRAP can be measured by western blotting. [9, 13–16]

1. Isolate the BMMs as described earlier under osteoclast culture (Subheading 3.1). Plate BMMs in 95 mm tissue culture plate (p100) at a density of 5×10^5 cells/p100mm plate. Add 50 ng/ml M-CSF to the plate.
2. Once the cells reach 80–90% confluence, remove the media and wash the cells twice with 10 ml of sterile DPBS. Add 10 ml serum free α-MEM to the p100 mm plates with cells and starve them for 2–3 h.
3. Post starvation treat the cells with 50 ng/ml RANKL for 0, 15, 30, and 60 min.
4. In addition, to analyze osteoclast marker, treat the BMMs in 95 mm tissue culture dish for 0, 1, 2, and 4 days.
5. At the end of treatment, transfer the plates to ice. Remove the media and wash the cells twice with ice cold DBPS. Aspirate all the DBPS from the plates and collect the cell lysates in 100–200 μl of 1× cell lysis buffer containing 1× protease and phosphatase inhibitor.
6. Determine protein concertation by BCA method.
7. Using equal amount of protein, perform sodium dodecyl sulfate–polyacrylamide gel electrophoresis (SDS-PAGE) and western blotting for p-IKK2 (1:1000), total-IKK2 (1:1000), TRAF6 (1:1000), p-p65(1:1000), total-p65 (1:1000), p-IκBα (1:10,000) and total IκBα (1:10,000) to measure the RANKL induced NF-κB activation in BMMs. Use Actin as loading control (1:10,000).
8. Further using the lysates from BMMs treated with RANKL for 0, 2, and 4 days, perform western blotting to measure the expression of NFATc1 (1:500), TRAP (1:500), and Cathepsin-K (1:500).

NF-κB Luciferase Reporter Assay

Another method to analyze NF-κB activation in response to RANKL or other stimulation is to measure NF-κB (RelA)-luciferase reporter activity. BMMs can be isolated from NF-κB (RelA)-luciferase reporter mice (Taconic, USA) or NF-κB-GFP-luciferase or NGL mice (The Jackson Laboratory, USA). The mice from Taconic carries transgenes with 6 U of NF-κB responsive elements and a luciferase cDNA. The NGL mice from Jackson lab expresses a NF-κB driven GFP-luciferase fusion protein. These mice model can be used to study the transcriptional regulation of NF-κB gene in response to various stimuli both in vitro and in vivo (RANKL in case of osteoclast differentiation) [13, 15].

1. Isolate the BMMs as described earlier under osteoclast culture from NF-κB reporter mice.
2. Culture the BMMs for 4 days in the same petri-plate. Keep adding 50 ng/ml M-CSF every alternate day.
3. On day 4, collect the adherent cells using TrypLE. Count and plate in 96-well plate at a density of 10,000 and 20,000 cells/well. Add 50 ng/ml M-CSF.
4. On day 5, treat the cells with 50 ng/ml RANKL for 0, 6, 12, and 24 h.
5. At the end of the treatment, lyse the cell using lysis buffer provided with luciferase assay kit. Measure the luciferase as per the kit instruction.
6. Use the same lysates to measure protein concentration using BCA assay. Normalize the luciferase activity with total protein content to measure the RANKL induced NF-κB activation.

3.2 Analyzing the Role of TAK1 During Osteoclastogenesis

TAK1 or MAP kinase TGFβ-activated kinase is an upstream activator of NF-κB and MAPK pathways and therefore is a key target for regulating osteoclast differentiation and function [17–20]. TAK1 deletion or inhibition results in a significant decrease in differentiation of osteoclast from BMMs [8, 9].

3.2.1 In Vivo Effect of TAK1 Deletion on Osteoclastogenesis and Bone Parameters

TAK1 floxed *(TAK1f/f)* mice were crossed with LysozymeM-cre (LysMcre) mice to generate heterozygous mice. These mice were further inter-bred to generate TAK1-null homozygous mice (LysMcre-TAK1-f/f or TAK1-cKO mice) which has complete deletion of TAK1 in the myeloid lineage or BMMs. Deletion of TAK1 leads to inhibition of NF-κB activation.

Micro-CT Analysis for Bone Parameters (*see* **Note 2**)

1. Bone density parameters are measured using micro-CT. To this end, isolate the long bones (femur and tibia) from 4- to 8-week-old WT and TAK1-cKO animals. Fix the bones in 10% neutral formalin for 16–24 h. After fixation wash the bone twice with DPBS and transfer them to 70% ethanol. The bones can be stored in 70% ethanol for extensive period of time.

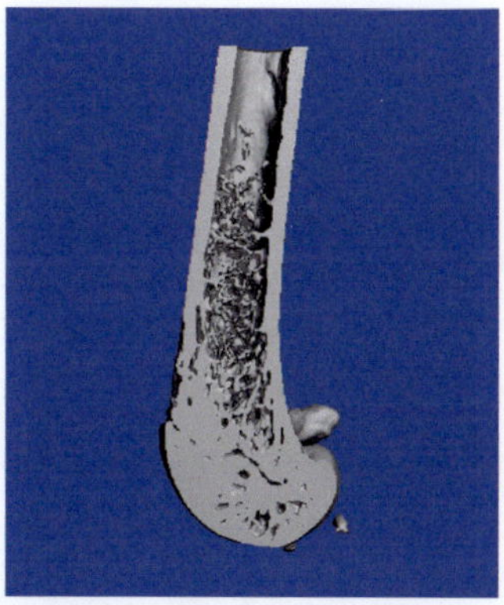

Fig. 2 Representative images for micro-CT analysis

2. Scan the mouse bones using micro-CT (μCT 40). Scan the distal metaphysis regions of femur to measure trabecular bone morphological parameters with the following settings: 55 kVp, 145 mA, standard resolution, 16.4-mm diameter, 16-mm voxel size, 300-ms integration time. Identify the area of interest just proximal to the growth plate and scan a height of 150 slices. Following the manufacturer's 3D analysis tools, measure bone mineral density (BMD), bone volume over total volume (BV/TV), trabecular thickness (Tb.Th), trabecular separation (Tb.Sp), and trabecular number (Tb.N). These analyses will show the change in the bone parameter for a TAK1-cKO mice compared with the WT mice (Fig. 2).

Histo-morphometric Analysis

1. For histo-morphometric analysis of bone sections, decalcify the fixed bones using 10% EDTA, pH 7.0, for 7–14 days with gentle rocking and daily replacement of solution. Dehydrate the decalcified bones in graded alcohol, clear through xylene, and embed in paraffin. Cut 5 μM-thick longitudinal sections of paraffin blocks and then stain them with hematoxylin and eosin or histochemical with tartrate-resistant acid phosphatase (TRAP) to determine osteoclasts numbers.

3.2.2 Ex Vivo Characterization of TAK1-Null BMMs for Osteoclastogenesis and NF-κB Activation

Retroviral Vector Mediated Transduction

1. PLAT-E cells are used as the packaging cells to generate retroviral particles. Plat-E cells are derived from HEK-293 cells, which are stably expressing retroviral structural proteins gag-pol and env for transient production of retrovirus. On Day 0, Plate 4 × 10^6 Plat-E cells per 96 mm tissue culture plate in DMEM with 10% FBS. Make three plates each for pMx-GFP, pMX-TAK1-WT, and pMX-TAK1-K63W (Inactive form of TAK1).
2. On day 1, transfect the Plat-E cells with pMx-GFP, pMX-TAK1-WT, and pMX-TAK1-K63W, using x-tremegene 9 or any other commercially available transfection reagent.

Specifically, in a 1.5 ml tube add 485 μl serum free DMEM, 15 μl of transfection reagent, and 5 μg of DNA (pMX-GFP, pMX-IKK2-WT, and pMx-IKK2-SSEE). Wait for 20 min. After 15–20 min, mix the transfection complex gently with a pipette and add it to the PLAT-E cells in a drop-wise manner. Make sure to cover the whole plate. Swirl the plate gently to spread the DNA evenly. Transfer the plate back to the CO_2 incubator.

3. Day 2, estimate the transfection efficiency of pMX-GFP under a UV-microscope. Normally the transfection efficiency is more than 90%. Remove the DMEM and add α-MEM with 10% FBS.
4. Day 3, collect the α-MEM having retroviral particles and filter it using a 0.45 μM syringe filter (first collection). Add fresh α-MEM with 10% FBS and collect it again after 24 h (Day 4, second collection).
5. The virus containing media can be stored on ice at 4 °C. Repeated freeze-thawing the virus reduces its transduction efficiency.
6. On day 3 (same day as first virus collection), sacrifice a WT and TAK1-cKO mice and isolate the BMMs and plate them in petri plates with 50 ng/ml M-CSF as described earlier in Subheading 3.1, Osteoclast culture) (*see* **Note 3**).
7. Day 4, collect all the non-adherent cells from the petri-plate as described earlier. Count the cells and plate them at a density of 5×10^6/p100 plate in retroviral particles for GFP, TAK1-WT, and TAK1-K63W containing α-MEM (**step 4**). Add polybrene (8 μg/ml) and 50 ng/ml M-CSF to the plate.
8. On day 5, collect all the floating and adherent cells using TrypLE or Accutase (Gibco). Centrifuge at 400 × *g* for 5 min and suspend the cells pellet in α-MEM with 10% FBS.

Osteoclast Differentiation, mRNA Expression, and Western Blot Analysis

1. At this stage, plate the cells for osteoclastogenesis, mRNA isolation, and protein lysates collection following the protocol described earlier in this chapter (Subheading 3).
2. After 4–5 days, fix and stain the OC for TRAP (refer Subheading 3.1, Osteoclast Culture). We will see a very significant decrease in the number of mature OC in TAK1-null BMMs. We will also observe that TAK1-WT but not TAK1-K63W overexpression rescues the osteoclast differentiation in TAK1-cKO BMMs.
3. Supporting the TRAP staining data, real-time PCR analysis will show that TAK1-WT expression corrected the mRNA

expression of osteoclast markers TRAP, CTSK, and NTAc1 in TAK1-cKO cells. Inactive form of TAK1 (TAK1-K63W) overexpression in TAK1-cKO cells will fail to rescue the mRNA expression.

4. The western blot analysis for p-IKK2, IKK2, p-p65, p65, p-IκBα, and IκBα also will further show decreases in NF-κB activation in the TAK1-null BMMs, which can be rescued using TAK1-WT exogenous expression but not TAK-K63W expression.

3.3 Analyzing Interaction of TRAF6 with NF-κB Components During Osteoclast Differentiation

Tumor necrosis factor receptor associated factors (TRAFs) are a family of proteins, which serves as adaptor proteins and mediates signal transduction from members of TNF receptor superfamily. TRAF6 is ubiquitin ligase, which in response to binding of RANKL to RANK receptor, catalyzes auto-K63 ubiquitination and in turn recruits TAK1/TAB2 complex. This further leads to formation of IKK2 complex and NF-κB activation [21, 22].

3.3.1 Analyzing TRAF6-K63-Ubiquitnatoin

1. Day 0, isolate the BMMs as described earlier under osteoclast culture from 4- to 8-week-old mice. Add 50 ng/ml M-CSF to the plate every alternate day.
2. Day 3, plate BMMs in 95 mm tissue culture plate (p100) at a density of 5×10^5 cells/p100mm plate. Once the cells reach around 90% confluence, treat them with RANKL for 0, 5, 15, 30, and 60 min.
3. At the end of treatment, wash the cells twice with ice cold DPBS and lyse the cells in 1 ml of lysis buffer containing protease and phosphatase inhibitor.
4. Save 200 μl of the total cell lysate for loading control.
5. Incubate equal amount of protein from the remaining lysates with 1 μg of anti-TRAF6 antibody for 2 h.
6. After 2 h, add protein-G beads (50ul) and incubate overnight at 4 °C on a shaker or notator.
7. Next day, wash the beads thrice with washing buffer containing protease and phosphatase inhibitor.
8. After washing, suspend the beads in protein loading buffer and process for western blotting for K63-Ub, TAK1, and other proteins.

3.4 ISGylation of NEMO to Modulate NF-κB Activity

NEMO is a key component of IKK-signaling complex. NEMO lacks any enzymatic activity but serves as a scaffold protein, which based upon the upstream signals recruits different substrate complex and thus regulates downstream signaling [1, 23]. NEMO undergoes different post-translational modification (PTM) such as ubiquitination, SUMOylation, and ISGylation. These PTMs enable nemo to serve as a signal-specific docking site to regulate distinctive downstream signaling [13, 24–29]. ISG15 (IFN-γ-stimulated gene 15) is a small ubiquitin-like intracellular protein modifier. ISGylation of NEMO has been shown to regulate RANKL induced NF-κB signaling during osteoclastogenesis [13].

3.4.1 ISGylation of NEMO (Fig. 3)

1. To study the ISGylation of NEMO during RANKL induced osteoclast differentiation, isolate the BMMs as described earlier under osteoclast culture from 4- to 8-week-old mice. Add 50 ng/ml M-CSF to the plate.
2. Next day, plate BMMs in 95 mm tissue culture plate (p100) at a density of 5×10^5 cells/p100mm plate. Add 50 ng/ml M-CSF and 50 ng/ml RANKL to the plate. Culture them for 2 days. These cells after 2 days are pre-osteoclast (pOC).
3. Starve the cell for 3 h. Post starvation, treat the pOC with RANKL for 0, 6, and 24 h.
4. At the end of treatment, process the cells for co-immunoprecipitation as described in Subheading 3.3.1.
5. Save 200 μl of the total cell lysate for loading control. Divide the remaining lysates from each condition into two fractions and incubate 1 μg of biotinylated-anti-ISG15 and anti-NEMO antibody, respectively, for 2 h (using biotinylation kit).

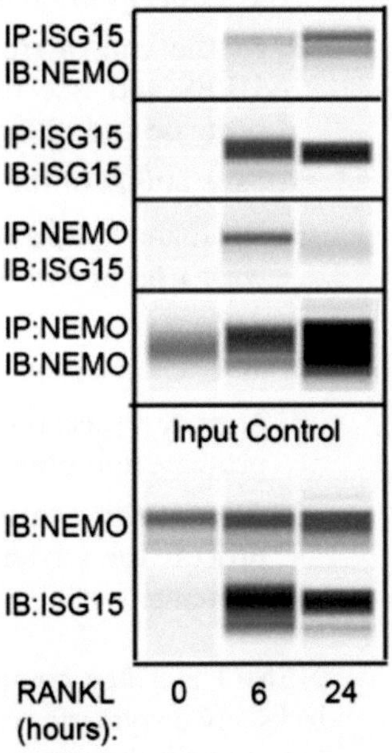

Fig. 3 RANKL induces ISGylation of NEMO. pOC cells were starved and treated with 50 ng/ml of RANKL for indicated time points. Immuno-blot from total cell lysates and ISG15- and NEMO-immuno-precipitates shows that NEMO interacts with ISG15

6. Perform immunoprecipitation using streptavidin magnetic beads according to manufacturer's protocol.
7. Next day, wash the beads thrice with washing buffer as described earlier and process for western blotting for ISG15 and NEMO.
8. Western blot analysis using total cell lysates fraction will show that RANKL increases total ISGylation as well as induces ISGylation of NEMO.

3.4.2 Immuno-Colocalization of ISG15 and NEMO in Pre-osteoclast (Fig. 4)

1. Isolate the BMMs as described earlier under osteoclast culture.
2. Culture the BMMs for 4 days in the same petri-plate. Keep adding 50 ng/ml M-CSF every alternate day.
3. On day 4, collect the adherent cells using TrypLE. Count and plate cells at a density of 10,000 cells on a glass coverslip in 24-well plate. Add 50 ng/ml M-CSF.
4. On day 5, treat the cells with 50 ng/ml RANKL for 24 h.
5. On day 6, fix the cells using 4% formaldehyde for 15 min. Wash the cells thrice with DPBS and proceed for immunofluorescence staining.
6. Block the slide using blocking buffer for 1 h followed by incubation with NEMO and ISG15 antibodies at 4 ° C overnight.
7. Next day, wash the coverslip with cells thrice with DPBS.
8. Incubate with secondary antibody for 1–2 h at room temperature. Wash again thrice with DPBS.

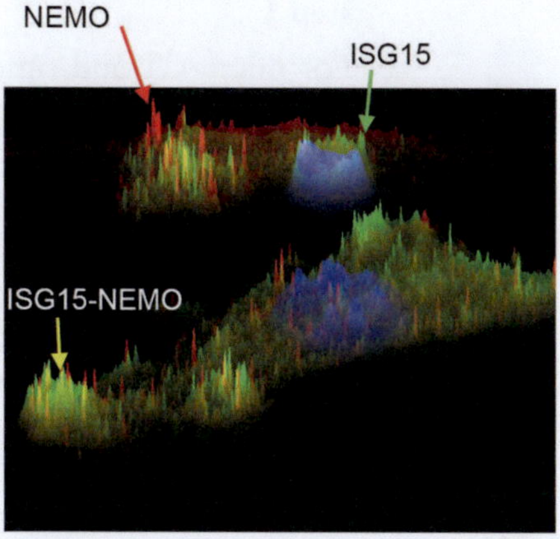

Fig. 4 Intensity plots of a dual immunofluorescence staining for NEMO (Red) and ISG15 (Green) in pre-osteoclasts at ×100 magnification demonstrating a colocalization of NEMO and ISG15 (Yellow) in NEMO WT. Nucleus is shown in blue color. Images were taken using a Nikon Ti microscope and the intensity plots were processed and analyzed by using a MetaMorph 7.0 software

9. Apply anti-fade solution with DAPI and transfer the coverslips with cells on a histological slide for imaging.

3.4.3 Generation of Retroviral Particles and Osteoclastogenesis

1. Generate pMX-GFP, pMX-NEMO, pMX-ISG15, pMX-NEMO-ISG15 retroviral vectors.
2. Transfect Plat-E cells with pMX-GFP, pMX-NEMO-, pMX-ISG15, and pMX-NEMO-ISG15, to generate retroviral particles as described earlier (sect. 3.2.2).
3. Transduce the wild-type BMMs with above viruses. To show the effect of interaction of ISG15 with NEMO, transduce the BMMs with NEMO with ISG15 (co-transduction) as well as with fusion constructs (pMX-NEMO-ISG15).
4. One day after transduction, collect adherent and non-adherent cells and plate the cells at different densities in 96-well plate for osteoclastogenesis in α-MEM with 10% FBS, 50 ng/ml M-CSF, and 50 ng/ml RANKL, as described earlier (in Subheading 3.1).
5. Fix and stain the cells for TRAP.
6. The TRAP staining will show that NEMO and ISG15 co-transduction as well as NEMO-ISG15 fusion results in decreased osteoclastogenesis.

4 Notes

1. Do not use tissue culture plates but use sterile bacterial culture petri-plate to plate the cells on day 0 of isolation (*see* Subheading "Differentiation of BMMs to Osteoclast (TRAP Staining)", **step 1**).
2. For micro-CT and other in vivo or in vitro analysis, animals with different age (older or younger than 4–8 weeks) can be used based on the experimental design.
3. The BMMs from TAK1-cKO mice do not survive for long due to absence of NF-κB and MAPK activity. To rescue the cell survival, the BMMs were cultured in the presence of TNF-α neutralizing antibody (0.2 μg/ml) throughout the experiment (*see* Subheading "Retroviral Vector Mediated Transduction", **step 6**).

Acknowledgments

This work was supported by R01 AR049192, AR054329 (NIH/NIAMS) and 85160 from the Shriners Hospital for Children to Y.A.

References

1. Abu-Amer Y (2013) NF-kappaB signaling and bone resorption. Osteoporos Int 24:2377–2386
2. Shiotani A, Takami M, Itoh K, Shibasaki Y, Sasaki T (2002) Regulation of osteoclast differentiation and function by receptor activator of NFkB ligand and osteoprotegerin. Anat Rec 268:137–146
3. Teitelbaum SL (2000) Bone resorption by osteoclasts. Science 289:1504–1508
4. Boyce BF, Yao Z, Xing L (2010) Functions of nuclear factor kappaB in bone. Ann N Y Acad Sci 1192:367–375
5. Franzoso G, Carlson L, Xing L, Poljak L, Shores EW, Brown KD, Leonardi A, Tran T, Boyce BF, Siebenlist U (1997) Requirement for NF-kappaB in osteoclast and B-cell development. Genes Dev 11:3482–3496
6. Abu-Amer Y, Darwech I, Otero J (2008) Role of the NF-kappaB axis in immune modulation of osteoclasts and bone loss. Autoimmunity 41:204–211
7. Hayden MS, Ghosh S (2004) Signaling to NF-kappaB. Genes Dev 18:2195–2224
8. Lamothe B, Lai Y, Xie M, Schneider MD, Darnay BG (2013) TAK1 is essential for osteoclast differentiation and is an important modulator of cell death by apoptosis and necroptosis. Mol Cell Biol 33:582–595
9. Swarnkar G, Karuppaiah K, Mbalaviele G, Chen TH, Abu-Amer Y (2015) Osteopetrosis in TAK1-deficient mice owing to defective NF-kappaB and NOTCH signaling. Proc Natl Acad Sci U S A 112:154–159
10. Karin M, Yamamoto Y, Wang QM (2004) The IKK NF-kappa B system: a treasure trove for drug development. Nat Rev Drug Discov 3:17–26
11. Ting AY, Endy D (2002) Signal transduction. Decoding NF-kappaB signaling. Science 298:1189–1190
12. Wong BR, Josien R, Lee SY, Vologodskaia M, Steinman RM, Choi Y (1998) The TRAF family of signal transducers mediates NF-kappaB activation by the TRANCE receptor. J Biol Chem 273:28355–28359
13. Adapala NS, Swarnkar G, Arra M, Shen J, Mbalaviele G, Ke K, Abu-Amer Y (2020) Inflammatory osteolysis is regulated by site-specific ISGylation of the scaffold protein NEMO. Elife 9:e56095
14. Chen TH, Swarnkar G, Mbalaviele G, Abu-Amer Y (2015) Myeloid lineage skewing due to exacerbated NF-kappaB signaling facilitates osteopenia in Scurfy mice. Cell Death Dis 6:e1723
15. Swarnkar G, Chen TH, Arra M, Nasir AM, Mbalaviele G, Abu-Amer Y (2017) NUMBL interacts with TAK1, TRAF6 and NEMO to negatively regulate NF-kappaB signaling during osteoclastogenesis. Sci Rep 7:12600
16. Swarnkar G, Shim K, Nasir AM, Seehra K, Chen HP, Mbalaviele G, Abu-Amer Y (2016) Myeloid deletion of nemo causes osteopetrosis in mice owing to upregulation of transcriptional repressors. Sci Rep 6:29896
17. Landstrom M (2010) The TAK1-TRAF6 signalling pathway. Int J Biochem Cell Biol 42:585–589
18. Lee SW, Han SI, Kim HH, Lee ZH (2002) TAK1-dependent activation of AP-1 and c-Jun N-terminal kinase by receptor activator of NF-kappaB. J Biochem Mol Biol 35:371–376
19. Mizukami J, Takaesu G, Akatsuka H, Sakurai H, Ninomiya-Tsuji J, Matsumoto K, Sakurai N (2002) Receptor activator of NF-kappaB ligand (RANKL) activates TAK1 mitogen-activated protein kinase kinase kinase through a signaling complex containing RANK, TAB2, and TRAF6. Mol Cell Biol 22:992–1000
20. Takaesu G, Surabhi RM, Park KJ, Ninomiya-Tsuji J, Matsumoto K, Gaynor RB (2003) TAK1 is critical for IkappaB kinase-mediated activation of the NF-kappaB pathway. J Mol Biol 326:105–115
21. Darnay BG, Besse A, Poblenz AT, Lamothe B, Jacoby JJ (2007) TRAFs in RANK signaling. Adv Exp Med Biol 597:152–159
22. Lamothe B, Webster WK, Gopinathan A, Besse A, Campos AD, Darnay BG (2007) TRAF6 ubiquitin ligase is essential for RANKL signaling and osteoclast differentiation. Biochem Biophys Res Commun 359:1044–1049
23. Yamamoto Y, Kim DW, Kwak YT, Prajapati S, Verma U, Gaynor RB (2001) IKKgamma / NEMO facilitates the recruitment of the IkappaB proteins into the IkappaB kinase complex. J Biol Chem 276:36327–36336
24. Cordier F, Grubisha O, Traincard F, Veron M, Delepierre M, Agou F (2009) The zinc finger of NEMO is a functional ubiquitin-binding domain. J Biol Chem 284:2902–2907
25. Hay RT (2004) Modifying NEMO. Nat Cell Biol 6:89–91
26. May MJ, Marienfeld RB, Ghosh S (2002) Characterization of the Ikappa B-kinase

NEMO binding domain. J Biol Chem 277:45992–46000

27. Rushe M, Silvian L, Bixler S, Chen LL, Cheung A, Bowes S, Cuervo H, Berkowitz S, Zheng T, Guckian K, Pellegrini M, Lugovskoy A (2008) Structure of a NEMO/IKK-associating domain reveals architecture of the interaction site. Structure 16:798–808

28. Schrofelbauer B, Polley S, Behar M, Ghosh G, Hoffmann A (2012) NEMO ensures signaling specificity of the pleiotropic IKKbeta by directing its kinase activity toward IkappaBalpha. Mol Cell 47:111–121

29. Sebban H, Yamaoka S, Courtois G (2006) Posttranslational modifications of NEMO and its partners in NF-kappaB signaling. Trends Cell Biol 16:569–577

Chapter 17

NF-κB Signaling in Ex-Vivo Mouse Intestinal Organoids

Kateryna Shostak, Caroline Wathieu, Sylvia Tielens, and Alain Chariot

Abstract

We describe here a protocol to assess NF-κB activation in ex-vivo organoids generated from mouse intestinal crypts. These structures are maintained in culture as crypt-villus forming organoids. These ex-vivo organoids maintain both self-renewal and multilineage differentiation overtime. We also describe the generation of ex-vivo organoids from *Apc*-mutated mouse intestinal crypts. Both wild-type and *Apc*-mutated organoids respond very well to NF-κB-activating signals such as TNFα but not to LPS. The kinetic of NF-κB activation in response to these signals in ex-vivo intestinal organoids is very similar to what we see in 2D cell lines. This protocol provides investigators a powerful tool to assess NF-κB activation in both healthy and transformed intestinal epitheliums maintained in culture as 3D structures.

Key words Ex-vivo organoids, Intestinal crypts, NF-κB, Cell culture

1 Introduction

The architecture of the intestinal epithelium includes multiple crypt-villus units covered by a single layer of post-mitotic epithelial cells [1]. This layer of cells is surrounded at its base by multiple epithelial invaginations, referred to as crypts of Lieberkühn. Each crypt-villus includes six differentiated cell types. These are lysozyme and defensin-producing Paneth cells located at the bottom of each intestinal crypt, absorptive enterocytes, Goblet and enteroendocrine cells which produce mucus or hormones, microfold (M) cells involved in mucosal immunity as well as rare Tuft cells (also referred to as Brush cells). The intestine is characterized by a very high regenerative potential. Indeed, a new intestine is generated in less than a week in rodents [2]. This intense self-renewal of the intestinal epithelium is driven by crypt base columnar (CBC) stem cells located at the bottom of each intestinal crypt. These cycling CBC cells express the Wnt target gene Leucine-rich repeat containing G-protein coupled receptor 5 (Lgr5) and undergo cell division every day to generate a pool of rapidly dividing transit-amplifying (TA) cells [3]. At one point, these cells stop proliferating

Guido Franzoso and Francesca Zazzeroni (eds.), *NF-κB Transcription Factors: Methods and Protocols*, Methods in Molecular Biology, vol. 2366, https://doi.org/10.1007/978-1-0716-1669-7_17,

to undergo cell differentiation in order to generate intestinal subtypes while progressing on the villi. Ex-vivo organoids can be generated from whole intestinal crypts or from single $Lgr5^+$ stem cells and will ultimately generate "mini-guts" in which all functional intestinal cell types can be found. These ex-vivo organoids can be passaged weekly and can be maintained in culture for many years as no genetic alterations would occur. Therefore, these ex-vivo organoids recapitulate stem cell self-renewal, maintenance as well as the hierarchical crypt-villus architecture [4]. Gene functions and cell fate determination are preserved, which makes these organoids a suitable experimental model for signal transduction studies.

The NF-κB family of transcription factors are critical in cell survival, proliferation, and inflammation. NF-κB activation, which occurs upon stimulation with a variety of signals including pro-inflammatory cytokines (TNFα, IL-1β, ...) and bacterial or viral molecular pathogens (LPS, dsRNA, ...), relies on the signal-induced and proteasome-dependent degradation of inhibitory proteins such as IκBα [5]. NF-κB heterodimers can then translocate into the nucleus in order to drive gene transcription of hundreds of target genes, including inhibitors of NF-κB such as A20 [6]. Assessing NF-κB activation by TNFα can be assessed by looking at IκBα degradation and/or at the phosphorylation of the NF-κB protein p65 and/or also at A20 protein levels by western blot.

Intestinal epithelial cells in ex-vivo organoids are responsive to most cytokines such as TNFα. All steps required for NF-κB activation indeed occur in ex-vivo organoids as seen in 2D-cell cultures. We will report here a detailed protocol to generate ex-vivo organoids from intestinal crypts and to assess NF-κB activation by cytokines in these cultures.

2 Materials

2.1 Reagents

1. VWR microscope slides.
2. 24-Well cell culture plate.
3. 50 and 15 ml conical centrifuge tubes.
4. 10 ml syringes.
5. Gavage needles.
6. Pipette tips; 5, 10, 25 ml sterile cell culture pipettes.
7. 70 mM cell strainer.
8. Phosphate buffered saline (PBS).
9. Matrigel matrix for organoids culture (Corning).
10. Fetal bovine serum (FBS).
11. Bovine serum albumin.

12. PBS.
13. Penicillin (100 I.U./mL) and streptomycin (100 μg/mL).
14. Sterile 1% FBS-PBS.
15. Ethylenediaminetetraacetic acid (EDTA).
16. 6 mM EDTA-PBS, pH 7.0.
17. IntestiCult OGM mouse basal medium and supplements (StemCell Technology).
18. Recombinant mTNFα (R&D Systems).
19. Recombinant mIL-17 (R&D Systems).
20. Trypsin.
21. Anti-IκBα, -HSP90, -A20, -p65 antibodies (Santa-Cruz).
22. Medium for *Apc*-mutated organoids: DMEM/F12 (Gibco), B27 supplement (Gibco), N2 supplement (Gibco), EGF (Cell Signaling), Murine Noggin (Peprotech).

2.2 Equipment

1. Cell culture hood NuAIR, NU-425-600E.
2. 37 °C, 5% CO_2 cell culture incubator (Binder).
3. Roller STUART (SRT9).
4. Brightfield inverted microscope Olympus CK40.
5. Centrifuges Eppendorf (5418R and 5810R).
6. Tools, pipettes.

3 Methods

3.1 Generation of Ex-Vivo Intestinal Organoids

1. Select 8- to 12-weeksold mice for the generation of ex-vivo organoids. Age of mice can vary and adjusted for investigator's needs (*see* **Notes 1** and **2**).
2. Carefully extract small intestine. Usually, we take the first 1/3 of the small intestine (10–15 cm of the duodenum) for organoids isolation. Clean from adjacent tissues (*see* **Note 3**).
3. Flush extracted small intestines with ice-cold PBS using gavage needle and 10 ml syringe, cut longitudinally and carefully open (internal side out). Remove mucus using a slide glass. Glass should not be pushed too hard since it could destroy intestinal crypts. Immediately place extracted intestines into 50 ml falcon tubes with 10 ml of ice-cold PBS and store on ice till next step. All subsequent steps are carried out under a cell culture hood.
4. Rinse tissues in PBS/Penicillin-Streptomycin, cut into 5 mm pieces, and place into a new 50 ml falcon tube with 10 ml ice-cold PBS/1% FBS. Use a 10 ml pipette to wash pieces of intestine with PBS/1% FBS by sucking up tissues up and down 8–10 times. If intestinal pieces are blocking the pipette, tap

pipette slightly at the bottom of the falcon tube. Immediately place the falcon tube on ice, let tissues settle by gravity for 30 s and carefully remove the PBS. Add again 10 ml of ice-cold PBS/1% FBS and carry out the washing step one more time. Repeat this step 7–10 times until PBS/1% FBS becomes clear. Make last wash only with ice-cold PBS (without FBS).

5. Carefully remove PBS from the last wash and add 25 ml of ice-cold PBS/6 mM EDTA. Place the falcon tube for 30 min at 4 °C with tender shaking or on a table top roller.
6. After this incubation, place the falcon tube on ice and wait for tissues to settle at the bottom. Carefully remove PBS/6 mM EDTA without disturbing the sediment.
7. Add 10 ml of ice-cold PBS without disturbing pellet and carefully remove (to remove remains of EDTA).
8. Add 10 ml of ice-cold PBS and wash tissue pieces with a 10 ml pipette (up and down around 3–4 times).
9. Wait again for tissue pieces to settle down. Collect PBS and transfer it into a 50 ml falcon tube placed on ice. This is Fraction 1. This fraction is usually discarded since it mainly contains remains of villi and debris. Keep this fraction on ice untill the end of the experiment.
10. Add 10 ml of ice-cold PBS to tissue pieces and wash it up and down 7–8 times with a 10 ml pipette. The falcon tube is maintained on ice. Wait again for tissue pieces to settle down. Collect PBS and transfer it into a 50 ml falcon tube placed on ice. This is Fraction 2. Fractions 3 and 4 are collected in the same 50 ml falcon tube as Fraction 2 using the same procedure.
11. Centrifuge the falcon tube with fractions 2–4 at 200 × *g* for 10 min at 4 °C.
12. Decant PBS and place pellet on ice and resuspend with 10 ml of ice-cold PBS.
13. Pass suspension through a 70 μm strainer into a new 50 ml falcon tube. Wash strainer with additional 5 ml of ice-cold PBS.
14. Transfer the suspension into a new 15 ml falcon tube.
15. Centrifuge the falcon tube at 200 × *g* for 10 min at 4 °C.
16. Decant PBS and resuspend with fresh 10 ml of ice-cold PBS.
17. Centrifuge the falcon tube at 200 × *g* for 5 min at 4 °C. The pellet is a mix of intestinal crypts (*see* **Note 4**).
18. Add 1 ml of Matrigel Matrix to the pellet. Resuspend carefully by trying to avoid the generation of bubbles. Keep tubes on ice (*see* **Note 5**).
19. Distribute the Matrigel with intestinal crypts as around 5–7 mm diameter drops into a pre-warmed 24-well dish.

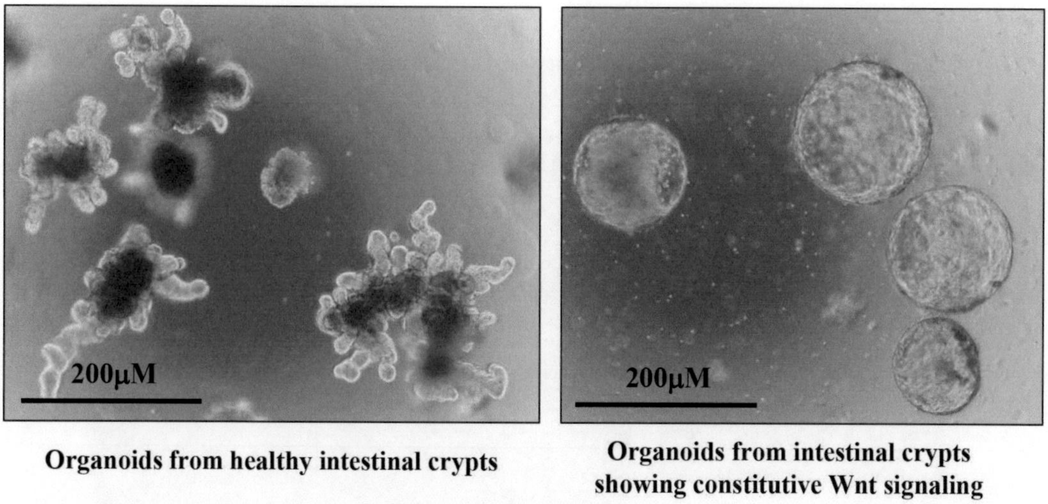

Fig. 1 Illustration of organoids generated with intestinal crypts from a healthy mouse or from a mouse showing constitutive Wnt signaling ($Apc^{+/Min}$ strain). On the left, ex-vivo organoids from the healthy mouse show differentiated cell types of the intestinal crypt while organoids on the right typically show a round shape

20. Check drops under a brightfield microscope. You should observe intact intestinal crypts.
21. Place the dish in a cell culture incubator for 15–30 min.
22. During this time, prepare the IntestiCult Medium according to the manufacturer's protocol.
23. Add 500 μl of this medium into each well.

This protocol is relevant for the isolation of intestinal crypts from healthy and *Apc*-mutated mice showing constitutive Wnt signaling [7, 8]. Note that a specific medium for *Apc*-mutated organoids has to be used (cf the list of reagents). Organoids should usually be passed in 6–8 days once they reach the shape and size of fully differentiated and grown organoids (Fig. 1, left panel). In contrast to ex-vivo organoids generated with intestinal crypts from healthy mice, organoids generated with intestinal crypts from *Apc*-mutated mice show a perfectly round shape and do not have too much of debris inside (Fig. 1, right panel) [8]. Immunofluorescence analyses carried out on *Apc*-mutated ex-vivo organoids indicate that some but not all cells are proliferating, as evidenced by anti-Ki67 and BrdU stainings (Fig. 2). Some cells are also undergoing cell apoptosis, as evidenced by anti-cleaved caspase 3 stainings (Fig. 2).

3.2 Passage and Expansion of Ex-Vivo Intestinal Organoids

1. To pass organoids, remove the medium from each well and wash Matrigel drops with ice-cold PBS (approximately 500 μl per well).

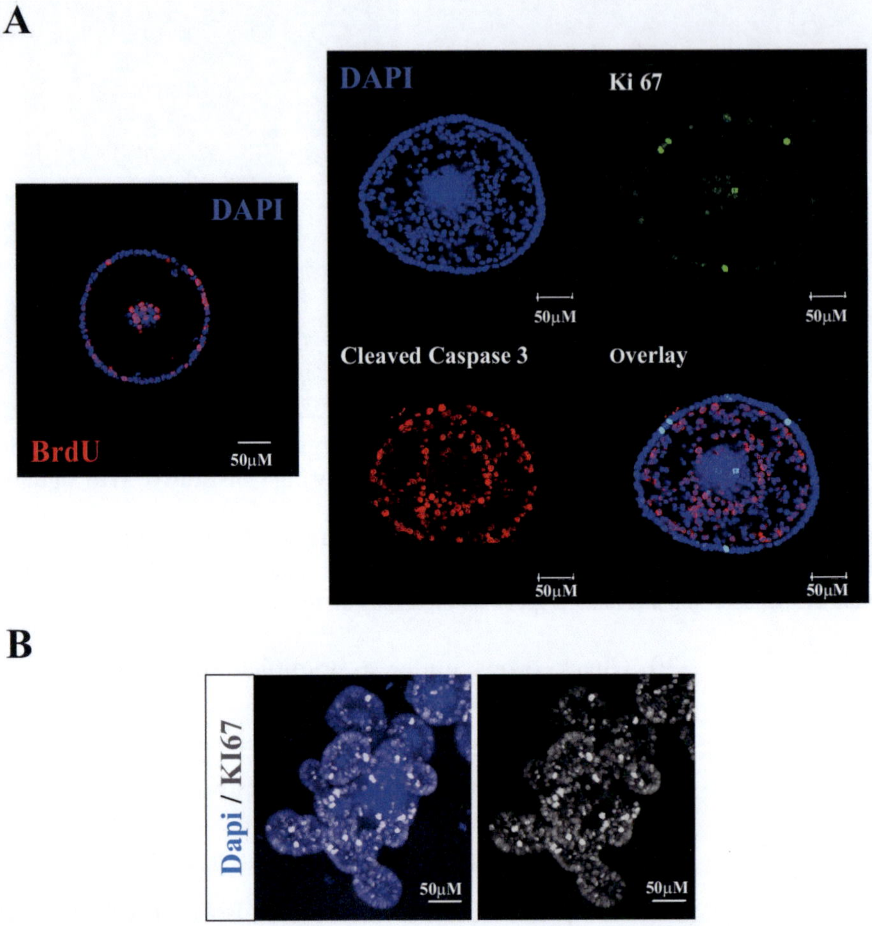

Fig. 2 Illustration of proliferative and dying cells in ex-vivo organoids. (**a**, **b**) Immunofluorescence was carried out on ex-vivo organoids generated with intestinal crypts from *Apc*$^{+/Min}$ mice (**a**) or from healthy mice (**b**). Proliferative and dying cells were stained (Ki67 and Cleaved Caspase 3 stainings, respectively)

2. Add 500 μl of fresh ice-cold PBS to each well. Using a 1 ml tip, resuspend Matrigel drops with PBS and collect into a 15 ml tube. Keep on ice. Mix well.
3. Add 10 ml of fresh ice-cold PBS and mix.
4. Centrifuge the falcon tube at 200 × *g* for 5 min at 4 °C.
5. Decant PBS. Remove remains with a 1 ml tip.
6. Add 200 μl of fresh PBS. Using a 200 μl tip, mechanically destroy/shred organoids by passing vigorously the pellet up and down around 20–30 times (*see* **Note 6**).
7. Add 10 ml of fresh ice-cold PBS and mix well.
8. Centrifuge the falcon tube at 200 × *g* for 5 min at 4 °C.

9. Decant PBS. Remove remains with a 1 ml or a 200 μl tip.
10. Add Matrigel Matrix (800 μl per 24-well dish). To reach a good "confluency" of organoids for treatment (*see* Fig. 1b), split organoids ¼, so that six wells of confluent organoids could be enough for a full 24-well plate.
11. Distribute Matrigel with intestinal crypts as around 5–7 mm diameter drops into a pre-warmed 24-well dish.
12. Place the dish in a cell culture incubator for 15–30 min.
13. During this time, prepare the IntestiCult Medium according to the manufacturer's protocol.
14. Add 500 μl of the IntestiCult Medium into each well.

3.3 Treatment of Ex-Vivo Intestinal Organoids with TNFα

1. Plate and grow organoids until you achieve a good "confluency" of well-differentiated ex-vivo organoids (Fig. 3a) (*see* **Note 7**).
2. Change the medium 1 day before the stimulation with any cytokine.
3. Treat ex-vivo organoids for the required time points once they reach the right confluency (Fig. 3a). As for concentrations, we've been using the same as for the stimulation of 2D cell cultures: To trigger NF-κB activation (as evidenced by IκBα degradation through the proteasome and by the induction of the NF-κB target gene A20) (Fig. 3b) [6], TNFα was used at 10 ng/ml (Fig. 3b). To induce cell death, TNFα was used at 100 ng/ml (Fig. 4). TNFα triggers some cell death at that concentration, as evidenced by positive PI stainings (Fig. 4). Intestinal organoids are also very well responsive to IL-17A (50–100 ng/ml) [9]. On the other hand, these organoids are poorly responsive to LPS, potentially because of the poor accessibility of TLR4.
4. At the end of each treatment, place dishes on ice, remove the medium, and wash each dish with ice-cold PBS.
5. Add fresh 500 μl of ice-cold PBS to each well. Using a 1 ml tip, resuspend Matrigel drops with PBS and collect into a 15 ml tube. Keep on ice. Mix well.
6. Add 10 ml of fresh ice-cold PBS and mix well.
7. Centrifuge the falcon tube at 200 × *g* for 5 min at 4 °C.
8. Decant PBS, add 1 ml of fresh ice-cold PBS, mix well and transfer the suspension into Eppendorf tubes.
9. Centrifuge the Eppendorf tubes at 2000 × *g* for 2 min at 4 °C. Remove PBS and Matrigel on top (transparent part, do not touch pellet).

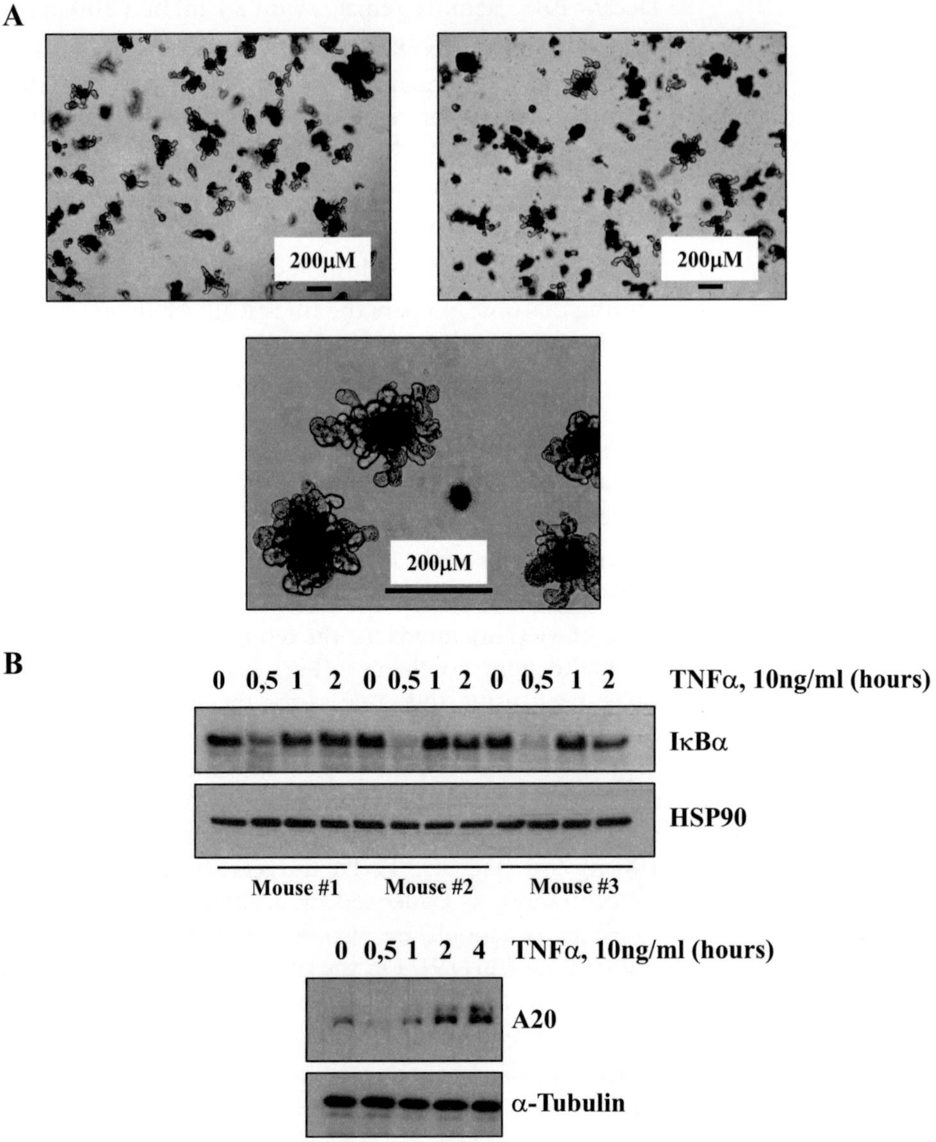

Fig. 3 TNFα-dependent NF-κB activation in ex-vivo organoids. (**a**) Illustration of ex-vivo organoids generated with intestinal crypts from healthy mice and maintained in culture at the right confluency. (**b**) Activation of NF-κB by TNFα. Ex-vivo organoids were treated or not with TNFα for the indicated periods of time and western blot analyses were conducted to assess IκBα degradation as well as A20 expression as readouts of NF-κB activation

10. Add 1 ml of fresh ice-cold PBS, vortex.
11. Repeat **step 9** until Matrigel is almost removed in order to obtain a clean organoids pellet. Proceed with lysis. Use the resulting extracts for western blot analyses.

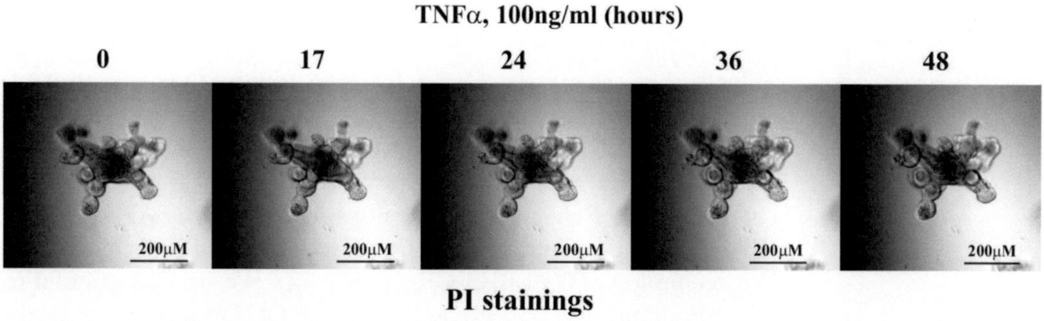

Fig. 4 TNFα-dependent cell death of ex-vivo organoids. Intestinal crypts from healthy mice were used to generate ex-vivo organoids which were treated or not with TNFα at the indicated periods of time. Cell death was visualized by PI stainings

4 Notes

1. For the selection of mice, too young mice have shorter intestine, which results in the generation of lower number of ex-vivo organoids. Selected mice are euthanized by cervical dislocation, according to the Ethical committee's recommendations.
2. We can also successfully generate ex-vivo organoids from other parts of the small intestine to meet investigator's needs. In case of comparison of two mice strains, the same parts of the small intestine are selected for experiments.
3. Once intestines are extracted, it is not recommended to keep them more than 30 minutes on ice since the yield will decrease due to cell death. It is also better not to process more than two mice simultaneously.
4. For the **step 17** (Subheading 3.1), once the centrifugation is done, the pellet of intestinal crypts is generated. If this pellet is still important and fluffy, repeat this step one more time.
5. For the **step 18** (Subheading 3.1), the volume of Matrigel to add depends on the size of the pellet and should be 6–8 times higher than the volume of the pellet.
6. For the **step 6** (Subheading 3.2), you could introduce some air bubbles to facilitate the mechanical disruption.
7. For the **step 1** (Subheading 3.3), in order to prepare ex-vivo organoids to be treated with cytokines, their size and confluency matters. Indeed, too small organoids may not have the adequate ratio of differentiated cell types, while too overgrown organoids may have too many dead cells and debris which could interfere with the signaling response to cytokines. To carry out western blot analyses and RNA quantifications, we plate minimum four and three wells for each experimental condition, respectively.

Acknowledgments

The authors thank the GIGA Imaging facility. Our laboratory is supported by Grants from the Belgian National Funds for Scientific Research (FNRS), from the Concerted Research Action Program (UBICOREAR) and Special Research Funds (FSR) at the University of Liege, the Belgian foundation against Cancer (FAF-F/2016/794), as well as from the Walloon Excellence in Life Sciences and Biotechnology (WELBIO-CR-2015A-02). We are also grateful to the "Foundation Leon Fredericq" of the CHU Liege for its financial support. A. Chariot and S. Tielens are Research Director and Senior Research Assistant at the FNRS, respectively.

References

1. Beumer J, Clevers H (2021) Cell fate specification and differentiation in the adult mammalian intestine. Nat Rev Mol Cell Biol 22:39–53
2. Leblond CP, Stevens CE (1948) The constant renewal of the intestinal epithelium in the albino rat. Anat Rec 100:357–377
3. Barker N, Van Es JH, Kuipers J, Kujala P, Van Den Born M, Cozijnsen M, Haegebarth A, Korving J, Begthel H, Peters PJ et al (2007) Identification of stem cells in small intestine and colon by marker gene Lgr5. Nature 449:1003–1007
4. Sato T, Clevers H (2013) Growing self-organizing mini-guts from a single intestinal stem cell: mechanism and applications. Science 340:1190–1194
5. Taniguchi K, Karin M (2018) NF-κB, inflammation, immunity and cancer: coming of age. Nat Rev Immunol 18:309–324
6. Krikos A, Laherty CD, Dixit VM (1992) Transcriptional activation of the tumor necrosis factor α-inducible zinc finger protein, A20, is mediated by κB elements. J Biol Chem 265:17971–17976
7. Sato T, Vries RG, Snippert HJ, Van De Wetering M, Barker N, Stange DE, Van Es JH, Abo A, Kujala P, Peters PJ et al (2009) Single Lgr5 stem cells build crypt-villus structures in vitro without a mesenchymal niche. Nature 459:262–265
8. Schwitalla S, Fingerle AA, Cammareri P, Nebelsiek T, Göktuna SI, Ziegler PK, Canli O, Heijmans J, Huels DJ, Moreaux G et al (2013) Intestinal tumorigenesis initiated by dedifferentiation and acquisition of stem-cell-like properties. Cell 152:25–38
9. Duong HQ, Nemazanyy I, Rambow F, Tang SC, Delaunay S, Tharun L, Florin A, Buttner R, Vandaele D, Close P et al (2018) The endosomal protein CEMIP links WNT signaling to MEK1–ERK1/2 activation in selumetinib-resistant intestinal organoids. Cancer Res 78:4533–4548

Chapter 18

Extracellular Flux Analysis to Investigate the Impact of NF-κB on Mitochondrial Respiration in Colorectal Carcinoma (CRC)

Daria Capece, Daniela Verzella, Federica Begalli, Jason Bennett, Daniel D'Andrea, Davide Vecchiotti, Francesca Zazzeroni, and Guido Franzoso

Abstract

The reprogramming of cell metabolism is a hallmark of cancer. NF-κB transcription factors coordinate the host defense responses to stress, injury, and infection. They also play a central role in oncogenesis, at least in part by regulating cell metabolism and the adaptation to energy stress conditions in various types of cancer, such as colorectal carcinoma (CRC). Here, we describe the XF Cell Mito Stress Test methodology aimed at characterizing the metabolic and bioenergetic profile of CRC cells following the silencing of the essential NF-κB subunit, RelA. This methodology may also be applied to other cancers to reveal novel core vulnerabilities of malignant cells.

Key words Extracellular Flux Analysis, NF-κB, Colorectal carcinoma, Cancer, Metabolism

1 Introduction

Metabolic reprogramming and energy stress adaptation play an important role in tumor progression and metastatic spread [1–4]. As such, the metabolic profiling of cancer cells is an increasingly more valuable approach to reveal novel core vulnerabilities of cancer cells, which could lead to better anticancer treatments. The advent of high-throughput respirometry and the availability of specific mitochondrial inhibitors have recently invigorated the development of methodologies that allow the measurement of bioenergetic parameters in intact cells [5–7]. Among these, extracellular flux (XF) analysis has become a mainstream method to measure the bioenergetic functions of cells and tissues. In particular, the Seahorse XFe96 respirometer can measure parameters such as the flux of oxygen (O_2), the oxygen consumption rate (OCR),

Guido Franzoso and Francesca Zazzeroni (eds.), *NF-κB Transcription Factors: Methods and Protocols*, Methods in Molecular Biology, vol. 2366, https://doi.org/10.1007/978-1-0716-1669-7_18,

the flux of protons (H^+), and the extracellular acidification rate (ECAR) in the medium immediately surrounding living cells, in real time and without the need to lyse cells or isolate mitochondria. This is significant because the OCR is a key readout of oxidative phosphorylation (OXPHOS), while ECAR reflects the rate of glycolysis (Fig. 1).

The Seahorse XFe96 system comprises a 96-well plate coupled with a sensor cartridge containing an equivalent number of individual sensor probes (Fig. 2). The tip of each probe is equipped with two separate polymer-embedded fluorophores that are sensitive to either O_2 or H^+. The quenching chemistry of the fluorophores in response to variations in O_2 and H^+ concentrations in the medium is responsible for the sensitivity of the probes. Both oxygen and protons fluxes are directly linked to the intensity of the fluorescence signals relative to the signals of a standard solution, and any changes in signal intensity can therefore be inferred to be proportional to changes in O_2 or H^+ concentrations, respectively. The rates of change in O_2 or H^+ concentration can be calculated by measuring the fluorescent signals over time [8].

During each measurement cycle, the sensor cartridge is lowered into the wells to create a transient microchamber of defined volume (Fig. 2). Drug delivery ports located on the cartridge itself then allow the automatic injection via air pressure of specific compounds. Fiber optic bundles are inserted into the probes. Light is forced into the sensor to excite the fluorophores. The resulting emitted light is then transmitted back through the fiber optic bundle and measured by the detector. The fluorescent emission is measured for a specified period of time (measurement time). Next, the slope of the linear decline (for O_2 concentration over time) or incline (for H^+ concentration over time) provides a basis for calculating the OCR (in pmol/min) or ECAR (mpH/min) value, respectively (Fig. 2) [9]. Gerencser et al. have reported the algorithm for calculating the OCR using XF respirometers [10].

The XF Cell Mito Stress Test is a comprehensive standard assay to assess key parameters of respiratory function by the measurement of the OCR [9]. The assay uses the following modulators of respiration to selectively target specific components of the electron transport chain (ECT): (a) Oligomycin, an inhibitor of ETC complex V, also known as ATP synthase; (b) a mitochondrial uncoupler, such as carbonyl cyanide-p-trifluoromethoxyphenylhydrazone (FCCP) or 2,4-dinitrophenol (2,4-DNP); and a combination of rotenone (an inhibitor of ETC complex I) and antimycin A (an inhibitor of ETC complex III). In brief, the first measurement is "basal OCR," which is measured before injection of the mitochondrial inhibitors. Once the basal measurements have been recorded, the inhibitors are serially injected to measure the following bioenergetic parameters: (1) basal respiration; (2) "ATP-linked respiration"; (3) "maximal respiration"; and (4) "non-

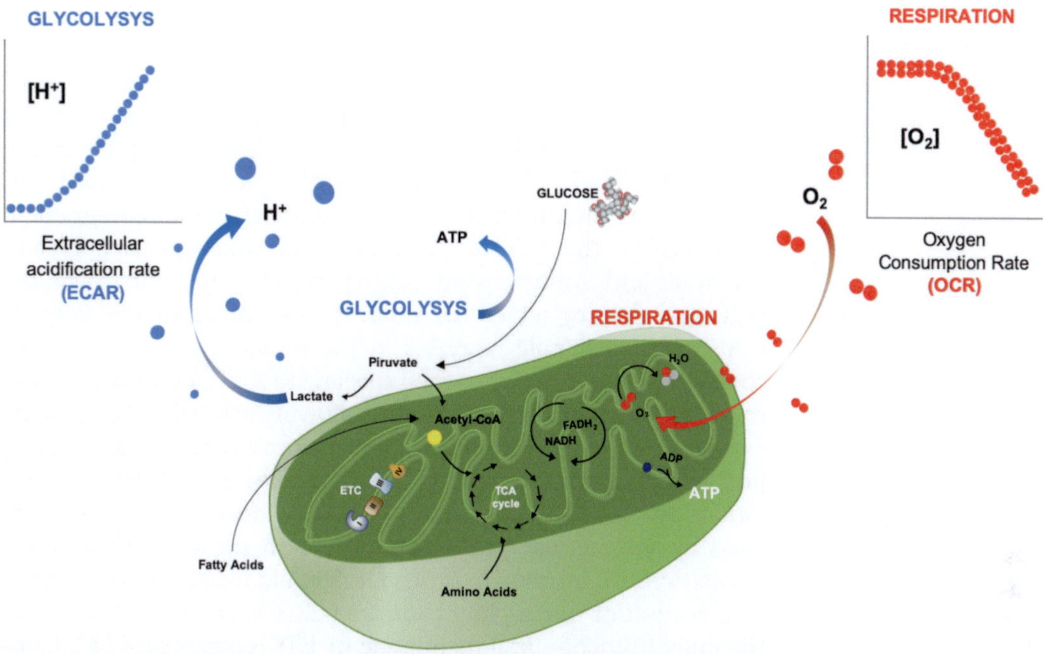

Fig. 1 XF analysis is the mainstream method for characterizing the bioenergetic and metabolic profile of a cell. XF analysis measures the flux of oxygen—e.g., OCR—as well as the flux of protons—e.g., ECAR. The OCR is largely the results of OXPHOS, while ECAR reflects the rate of glycolysis

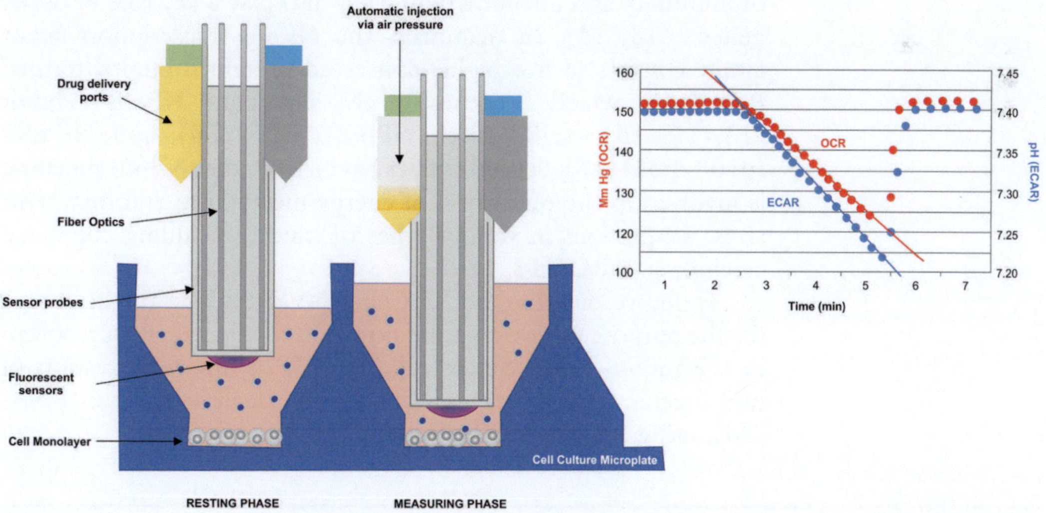

Fig. 2 Schematic representation of a typical Seahorse XFe96 respirometer measurement cycle. During each measurement cycle, the sensor cartridge is lowered onto the wells to create a transient microchamber. Drug delivery ports located on the cartridge itself allow the automatic injection of specific compounds. Fiber optic bundles are inserted into the probes, and light is forced into the sensor to excite the fluorophores. Changes in fluorescence signals are proportional to changes in O_2 or H^+ concentrations. The slope of the linear decline (for O_2 concentration over time) or incline (for H^+ concentration over time) provides a basis for extrapolating the OCR value (expressed in pmol/min) or ECAR value (expressed in mpH/min), respectively

mitochondrial respiration." Specifically, by blocking ATP synthase, oligomycin allows the measurement of "ATP-linked respiration" [11]. An increase in ATP-linked OCR indicates a surge in ATP demand, while a decrease in this parameter indicates a reduced ATP demand, a lack of substrate availability, and/or an impairment in OXPHOS [11]. The mitochondrial uncoupler (FCCP or 2,4-DNP) is then used to collapse the inner membrane gradient and so allow the ETC to function at maximal capacity. Finally, rotenone and antimycin are added together to shut down the ETC and thereby unveil the residual, non-mitochondrial respiration, which generally increases in the presence of mitochondrial stressors such as reactive oxygen species (ROS) [11]. In addition to these bioenergetic parameters, the following other functions can be extrapolated: (5) "proton leak," which is calculated by subtracting non-mitochondrial respiration from the value of ATP-linked respiration; and (6) "spare respiratory capacity" (SRC), which is calculated by subtracting basal respiration from the value of maximal respiration. An increase in proton leak could be due to for instance an elevation of uncoupling proteins (UCP) activity or damage to the inner mitochondrial membrane or ETC complexes [11]. Conversely, the SPR reflects the respiratory reserve available for ATP production to support cell survival during energy stress conditions or increased metabolic demand [11, 12]. A typical OCR graph obtained by using a Cell Mito Stress Test is depicted in Fig. 3.

NF-κB transcription factors are central coordinating regulators of immunity and inflammation. They also play a key role in oncogenesis [13, 14]. In mammals, the NF-κB transcription factor family consists of five highly conserved protein subunits, termed RelA/p65, which is generally the dominant NF-κB subunit [13, 15, 16], RelB, c-Rel, NF-κB1(p105/p50), and NF-κB2 (p100/p52) [17]. Several studies have shown that NF-κB signaling is involved in the regulation of energy metabolism and metabolic stress adaptations in various types of cancer, including colorectal carcinoma (CRC) [13, 18–22].

Here, we describe the XF Cell Mito Stress Test methodology for the purpose of investigating how NF-κB affects cell metabolism in the mouse colon cancer cell line, CT26, upon expression of non-specific (ns) or RelA-specific (RelA) small-hairpin (sh)RNAs [20], using a Seahorse XFe96 respirometer.

2 Materials

2.1 Equipment

1. Seahorse XFe96 Extracellular Flux analyzer.
2. Non-CO_2 incubator.
3. Seahorse XFe96 FluxPak, including XF96 Cell Culture Microplate and XFe96 sensor cartridge.

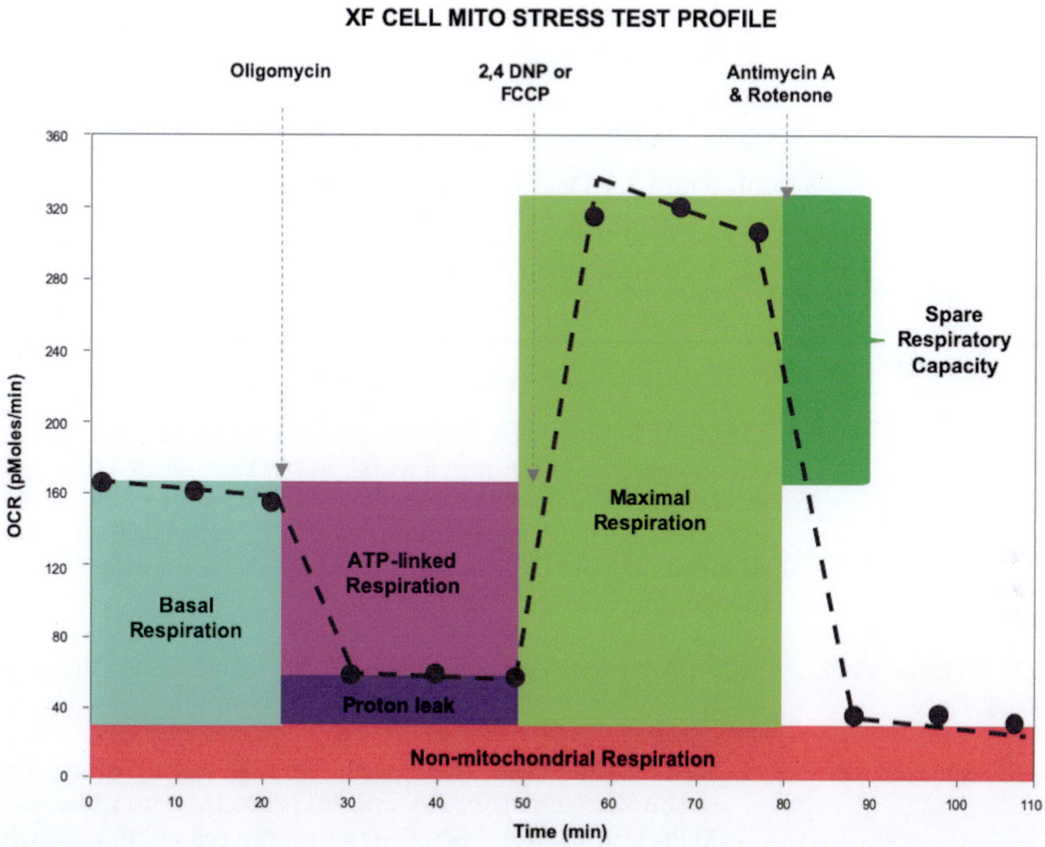

Fig. 3 Typical XF Cell Mito Stress Test profile, showing the key parameters of mitochondrial function. Basal OCR is measured before the injection of the mitochondrial inhibitors. Once the basal measurements are recorded, specific drugs are serially injected onto the microplate to measure the following bioenergetic parameters: (1) basal respiration; (2) ATP-linked respiration; (3) maximal respiration; and (4) non-mitochondrial respiration. Additionally, (5) the proton leak value can be calculated by subtracting non-mitochondrial respiration from the ATP-linked respiration value, whereas (6) the spare respiratory capacity (SPR) can be calculated by subtracting basal respiration from the maximal respiration value

4. Cell Counter.
5. Water Bath.

2.2 Reagents

1. CT26 colon cancer cell line expressing ns-shRNAs or RelA-shRNAs (*see* **Note 1**).
2. Cell culture growth medium: RPMI-1640 medium without glucose, 10% FBS, 150 U/mL penicillin, 200 U/mL streptomycin, 10 mM HEPES buffer, 1 mM sodium pyruvate.
3. 1× Phosphate Buffer Saline (PBS).
4. Seahorse XF calibrant solution (Agilent).
5. XF RPMI assay medium: RPMI medium, pH 7.4, 1 mM pyruvate, 2 mM glutamine, and 10 mM D-glucose.

6. 10 mM D-Glucose.
7. 2 mM Glutamine.
8. 1 mM Sodium Pyruvate.
9. 1 mM Oligomycin A stock solution.
10. 5 mM 2,4-Dinitrophenol stock solution.
11. 1 mM Rotenone stock solution.
12. 1 mM Antimycin A stock solution.

3 Methods

The XF Cell Mito Stress Test is carried out over a two-day period. Accordingly, the experimental workflow has been divided into two sections: (1) The first describes the steps to be undertaken on the day before the assay (day 0) (Subheading 3.1); (2) while the second describes the steps to be undertaken on the day of the assay (day 1) (Subheading 3.2) (Fig. 4).

3.1 Seahorse XFe96 Assay: Day 0

1. Turn on the Agilent Seahorse XFe96 Extracellular Flux analyzer and let it warm up overnight (*see* **Note 2**).

3.1.1 Seeding the Cells

1. Seed CT26 cells (80μl/well, 25,000 cells) expressing ns-shRNAs (sh-ns) or RelA-shRNAs (sh-RelA) onto a Seahorse XF96 Cell Culture Microplate using the cell culture growth medium (*see* **Note 3**).
2. Incubate the Microplate at 37 °C, 5% CO_2 overnight. Monitor the growth and health of the cells using an inverted microscope.

3.1.2 Hydrating the XFe96 Sensor Cartridge

1. Aliquot at least 20 mL of XF Calibrant into a 50 mL conical tube and place the tube in a non-CO_2 37 °C incubator overnight.
2. Open the XFe96 Extracellular Flux Assay Kit and remove the green sensor cartridge from the utility plate. Fill each well of the utility plate with 200μL of sterile, tissue culture-grade water and place the sensor cartridge back into the utility plate, submerging the sensors into water.
3. Place the cartridge at 37 °C in a non-CO_2 incubator overnight (*see* **Note 4**).

3.1.3 Designing the Experiments

1. Launch the Wave 2.6 software and open the assay template listed as "XF Mito Stress Assay."
2. Prepopulated information relating to the injection strategy, cell pre-treatments, assay media, and cell types are provided in the "Group Definitions" menu. Modify the available experimental

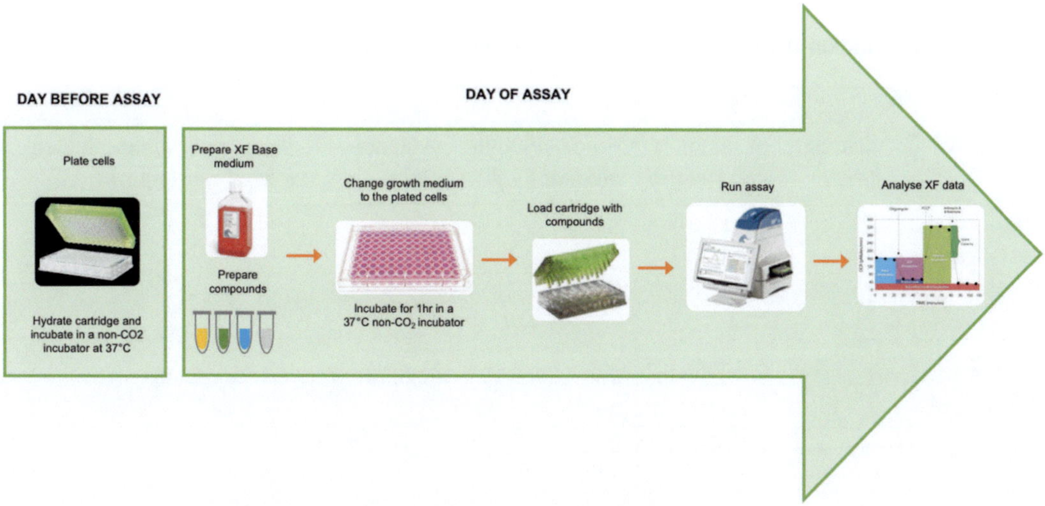

Fig. 4 Schematic representation of the XF Mito Stress Test workflow

parameters as follows: (a) In "Injection strategy," add the concentration of each drug used in the assay (i.e., Port A: 1.8μM Oligomycin A; Port B: 60μM 2,4-DNP; Port C: 0.6μM Rotenone/Antimycin A). (b) In "Pre-treatments," select "Control" and add "sh-ns"; then select "Experimental" and add "sh-RelA." In "Assay Media," add "XF RPMI assay medium, pH 7.4." In "Cell Type," delete the default entry and add "CT26," specifying the cell seeding density for your own records.

3. Once the experimental groups have been named, select "Generate Groups"; the software will then automatically create two unique assay groups: CT26 sh-ns and CT26 sh-RelA.
4. Select the "Plate Map" menu to assign experimental groups to the microplate map by first choosing the group name in the list provided and then clicking individual experimental points on the microplate map (*see* **Note 5**).
5. Select the "Instrument Protocol" menu to view the default instrument protocol (*see* **Note 6**).
6. Click "Run Assay" in the "Functions" ribbon to save the template file and start the assay.

3.2 Seahorse XFe96 Assay: Day 1

3.2.1 Preparing the Assay Medium and the Compound Working Solutions

1. Warm up the Seahorse XF RPMI medium to 37 °C in a water bath.
2. Remove the Seahorse XF Cell Culture Microplates from the 37 °C CO_2 incubator and examine the cells using an inverted microscope to confirm their confluence.
3. Remove the cell growth medium from the cell culture microplate and add 180μL of pre-warmed assay medium.

Table 1
Compound preparation instructions for loading onto the XFe/XF96 sensor cartridges[a]

	Stock solution (mM)	Final well (μM)	Stock solution volume (μL)	Media volume (μL)	10× (Port) (μM)	Volume added to port (μL)
Port A: Oligomycin	1	1.8	54	2946	18	20
Port B: 2,4 DNP	5	60	360	2640	600	22
Port C: Rot/AA	1	0.5	15	2985	5	25

[a]The starting volume of assay medium for the cell microplates is 180μL per well

4. Place the microplate into a 37 °C non-CO_2 incubator for 1 h prior to the assay.
5. Prepare 3 mL of working solutions in the assay medium for each compound by using the volumes specified in Table 1 (*see* **Note 7**).
6. Load the solutions onto the injection ports of the sensor cartridge by using the guides provided.

3.2.2 Running the Assay

1. Browse the saved design file and then open it. Select the "Review and Run" tab and then choose "Start Run."
2. Place the loaded sensor cartridge with the calibrant microplate into the instrument, and then select "I'm ready" (*see* **Note 8**).
3. Following calibration and equilibration of the cell culture microplate, when prompted select "I'm ready."
4. Remove the calibration utility plate and load the cell culture microplate and select "I'm ready" to run the assay.

3.3 Data Analysis

1. The Seahorse Mito Stress Test assay can be analyzed by using the Seahorse XF Mito Stress Test Report Generator, which automatically calculates the main Seahorse XF Cell Mito Stress Test parameters from the Wave data that were exported to Excel.
2. The Seahorse XF Report Generator can be installed together with the Wave 2.6 software or directly from the Agilent Cell Analysis website (*see* **Note 9**).

4 Notes

1. Production of high-titre lentiviral preparations in HEK-293T cells and lentiviral infections of CT26 cells may be carried out as described in [20].
2. The Agilent Seahorse XFe96 instrument needs to warm up for at least 5 h. This step is required to allow the instrument and its program to stabilize at 37 °C, since a stable 37 °C temperature of the XF instrument is required for accurate Seahorse readings.
3. It is important to predetermine the optimal seeding density for each cell line used. The optimal cell seeding density may vary widely depending on the cell line, although this is typically in a range between 5×10^3 and 4×10^4 cells per well. The seeding density must be determined empirically for each cell line. After seeding the cells, allow the plate to rest in the tissue culture hood at room temperature for at least 1 h. This step facilitates an even distribution of the cells and reduces edge effects, which can be particularly pronounced for certain cell types. In all cases, empirically monitor the cell adherence using an inverted microscope.
4. The maximum time of incubation of the cartridge in the non-CO_2 incubator is 72 h. If for any reason the cartridge is hydrated for more than 24 h, then wrap it up in parafilm to avoid evaporation. In any case, the hydration period should be no less than 4 h.
5. The number of experimental replicates should be no less than 5.
6. The default instrument protocol does not require modifications. However, the name of the protocol command, the number of measurements recorded before or after injection, and the length of time for each measurement can be changed. Modifying the instrument protocol settings may affect how the data are acquired during the assay. For the CT26 cell line, the default instrument protocol is recommended for the analyses.
7. In our system, 2,4-DNP was used instead of FCCP because it demonstrated less toxicity.
8. The calibration takes approximately 15–30 min. Remove the cartridge lid and verify the correct microplate orientation.
9. https://www.agilent.com/en/products/cell-analysis/cell-analysis-software/data-analysis/seahorse-xf-cell-mito-stress-test-report-generators.

Acknowledgments

This work was supported in part by Medical Research Council (MRC) Biomedical Catalyst grant MR/L005069/1, Bloodwise project grant 15003, and NIHR Imperial Biomedical Research Centre (BRC) Push for Impact Award WIIS_P81135 to G.F. Infrastructure support for this research was provided in part by the NIHR Imperial BRC.

References

1. Cairns RA, Harris IS, Mak TW (2011) Regulation of cancer cell metabolism. Nat Rev Cancer 11(2):85–95
2. Pavlova NN, Thompson CB (2016) The emerging hallmarks of cancer metabolism. Cell Metab 23(1):27–47
3. Boroughs LK, DeBerardinis RJ (2015) Metabolic pathways promoting cancer cell survival and growth. Nat Cell Biol 17(4):351–359
4. Kreuzaler P, Panina Y, Segal J, Yuneva M (2019) Adapt and conquer: metabolic flexibility in cancer growth, invasion and evasion. Mol Metab 33:83–101
5. van der Windt GJW, Chang CH, Pearce EL (2016) Measuring bioenergetics in T cells using a seahorse extracellular flux analyzer. Curr Protoc Immunol 113:3.16B.1–3.16B.14
6. Dranka BP, Benavides GA, Diers AR, Giordano S, Zelickson BR, Reily C, Zou L, Chatham JC, Hill BG, Zhang J, Landar A, Darley-Usmar VM (2011) Assessing bioenergetic function in response to oxidative stress by metabolic profiling. Free Radic Biol Med 51 (9):1621–1635
7. Salabei JK, Gibb AA, Hill BG (2014) Comprehensive measurement of respiratory activity in permeabilized cells using extracellular flux analysis. Nat Protoc 9(2):421–438
8. Koopman M, Koopman M, Michels H, Dancy BM, Kamble R, Mouchiroud L, Auwerx J, Nollen EA, Houtkooper RH (2016) A screening-based platform for the assessment of cellular respiration in *Caenorhabditis elegans*. Nat Protoc 11(10):1798–1816
9. https://www.agilent.com/en/products/cell-analysis/seahorse-analyzers/seahorse-xfe96-analyzer
10. Gerencser AA, Neilson A, Choi SW, Edman U, Yadava N, Oh RJ, Ferrick DA, Nicholls DG, Brand MD (2009) Quantitative microplate-based respirometry with correction for oxygen diffusion. Anal Chem 81(16):6868–6878
11. Chacko BK, Kramer PA, Ravi S, Benavides GA, Mitchell T, Dranka BP, Ferrick D, Singal AK, Ballinger SW, Bailey SM, Hardy RW, Zhang J, Zhi D, Darley-Usmar VM (2014) The Bioenergetic Health Index: a new concept in mitochondrial translational research. Clin Sci (Lond) 127(6):367–373
12. Zhang L, Romero P (2018) Metabolic control of CD8(+) T cell fate decisions and antitumor immunity. Trends Mol Med 24(1):30–48
13. Taniguchi K, Karin M (2018) NF-kappaB, inflammation, immunity and cancer: coming of age. Nat Rev Immunol 18(5):309–324
14. DiDonato JA, Mercurio F, Karin M (2012) NF-kappa B and the link between inflammation and cancer. Immunol Rev 246:379–400
15. Schwitalla S, , Fingerle AA, Cammareri P, Nebelsiek T, Göktuna SI, Ziegler PK, Canli O, Heijmans J, Huels DJ, Moreaux G, Rupec RA, Gerhard M, Schmid R, Barker N, Clevers H, Lang R, Neumann J, Kirchner T, Taketo MM, van den Brink GR, Sansom OJ, Arkan MC, Greten FR (2013) Intestinal tumorigenesis initiated by dedifferentiation and acquisition of stem-cell-like properties. Cell 152(1–2):25–38
16. Hayden MS, Ghosh S (2004) Signaling to NF-kappaB. Genes Dev 18(18):2195–2224
17. Capece D, Verzella D, Di Francesco B, Alesse E, Franzoso G, Zazzeroni F (2019) NF-kappaB and mitochondria cross paths in cancer: Mitochondrial metabolism and beyond. Semin Cell Dev Biol 98:118–128
18. Ishak Gabra MB, Yang Y, Lowman XH, Reid MA, Tran TQ, Kong M (2018) IKKbeta activates p53 to promote cancer cell adaptation to glutamine deprivation. Oncogenesis 7(11):93
19. Wang X, Liu R, Qu X, Yu H, Chu H, Zhang Y, Zhu W, Wu X, Gao H, Tao B, Li W, Liang J, Li G, Yang W (2019) Alpha-ketoglutarate-activated NF-kappaB signaling promotes compensatory glucose uptake and brain tumor development. Mol Cell 76(1):148–62.e7

20. Mauro C, Leow SC, Anso E, Rocha S, Thotakura AK, Tornatore L, Moretti M, De Smaele E, Beg AA, Tergaonkar V, Chandel NS, Franzoso G (2011) NF-kappaB controls energy homeostasis and metabolic adaptation by upregulating mitochondrial respiration. Nat Cell Biol 13(10):1272–1279
21. Tornatore L, Thotakura AK, Bennett J, Moretti M, Franzoso G (2012) The nuclear factor kappa B signaling pathway: integrating metabolism with inflammation. Trends Cell Biol 22(11):557–566
22. Capece D, D'Andrea D, Begalli F, Goracci L, Tornatore L, Alexander JL, Di Veroli A, Leow SC, Vaiyapuri TS, Ellis JK, Verzella D, Bennett J, Savino L, Ma Y, McKenzie JS, Doria ML, Mason SE, Chng KR, Keun HC, Frost G, Tergaonkar V, Broniowska K, Stunkel W, Takats Z, Kinross JM, Cruciani G, Franzoso G (2021) Enhanced triacylglycerol catabolism by Carboxylesterase 1 promotes aggressive colorectal carcinoma. J Clin Invest 20:137845. https://doi.org/10.1172/JCI137845. Online ahead of print

Chapter 19

Conditional Knockout Mouse Models to Study the Roles of Individual NF-κB Transcription Factors in Lymphocytes

Emma J. Adams, Nilushi S. De Silva, and Ulf Klein

Abstract

The NF-κB signal transduction pathway has crucial functions in cell growth, survival, and the development of lymphocytes and other immune cells. Upon activation of the pathway, five distinct NF-κB transcription factor subunits that occur as homodimers or heterodimers comprise the downstream mediators that transcribe NF-κB target genes. A major quest in NF-κB research is to understand the biology of the separate subunits. However, determining the functions of the individual subunits using constitutional knockout mice is often hampered by the marked cell type and/or developmental stage-specific activation of the NF-κB pathway. To overcome these problems, we and others have generated loxP-flanked alleles of the genes encoding the different NF-κB subunits that upon crossing to suitable Cre-expressing mouse lines can be conditionally deleted in the desired cell type or developmental stage. We here describe the basic characteristics of conditional NF-κB subunit alleles *rel* (encoding c-REL), *rela* (RELA), *relb* (RELB), and *nfkb2* (NF-κB2) generated in our laboratory that are available to the research community through a repository, and provide basic methods to study the consequences of tissue-specific ablation of NF-κB transcription factors in lymphocytes.

Key words Nuclear factor-κB, NF-κB, NF-κB transcription factors, NF-κB subunits, Conditional alleles, Transgenic animal models, Cre/loxP

1 Introduction

Activation of the NF-κB signal transduction pathway ultimately leads to the nuclear translocation of five NF-κB transcription factor subunits, c-REL, RELA, RELB, p50 (NF-κB1), and p52 (NF-κB2), which transcribe NF-κB target genes as heterodimers or homodimers [1, 2]. Ever since their discovery, a major focus of NF-κB research has been to understand the biology of the distinct subunits. The analysis of whole-body knockout mice of the individual NF-κB subunits could provide important insights into their biological function in lymphocytes and immune cells [3–11]. However, with regard to the analysis of adaptive immune responses, there are two caveats associated with the constitutional knockout

Guido Franzoso and Francesca Zazzeroni (eds.), *NF-κB Transcription Factors: Methods and Protocols*, Methods in Molecular Biology, vol. 2366, https://doi.org/10.1007/978-1-0716-1669-7_19,

approach. First, during the lifetime of a lymphocyte, NF-κB activation can occur at different developmental stages, including antigen-activation and at later differentiation stages, for instance during the differentiation of an antigen-activated B cell in a plasma cell. Thus, the employment of knockout mice may allow only the study of the first stage. Second, NF-κB activation occurs in the various immune cell types that interact with each other during an immune response, e.g., T cell–B cell interactions; this circumstance makes it impossible to determine which cell type is inherently responsible for the observed knockout phenotype. To overcome the problems associated with cell type and developmental stage-specific NF-κB activation, loxP-flanked alleles of the NF-κB subunit genes were generated that by crossing to a Cre-expressing mouse line allowed the conditional ablation of the respective NF-κB subunit in the desired cell type or stage [12, 13].

Our laboratory has generated conditional NF-κB subunit alleles for *rel* (encoding c-REL) [12], *rela* (RELA) [12], *relb* (RELB) [13], and *nfkb2* (NF-κB2) [13] in order to determine the specific roles of the individual subunits in B cells of various developmental stages, including B-cell maintenance, B-cell activation and differentiation in T cell-dependent immune responses [12–15]. (Conditional *rela* and *relb* alleles have been generated also by others [16–18].) In summary, the results from our studies demonstrated that the canonical subunits c-REL and RELA were jointly required for mature B-cell maintenance [15]. Likewise, the alternative subunits RELB and NF-κB2 were jointly essential for this process in a non-redundant fashion with c-REL/RELA [13]. Following antigen-dependent B-cell activation, c-REL [12] and RELB/NF-κB2 [14] were required in a non-redundant fashion for the maintenance of the germinal center B-cell reaction, whereas the RELA subunit was dispensable at this stage and instead was found to control germinal center B cell to plasma cell differentiation [12]. NF-κB subunit-specific roles have been identified also within the T-cell lineage [19–21]. Overall, these findings may provide the rationale for more specific therapeutic targeting of aberrant NF-κB activity by focusing on the individual NF-κB subunits [22] that are constitutively activated in lymphomas corresponding to different B-cell developmental stages [23].

The purpose of this article is twofold. First, we describe the basic characteristics of the conditional NF-κB subunit alleles *rel* (encoding c-REL), *rela* (RELA), *relb* (RELB), and *nfkb2* (NF-κB2) generated in our laboratory that are available to the community through a repository. Since it is becoming increasingly clear that the separate NF-κB subunits have distinct functions in many different cell types and cell states (healthy and diseased), the availability of conditional NF-κB subunit alleles provides a means to determine the subunit-specific roles in the respective cells of interest for which suitable Cre lines are available.

Second, we describe basic methods to study the consequences of the tissue-specific ablation of NF-κB transcription factors in lymphocytes. While the protocols have been devised to study lymphocytes, some of the methods can easily be modified to the particular cell or tissue type of interest, and the protocols for tissue preparation and immunohistochemistry (IHC) may be directly applicable to most animal tissue types.

1.1 Conditional Knockout NF-κB Subunit Alleles

We have generated conditional alleles of the genes encoding the NF-κB subunits c-REL (*rel*), RELA (*rela*), RELB (*relb*), and NF-κB2 (*nfkb2*) [12, 13]. All of the four conditional alleles were constructed such that the promoter region and the exon containing the translational start codon are flanked by loxP sites (Fig. 1), allowing for Cre-mediated deletion of the intervening region and generation of a null allele of the respective NF-κB subunit. In addition, concomitant with the deletion, the expression of an enhanced green fluorescent protein (eGFP) minigene is activated, which allows deleted cells to be traced by flow cytometry or staining for eGFP in histological sections.

The loxP deletion can also be detected by a PCR strategy, which is common for all four transgenic NF-κB subunit mice due to the localization of the primers in the eGFP minigene and the 3′ loxP site (Fig. 1). The deletion PCR strategy makes use of the circumstance that a novel primer-binding site is generated as a consequence of the loxP recombination at the 3′ loxP site. While the 5′ anchor primer is the same in the PCRs detecting the unrearranged and rearranged locus, the 3′ primers are different; however, since they localize in the same region, the size of the PCR product is the same, necessitating two separate PCR reactions. The deletion PCR strategy is not only suitable for determining the presence of an NF-κB subunit gene deletion in a tissue of interest (although as described in the Methods, it is not quantitative), but also to test for possible germline deletion of the conditional NF-κB subunit genes as the result of the occasional activity of some Cre recombinases in cells of the germline. In our experience with Cγ1-Cre mice, germline deletion of the loxP-flanked gene occurs in a small fraction of mice, necessitating analysis for potential germline deletion along with genotyping. The deletion PCR is sensitive and unambiguously detects pups with an NF-κB subunit allele deleted in the germline.

Moreover, frt sites have been introduced into the targeting vector that flank the eGFP cassette, the loxP-flanked promoter region, and the promoter driving eGFP expression upon Cre-recombination. Therefore, upon crossing the conditional NF-κB subunit lines to a mouse that expresses the FLP recombinase in the germline, a constitutional knockout (null) allele without eGFP expression can be generated (Fig. 1).

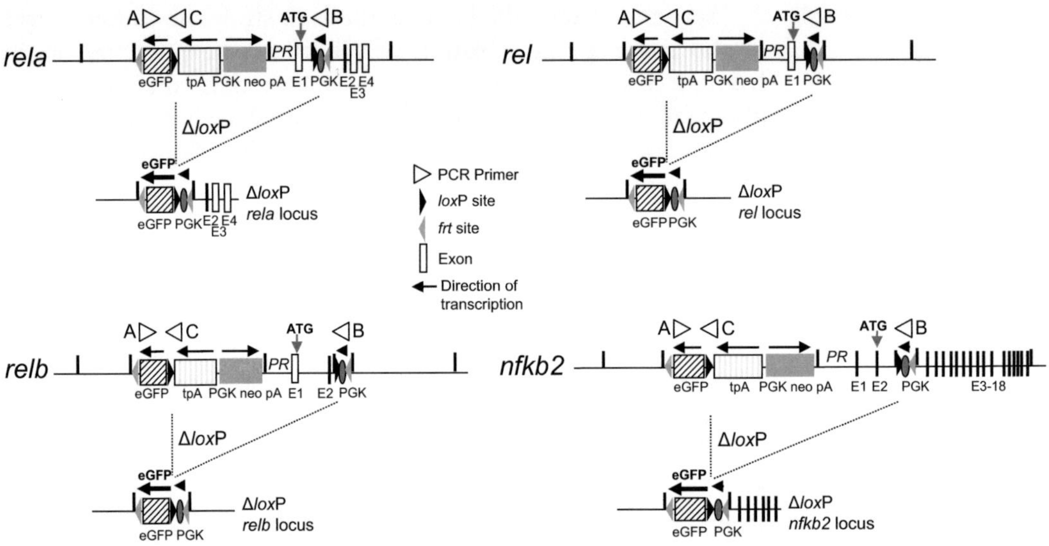

Fig. 1 Conditional knockout alleles of NF-κB subunits. The conditional alleles with their exons, promoter regions (PR) and translational start site (ATG) encoding RELA (*rela*), RELB (*relb*), c-REL (*rel*) and NF-κB2 (*nfkb2*) are depicted alongside the transgenic elements eGFP (eGFP minigene coding region), tpA (triple poly A site), PGK neo pA (neomycin resistance gene), PGK (PGK promoter), loxP sites and frt sites. Arrows denote the orientation of transcription. The thick arrow above the eGFP minigene indicates expression of eGFP following loxP deletion. The sites and orientations for the primers of the deletion PCR (A–C) are indicated by triangles above each locus. See text for further details

2 Materials

2.1 PCR

1. Primers to detect deletion (for location in locus, *see* A, B, C Fig. 1) from purified DNA.

   Primer A: eGFP_Fw: 5′- CAG GATGTTGCCGTCCTCCTTG-3′.

   Primer B: loxP_Hygro_Rv: 5′- CCTGCAGCCAA TAAGCTTATAAC-3′.

   Primer C: loxP_eGFP_Rv: 5′- TATCATGTCTGGAT CAATTCATAAC-3′.

   Expected sizes: A + B, 470 bp; A + C, 470 bp.
2. Taq polymerase.
3. Taq polymerase buffer.
4. $MgCl_2$.
5. Nucleotides.

2.2 Immunization with Sheep Red Blood Cells (SRBC)

1. Sterile phosphate-buffered saline (PBS).
2. Trypan Blue.
3. 40% SRBCs in Alsever's solution.
4. 50-ml Falcon tube tubes.

5. Refrigerated centrifuge.
6. 1-ml syringes.
7. 25G1 needles.

2.3 Sample Generation from Mouse Tissues Following Immunization

1. PBS-0.5% BSA.
2. 40-μm cell strainer.
3. 50-ml Falcon tube
4. Plunger of a 5-ml syringe.
5. RBC lysis buffer (e.g., Red Blood Cell Lysis HybriMax; Sigma Aldrich); alternatively, prepare lysis buffer: 0.15 M ammonium chloride (NH_4Cl), 8.29 g/l; 10 mM potassium bicarbonate ($KHCO_3$), 1 g per l; EDTA 0.1 mM, 0.037 g/l; 1 l H_2O; filter solution with 0.45 μm filter, store at 4 °C.
6. 15-ml Falcon tubes.
7. Refrigerated centrifuge.
8. Optional: cell counter.

2.4 Flow-Cytometric Analysis

1. Antibodies against lymphocyte surface antigens (e.g., B220, CD19, CD3) as conjugates (e.g., PE, PerCP, APC).
2. Fluorescent-activated cell analyzer.
3. 1.5-ml collection tubes.
4. PBS-0.5% BSA.
5. Refrigerated centrifuge.

2.5 Preparation of Tissue for Immuno-histochemistry (IHC)

1. Paraformaldehyde in PBS (4%).
2. Paraffin.
3. Microtome.
4. Histology slides.

2.6 IHC of Lymphoid Tissue

1. Xylene.
2. Ethanol.
3. Heat plate.
4. 10 mM sodium citrate buffer with 0.05% Tween 20, pH 6: Add 2.94 g Trisodium citrate (dehydrate) to 1 l distilled water. Mix to dissolve and adjust pH to 6.0 with 1 N HCL. Add 0.5 mL Tween 20 and mix well. Store at RT for 2 months or at 4 °C for longer storage.
5. Microwave.
6. 1 mM EDTA, pH 7.5.
7. Tris-buffered saline (TBS) 10×: Dissolve 24 g Tris base and 88 g NaCl in 900 ml distilled water. Adjust pH to 7.6 with HCl (hydrochloric acid). Add distilled water to a final volume of 1 l.

8. TBS-Tween 20 (TBS-T): Mix 100 ml TBS with 900 ml distilled water. Add 1 ml Tween 20.
9. 3% or 2% bovine serum albumin (BSA) in TBS-T: Add 15 g BSA or 10 g BSA, respectively, in 500 ml TBS-T.
10. Hydrophobic Barrier Pen.
11. Antibodies against lymphocyte antigens (*see* **Note 1**) and horse radish peroxidase (HRP) developing systems (for the latter, *see* **item 14**).
12. 0.1 M Tris-buffer, pH 9.5: Add 12.1 g Tris base in 900 ml ddH_2O; pH to 9.5 with HCL, add ddH_2O to 1 l.
13. Alkaline phosphatase (AP)-conjugate development buffer (NBT/BCIP buffer): Add 4.76 g $MgCl_2$ (to make 0.05 M) and 5.84 g NaCl (to make 0.1 M) and make up 1 l with 0.1 M Tris-buffer pH 9.5.
14. DAKO EnVision dual link HRP system (Cambridge Biosciences).
15. 1 g/l Hematoxylin.
16. Glycerol gelatin (e.g., pre-prepared from Sigma).
17. 30%Hydrogen peroxide (diluted to 3% H_2O_2 in PBS).

2.7 Acquisition of Blood Samples and Serum Preparation

1. Razor blade or scalpel.
2. 1- or 2-ml syringes.
3. 25G1 needles.
4. 1.5-ml collection tubes.
5. Centrifuge.

2.8 ELISA

1. 96-well immune plates.
2. 96-well culture plates.
3. Parafilm.
4. PBS.
5. 2% fetal bovine serum (FBS) in PBS.
6. Wash buffer: PBS containing 0.05% Tween-20.
7. Capture antibody: Goat anti-mouse IgG1 (Southern Biotech).
8. Isotype control mouse IgG1-unconjugated (Southern Biotech).
9. Detection antibody: AP-conjugated anti-mouse Ig (e.g., IgM, IgG1) (Southern Biotech).
10. PNPP (p-nitrophenylphosphate; Southern Biotech).
11. Substrate buffer (500 ml): 24.5 mg $MgCl_2{\cdot}H_2O$ + 48 ml diethOlamine; add ddH_2O to a volume of 500 ml. Adjust pH to 0.8 with 5 N HCl.
12. Microplate reader.

3 Methods

3.1 PCR to Detect Cre-Mediated Deletion of Transgenic NF-κB Subunit Alleles

1. Use 1–2 μl of genomic DNA from DNA isolated by standard protocols from animal tissues in which the deletion is to be analyzed, or tails or ear punches from pups in order to determine potential germline deletion.
2. Set up two separate standard PCR reactions with primer pairs A + B and A + C (*see* **Note 2**). The reason for the separate PCRs is that the length of the PCR products is 470 bp in both cases. Perform the PCR reaction with an annealing temperature of 58 °C for 33 cycles, analyze on a 1.5% agarose gel. The $MgCl_2$ concentrations in the respective PCR reactions using Taq polymerase from Sigma are 1.5 mM for the A + B pair and 2.5 mM for A + C.

3.2 Primary and Secondary Immunization with SRBC

3.2.1 Preparation of SRBCs

1. (Work is performed in a biosafety hood.) To calculate the SRBCs required for the immunization of one mouse, take 100 μl of the commercially available 40% SRBC solution in Alsever's and dilute up to 1:1000 by serial dilution into 1 ml PBS. Mix 50 μl diluted SRBC suspension and 50 μl Trypan blue and transfer to cell counting chambers or automated cell counter. Determine live SRBCs and calculate the dilution required to obtain 5×10^8 SRBC per 200 μl injection volume per mouse, i.e., 2.5×10^6 per μl.
2. Calculate volume of the 40% SRBC in Alsever's solution for 8 ml (which should give sufficient SRBCs for many injections) and wash with PBS in a 15 ml Falcon tube; spin down for 5 min at 800 × *g* (2000 rpm in large table-top centrifuge) in a refrigerated centrifuge.
3. Repeat washing with PBS and centrifugation steps until the supernatant is colorless and/or not turbid anymore.
4. After the final centrifugation, resuspend SRBCs in 8 ml PBS and keep on ice.

3.2.2 Immunization

1. Primary immunization: In the animal facility, inject 200 μl SRBCs per animal intraperitoneally.
2. Secondary immunization: After at least 3 weeks, repeat **steps 1–4** (Subheading 3.2.1) and **step 1** (Subheading 3.2.2).

3.3 Sample Generation from Mouse Tissues Following Immunization for Flow Cytometry

1. Isolate mononuclear cells from lymphoid tissue (e.g., spleen, lymph node) in PBS-0.5% BSA (refrigerated) at day 14 (*see* **Note 3**) by mincing the tissue using a 40-μm Falcon cell strainer placed on top of a 50-ml Falcon tube and a plunger from a 5-ml syringe to mince the lymphoid tissue with occasional flushing with PBS-0.5% BSA.

2. Spin down the cell suspension in a refrigerated centrifuge in a 15-ml Falcon tube and resuspend in RBC lysis buffer for 3–5 min, then add 10 ml PBS-0.5% BSA and spin; after resuspending in 10 ml, let the debris (sometimes referred to as "ghosts") resulting from the lysed red blood cells settle and transfer supernatant into a new 15-ml tube, count cells (e.g., using automated cell counter or standard counting method). This fraction is used for the analytical flow cytometry analysis and can also be used for fluorescent-activated cell sorting (FACS) of a population of interest.

3.4 Flow-Cytometric Analysis of Cell Subsets from Immunized Mice

1. Use 500,000 to two million cells per staining (*see* **Notes 4** and **5**). Prepare cell mixture such that the desired number of cells is in a volume of 50 μl.
2. To set the compensation on the flow analyzer, prepare either tubes for single stains of each fluorochrome in addition to a sample without staining, or use compensation beads with the respective fluorochromes. Take into consideration that cells of the conditional NF-κB subunit mice may be eGFP-positive when crossed to a Cre-expressing mouse line, implying that the "FITC channel" (emission spectrum 500–550 nm; "green channel") cannot be used for other antibodies. Therefore, cells from those mice are not suitable for setting the compensation; rather, cells from a Cre-only mouse that contains wild-type NF-κB subunit genes is the ideal control (both as biological experimental littermate control and for setting compensation).
3. Dilute the antibodies coupled to fluorochromes used in the experiment in a volume of 50 μl (when making a master-mix, prepare for one more staining, or two more if many samples are being stained), which is subsequently added to the 50 μl cell suspension to have 100 μl total staining volume. Thus, the amount or concentration of antibodies used should be calculated for 100 μl staining volume; for many commercially available antibody conjugates, the suggested staining dilution is 1:100; however, the best antibody concentration should be titrated individually for each antibody.
4. Stain cell suspensions for 15 min on ice, flicking the tube intermittently once or twice.
5. Wash cells by adding 1 ml PBS-0.5% BSA and spin down at 290 × *g* (1200 rpm) for 5 min in a refrigerated benchtop centrifuge (4 °C). Resuspend the cell pellet by flicking the tube before adding 500 μl PBS-0.5% BSA.
6. If a streptavidin-biotin staining system is being used, or if the primary antibody is detected by a secondary antibody coupled to a fluorochrome, repeat **steps 3–5**.

7. Before analyzing the cells in a flow cytometer, pass the cell suspension through a 40-μm cell strainer to exclude debris that may clog the tubing of the instrument.
8. Analyze the cell suspensions on a flow cytometer after the compensation has been set.

3.5 Preparation of Tissue for IHC

1. Dissect tissue of interest, e.g., spleen or lymph nodes; keep the tissue in PBS-0.5% BSA until further processing. Optional: If tissue needs to be weighed, briefly dry off buffer with a Kim-wipe tissue and use weighing boat or paper.
2. Place tissue in 5 ml paraformaldehyde overnight.
3. Remove paraformaldehyde and fill tube with dH_2O; invert several times.
4. Remove dH_2O and add 5 ml 70% Ethanol (*see* **Note 6**).
5. Send tissue to histology core facility for embedding in paraffin and cutting of sections with 2–5 μm in thickness for mounting onto histology slides. Ask for a hematoxylin & eosin (H&E)-stained slide, which is useful for evaluating the quality of the section as well as tissue morphology.

3.6 IHC of Lymphoid Tissue

3.6.1 Deparaffinization

1. Heat slides at ~60 °C for 20 min on a heat plate (Fig. 2a).
2. Deparaffinize tissue sections on the slides with xylene for a total of 15 min in a coplin jar.
3. Dehydrate with 100% ethanol for 15 min in a coplin jar, wash out ethanol with dH_2O for 5 min by changing the dH_2O three times in a coplin jar.

3.6.2 Antigen Retrieval

1. Immerse slides and holder in 800 ml of 10 mM sodium citrate buffer with 0.05% Tween 20.
2. Boil in microwave at maximum power until solution boils.
3. Boil for an additional 15 min at a reduced power so that the solution continues to simmer. Cool at room temperature (RT) for 30 min to 1 h.

3.6.3 IHC

1. Briefly wash slides in TBS-T.
2. Incubate slides in 3% H_2O_2 in PBS for 20 min and place back into TBS-T.
3. Clip slides into IHC rack (Fig. 2b; *see* **Note 7**) making sure TBS-T covers specimen when pressure applied.
4. Check running of liquid by applying 1 ml TBS-T per slide.
5. Blocking: Add 1–2 ml 3% BSA in TBS-T to slides in IHC rack for a period of 45 min to 1 h.
6. Add 200 μl antibody in 3% BSA/TBS-T. In our example staining, the primary antibody is an anti-BCL6 rabbit polyclonal

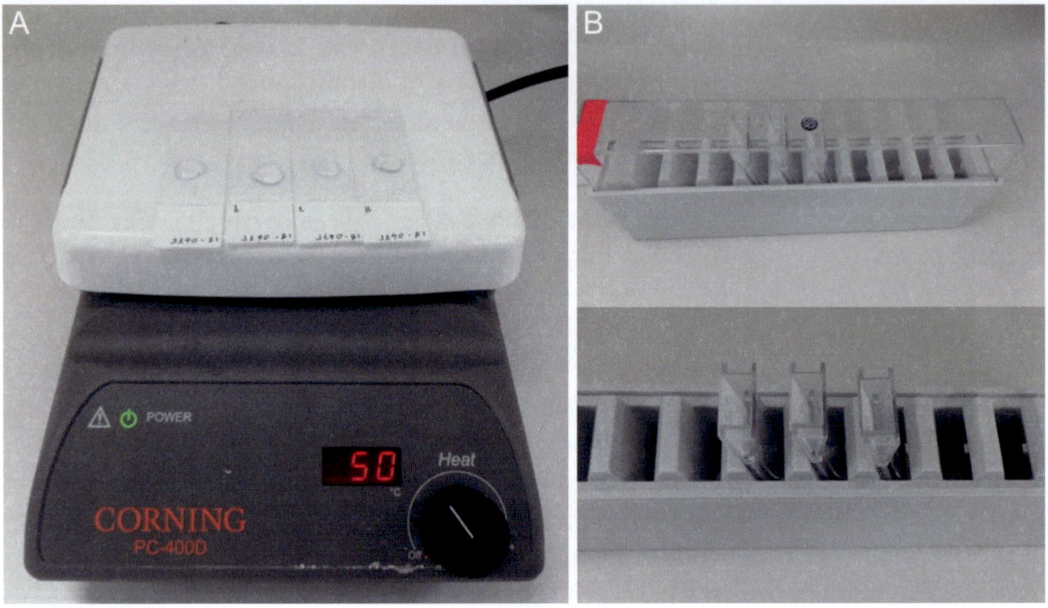

Fig. 2 Heat plate and IHC rack. (**a**) Example of a heat plate. (**b**) Example of an IHC rack with and without lid

antibody (Proteintech) at 1:100 (*see* **Note 8**); leave at 4 °C overnight (e.g., in cold room).

7. Wash 3× 20 min in TBS-T.
8. Apply 200 μl of secondary antibody (anti-rabbit HRP; comes with the DAKO EnVision dual link HRP system) in appropriate concentration and incubate in humidity chamber (note that the IHC rack with a lid on can be used as humidity chamber) for 45 min at RT.
9. Wash 3× 20 min in TBS-T.
10. Remove slides from slide rack, dry with a Kimwipe tissue, draw around the specimen with hydrophobic IHC pen.
11. Preheat heat plate to ~30 °C.
12. Focus light microscope (for example on a previously stained slide to not dry out the slides while staining).
13. Develop using, e.g., the very sensitive (and thus best suitable for BCL6, a transcription factor that is not highly expressed) DAKO EnVision dual link HRP system according to manufacturer's instructions; pipette 1 ml substrate buffer into a 15-ml falcon tube, add one drop DAB+, vortex; put slide on heat plate set at ~30 °C and pipette 100–200 μl of DAB+ reagent onto tissue section.
14. Determine levels of staining under microscope, when optimum staining reached, quench in TBS-T. Note that if no secondary staining is planned, proceed to hematoxylin counter stain and slide mounting.

15. Secondary stain: repeat **steps 3–5** from above.
16. Add 200 μl antibody in 3% BSA/TBS-T. In our example, the antibodies are human-adsorbed goat anti-mouse IgG1 or IgM polyclonal antibodies, both of which are AP-conjugated; incubate for 1 h in humidity chamber (or IHC rack as in **step 8**) at RT (*see* **Note 9**).
17. Wash 3× 20 min in TBS-T.
18. Remove slides from slide rack, dry with a tissue (Kimwipe), preheat heat plate to ~30 °C, focus microscope.
19. Develop with AP-substrate; prepare developing solution and use within 5 min of adding the NBT/BCIP stock solution. Specifically, to every 1 ml NBT/BCIP add 20 μl of NBT/BCIP stock solution; vortex solution.
20. Place slide on heat plate and pipette 100–200 μl of AP-solution, check development under microscope; when optimum staining reached, quench in TBS-T.

3.6.4 Counterstain and Mounting

1. Counterstain with hematoxylin for 15 s, then wash in pot with running water facing away from mounted tissue.
2. Heat glycerol gelatin so that it is liquid without forming bubbles. This can be achieved by heating for ~5–8 s in the microwave, or 30 min in a 45 °C waterbath.
3. Heat dH_2O in microwave in coplin jar, full power ~30 s.
4. Dip slides in heated water and dry slide off with a Kimwipe tissue.
5. Drop glycerol gelatin onto slide in diagonal line, take coverslip and lay down from one end, making sure not to trap bubbles.
6. Leave to dry at RT overnight.

3.7 Acquisition of Blood Samples and Serum Preparation

1. Bleed mice by a method sanctioned by the animal facility 7 days before the immunization and at day 14 after primary immunization and at least day 7 after secondary immunization; take a few drops of peripheral blood, although one drop (~40 μl) is sufficient (*see* **Note 10**). After the final time-point, blood can be withdrawn from the heart of the sacrificed animal. When using syringes for blood collection (1 or 2 ml-syringe with 25G1 needle), transfer into 1.5-ml reaction tube.
2. Keep the reaction tubes at RT for the remainder of the day and at 4 °C overnight to allow clotting (*see* **Note 11**).
3. Spin down for 5 min at 12,000 × *g* in a benchtop centrifuge.
4. Transfer supernatant into new 1.5-ml reaction tube, ideally labeled with mouse number, cohort, and date of experiment.
5. Repeat the process to remove any clotted material and store in appropriate box at −20 °C until use.

3.8 ELISA

1. For analysis of total serum immunoglobulin (Ig) in SRBC-immunized mice by ELISA, coat 96-well (flat bottom) cell culture plates with 50 μl per well of 2.5 μl/ml capture antibody diluted in PBS in all wells. (For the antibody response to a model hapten, for T-independent immunization, and for detection of antigen-specific cells by ELISPOT, *see* **Note 12**)
2. Seal the plates with Parafilm and incubate overnight at 4 °C. Discard the solution and wash the wells 3× with 200 μl wash buffer.
3. Apply 200 μl of PBS-2% FBS per well to block unspecific binding, seal the plates with Parafilm and incubate at RT for at least 1 h.
4. Standard curves in rows 1 and 2 of each plate are generated by a 1:2 serial dilution starting with 50 μl of 400 ng/ml isotype control IgG1-unlabeled. Specifically, the standard curve should be 400–200–100–50–25–12.5–6.25–3.125–0 ng/ml per well.
5. Generate 1:2 serial dilutions in PBS-2% FBS of each serum sample in separate 96-well plates with 50 μl of each serum sample dilution per well, starting with a 1:8000 dilution and ending with a 1:256,000 dilution (*see* **Note 13**). Plate every condition in duplicate. Seal plates and incubate for 2 h at RT.
6. Discard the samples and wash the wells 3× with 200 μl wash buffer.
7. Add 50 μl of detection antibody (anti-IgG1-AP or anti-IgM-AP) diluted 1:2000 in PBS-2% FBS to each well. Seal the plates and incubate for 1 h at RT.
8. Discard the samples and wash the wells 5× with 200 μl wash buffer.
9. For detection of anti-IgG1-AP or anti-IgM-AP, dissolve one PNPP 5 mg tablet in 5 ml substrate buffer and apply 50 μl to each well; incubate at RT for 10–30 min, depending on the intensity of color development (*see* **Note 14** for HRP detection, which constitutes an alternative to AP detection).
10. Read the plates at 405 nm OD. Save the "raw data" as well as the "blanked data" and "calculated concentrations" in an Excel data sheet.
11. For the analysis of the amount of secreted IgG or IgM, plot the standard curve and include all sample values of the Ig coated plate that lie within the linear range (standard curve is 4 parameter logistic regression) of the standard curve, indicated by similar concentrations at different sample dilutions. An average of at least two different dilutions in duplicate is recommended for analysis.

4 Notes

1. In our example, we used anti-BCL6 rabbit polyclonal antibody (Proteintech) and human-adsorbed goat anti-mouse IgG1 or IgM polyclonal antibodies, both of which are AP-coupled (Southern Biotech).
2. In the case of using the deletion PCR to test for potential germline deletion of the NF-κB subunit allele, the A + C PCR reaction is not required if previously performed genotyping PCRs have verified sufficient DNA quantity and quality. It is sufficient to perform only the A + B PCR reaction that detects the deletion, and which in our experience is very sensitive even with low amounts of DNA.
3. Day 14 is a suitable day to analyze the primary antibody response since the germinal center reaction is at its peak. In the literature, one also finds days 10 or 12 as time-points for the analysis of the primary response. The secondary response after a repeat immunization is usually analyzed at day 7 since the memory B-cell response is faster than the primary B-cell response, but day 14 and even later days for analysis are possible as well.
4. We use two million cells per staining when our intention is to detect a rare cell population such as plasma cells in spleen that comprise less than 0.5% of splenic mononuclear cells in wild-type mice, and accordingly acquire more events on the flow analyzer. Generally, even less than 100,000 cells can be used in analytical flow cytometry stainings if the material from the lymphoid tissue is limited or if the main purpose is to collect the tissue for other downstream analysis, such as FACS sorting of specific cell populations, IHC on tissue sections or DNA, RNA or protein analyses. For example, with regard to the analysis of spleen after immunization, following weighing of the tissue, we use a scalpel to cut the spleen into two pieces, one measuring 1/3 and the other 2/3 of the organ.
5. When staining with antibodies for multiple antigens, using a panel of different fluorochromes, it is highly suggested to use the same amount of cells in all tubes of a given experiment, because differing numbers of cells can affect the staining intensities of certain fluorochromes, which may have an impact on the compensation and the potential consequence that samples among an experiment cannot be compared with each other.
6. Tissue can be stored in 70% ethanol for a long time (years); however, the quality of the sections deteriorates if the tissue is not processed within the first several months.

7. As an alternative to using an IHC rack, slides can be stained by applying the staining solution onto the tissue section and incubating in a humid chamber box.
8. For antibody concentration, refer to data sheet and/or determine the optimal concentration and staining conditions empirically.
9. Mind what species the antibodies are deriving from. For two consecutive stainings with two different primary antibodies, the species the antibodies were raised in needs to be different if secondary antibodies are used for development, such as in the case of BCL6 in our example.
10. Important: no addition of heparin to the collection tube or syringe as the blood needs to coagulate.
11. The serum can be isolated a day or two later if the tube is kept in the fridge.
12. A protocol that describes an approach to study the antibody response to the model hapten 4-hydroxy-3-nitrophenyl-acetyl (NP) by measuring NP-specific antisera by ELISA is described in ref. 24. This publication in addition describes other approaches, including ELISPOT analysis and immunoglobulin variable region gene sequence analysis, to study somatic hypermutation and affinity maturation using the NP system.
13. All samples with values outside of the linear range should be re-analyzed at a higher serial dilution.
14. For detection of anti-IgG1-HRP or anti-IgM-HRP, add 50 μl TMB substrate (prepared according to the manufacturer's instructions) to each well and incubate at RT for 5 min. Read the plates at 650 nm OD. Both AP and HRP detection methods are equally suitable for the serum Ig ELISA (HRP color development is markedly faster than AP color development).

Acknowledgments

We thank Qiong Shen and Riccardo Dalla-Favera for their involvement in the generation of the *rel* and *rela* alleles.

References

1. Hayden MS, Ghosh S (2004) Signaling to NF-κB. Genes Dev 18:2195–2224
2. Vallabhapurapu S, Karin M (2009) Regulation and function of NF-κB transcription factors in the immune system. Annu Rev Immunol 27:693–733
3. Gerondakis S, Siebenlist U (2010) Roles of the NF-κB pathway in lymphocyte development and function. Cold Spring Harb Perspect Biol 2:a000182
4. Kaileh M, Sen R (2012) NF-κB function in B lymphocytes. Immunol Rev 246:254–271
5. Köntgen F, Grumont RJ, Strasser A, Metcalf D, Li R, Tarlinton D, Gerondakis S (1995) Mice lacking the c-rel proto-oncogene exhibit defects in lymphocyte proliferation,

humoral immunity, and interleukin-2-expression. Genes Dev 9:1965–1977
6. Beg AA, Sha WC, Bronson RT, Ghosh S, Baltimore D (1995) Embryonic lethality and liver degeneration in mice lacking the RelA component of NF-κB. Nature 376:167–170
7. Sha WC, Liou HC, Tuomanen EI, Baltimore D (1995) Targeted disruption of the p50 subunit of NF-kappa B leads to multifocal defects in immune responses. Cell 80:321–330
8. Tumang JR, Owyang A, Andjelic S, Jin Z, Hardy RR, Liou ML, Liou HC (1998) c-Rel is essential for B lymphocyte survival and cell cycle progression. Eur J Immunol 28:4299–4312
9. Weih F, Carrasco D, Durham SK, Barton DS, Rizzo CA, Ryseck RP, Lira SA, Bravo R (1995) Multiorgan inflammation and hematopoietic abnormalities in mice with a targeted disruption of RelB, a member of the NF-kappa B/Rel family. Cell 80:331–340
10. Franzoso G, Carlson L, Poljak L, Shores EW, Epstein S, Leonardi A, Grinberg A, Tran T, Scharton-Kersten T, Anver M, Love P, Brown K, Siebenlist U (1998) Mice deficient in nuclear factor (NF)-kappa B/p52 present with defects in humoral responses, germinal center reactions, and splenic microarchitecture. J Exp Med 187:147–159
11. Caamano JH, Rizzo CA, Durham SK, Barton DS, Raventos-Suarez C, Snapper CM, Bravo R (1998) Nuclear factor (NF)-kappa B2 (p100/p52) is required for normal splenic microarchitecture and B cell-mediated immune responses. J Exp Med 187:185–196
12. Heise N, De Silva NS, Silva K, Carette A, Simonetti G, Pasparakis M, Klein U (2014) Germinal center B cell maintenance and differentiation are controlled by distinct NF-κB transcription factor subunits. J Exp Med 211:2103–2118
13. De Silva NS, Silva K, Anderson MM, Bhagat G, Klein U (2016) Impairment of mature B cell maintenance upon combined deletion of the alternative NF-κB transcription factors RELB and NF-κB2 in B cells. J Immunol 196:2591–2601
14. De Silva NS, Anderson MM, Carette A, Silva K, Heise N, Bhagat G, Klein U (2016) Transcription factors of the alternative NF-κB pathway are required for germinal center B-cell development. Proc Natl Acad Sci U S A 113:9063–9068
15. Milanovic M, Heise N, De Silva NS, Anderson MM, Silva K, Carette A, Orelli F, Bhagat G, Klein U (2017) Differential requirements for the canonical NF-κB transcription factors c-REL and RELA during the generation and activation of mature B cells. Immunol Cell Biol 95:261–271
16. Algül H, Treiber M, Lesina M, Nakhai H, Saur D, Geisler F, Pfeifer A, Paxian S, Schmid RM (2007) Pancreas-specific RelA/p65 truncation increases susceptibility of acini to inflammation-associated cell death following cerulein pancreatitis. J Clin Invest 117:1490–1501
17. Luedde T, Heinrichsdorff J, de Lorenzi R, De Vos R, Roskams T, Pasparakis M (2008) IKK1 and IKK2 cooperate to maintain bile duct integrity in the liver. Proc Natl Acad Sci U S A 105:9733–9738
18. Powolny-Budnicka I, Riemann M, Tänzer S, Schmid RM, Hehlgans T, Weih F (2011) RelA and RelB transcription factors in distinct thymocyte populations control lymphotoxin-dependent interleukin-17 production in γδ T cells. Immunity 34:364–374
19. Oh H, Grinberg-Bleyer Y, Liao W, Maloney D, Wang P, Wu Z, Wang J, Bhatt DM, Heise N, Schmid RM, Hayden MS, Klein U, Rabadan R, Ghosh S (2017) An NF-κB transcription-factor-dependent lineage-specific transcriptional program promotes regulatory T cell identity and function. Immunity 47:450–465
20. Grinberg-Bleyer Y, Oh H, Desrichard A, Bhatt DM, Caron R, Chan TA, Schmid RM, Klein U, Hayden MS, Ghosh S (2017) NF-κB c-Rel is crucial for the regulatory T cell immune checkpoint in cancer. Cell 170:1096–1108
21. Grinberg-Bleyer Y, Caron R, Seeley JJ, De Silva NS, Schindler CW, Hayden MS, Klein U, Ghosh S (2018) The alternative NF-κB pathway in regulatory T cell homeostasis and suppressive function. J Immunol 200:2362–2371
22. Perkins ND (2012) The diverse and complex roles of NF-κB subunits in cancer. Nat Rev Cancer 12:121–132
23. Kennedy R, Klein U (2018) Aberrant activation of NF-κB signaling in aggressive lymphoid malignancies. Cells 7:189
24. Heise N, Klein U (2017) Somatic hypermutation and affinity maturation analysis using the 4-hydroxy-3-nitrophenyl-acetyl (NP) system. Methods Mol Biol 1623:191–208

Chapter 20

Generation and Surgical Analysis of Genetic Mouse Models to Study NF-κB-Driven Pathogenesis of Diffuse Large B Cell Lymphoma

Bernard D. Maybury, Yolanda Saavedra-Torres, Thomas J. A. Snoeks, Jude Fitzgibbon, and Dinis P. Calado

Abstract

Enforced activation of NF-κB signaling can be achieved by constitutive NF-κB-inducing kinases, IKK2 and NIK, or via lymphoma-associated mutants of MYD88, CARD11, and CD79B. In order to model Diffuse Large B Cell Lymphoma (DLBCL) in mice, conditional alleles for these proteins are combined with alleles targeting Cre recombinase expression in mature B cells. However, unopposed NF-κB signaling promotes plasmablast differentiation, and as a consequence the model system must be complemented with further mutations that block differentiation, such as *Prdm1*/BLIMP1 inactivation or overexpression of BCL6. Here, we describe the currently available tools for DLBCL models in mice and their relative advantages and drawbacks. Furthermore, we describe methods to monitor lymphomagenesis, using ultrasound tomography of the spleen, and the technique of partial splenectomy surgery with recovery. These powerful techniques allow paired comparison of individual lymphoma cases before and after interventions, including therapies, and to study the evolution of lymphoma over time. NF-κB activation also promotes widespread nodal involvement with lymphoma and we describe the post-mortem dissection of major nodal groups.

Key words NF-κB, Diffuse large B cell lymphoma, Relapse/refractory, Therapy, Surgery, Partial splenectomy, Spleen ultrasound, Lymphoma evolution

1 Introduction

The NF-κB family is a group of heterodimeric transcription factors which are pivotal in the survival and activation of B cells, in response to signaling through the B cell receptor (BCR), TNF family members (CD40, BAFF), and Toll-like receptors (TLR) [1, 2] (Fig. 1). There are two characterized pathways, the canonical and the non-canonical. Activation of the canonical NF-κB pathway involves the translocation, from the cytoplasm to the nucleus, of heterodimers of p50 plus either RelA, RelB, or cRel, whereas activation of the non-canonical pathway leads to nuclear

Guido Franzoso and Francesca Zazzeroni (eds.), *NF-κB Transcription Factors: Methods and Protocols*, Methods in Molecular Biology, vol. 2366, https://doi.org/10.1007/978-1-0716-1669-7_20,

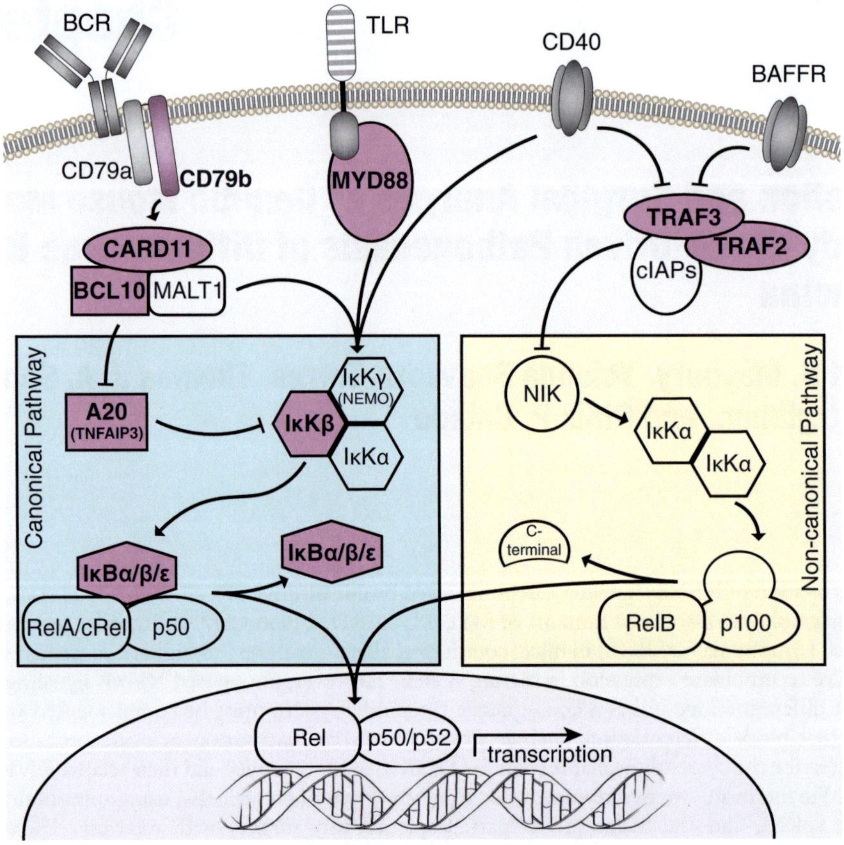

Fig. 1 A summary of canonical and non-canonical NF-κB signaling in B cells, highlighting in **bold** the negative regulators (A20, IκBα, IκBβ, IκBε, and TRAF2/3) and positive regulators (MYD88, CD79b, CARD11, BCL10, IκKβ) which are frequently mutated or lost in DLBCL. Physiologically, NF-κB signaling is determined by surface receptors binding with their cognate ligands, but in DLBCL signaling is often autonomous due to mutations in the pathway

translocation of p52/RelB heterodimers. These heterodimers bind specific response elements on DNA and recruit other proteins to activate transcription. At rest, p50/Rel heterodimers are sequestered in cytoplasm by binding to IκB family proteins, but activation of the IκB kinase (IKK) complex leads to phosphorylation, ubiquitination, and proteasome-mediated degradation of IκB proteins. The isolated p50/Rel heterodimer is then free to enter the nucleus. In the canonical pathway, p50 is constitutively cleaved from its precursor protein p105, whereas for the non-canonical transcription factor the rate-limiting step is the cleavage of p52 from its precursor p100. Activation of the non-canonical pathway is controlled by NF-κB-inducing kinase (NIK) which via a different IKK complex leads to phosphorylation and processing of p100 to form the active p52/RelB heterodimer.

Table 1
Proportion of DLBCL cases with a mutation in NF-κB pathway genes

	Chapuy et al. [5], %	Reddy et al. [9], %	Lacy et al. [6], %
MYD88	18	17	18
CD79B	14	5	9
CARD11	11	11	11
BCL10	5	3	4
TNFAIP3	9	6	10 inc. deletions
NF-κBIA	5	3	6
NF-κBIE	3	4	5

Diffuse large B cell lymphoma (DLBCL) is a biologically heterogenous group of conditions which present clinically as a high-grade lymphoma. The large majority, if not all cases derive from germinal center B cells. The majority of cases can be ascribed to either the germinal center-like (GCB) or activated B cell-like (ABC) groups based on transcriptional profiling [3]. NF-κB pathway activation is very common in diffuse large B cell lymphoma (DLBCL), particularly in the ABC subtype. Comparing DLBCL biopsies with healthy germinal center B cells shows gene set enrichment for the NF-κB signature in >95% of ABC and 47% of GCB case, and immunohistochemistry investigating nuclear localization of the p50 or p52 proteins demonstrates NF-κB activation in 61% of ABC and 30% of GCB biopsies [4]. Mutations in the BCR signaling pathway and other NF-κB-activating pathways are recurrent in DLBCL (Table 1) and particularly enriched in the ABC subtype (Table 2). In a DLBCL classification using unsupervised clustering based on whole exome sequencing [5], several subgroups are enriched for mutations that lead to NF-κB activation, but each of those subgroups is dominated by a distinct set of genes within the signaling pathway (Table 3). Those subgroups have been independently reproduced with minor variations [6]. In vitro experiments using ABC-DLBCL-derived cell lines have demonstrated dependence on CARD11, MALT1, BCL10, IκBKB, and NF-κBIA for cell survival [7, 8].

Mouse studies of enforced NF-κB activation in B cells have depended on the use of Cre-loxP technology, mostly using Cre recombinase expression from the *Cd19*, *Cd79a ('mb1')*, *Ighg1 ('cg1')*, and *Aicda* (AID) loci to target recombination in mature or germinal center B cells (Table 4) (*see* **Note 1**). NF-κB signaling can be activated by introducing mutant upstream signaling proteins, such as MYD88 and CARD11 which are constitutively active,

Table 2
Proportion of DLBCL cases within the ABC and GCB subtypes with mutations in NF-κB pathway genes

	ABC, %	GCB, %
MYD88	40	8
CD79B	25	2
CARD11	15	15
BCL10	10	4
TNFAIP3	14	18
NF-κBIA	1	7
NF-κBIE	6	10
TRAF2 inc. deletions	3	3
TRAF3 inc. deletions	16	10

Data from Schmitz et al. [10]

Table 3
Proportion of DLBCL cases carrying mutations in NF-κB pathway genes within clusters identified by Chapuy et al. [5]

	C1, %	C2, %	C3, %	C4, %	C5, %
MYD88	23	14	4	6	44
CD79B	9	6	5	10	47
CARD11	7	3	16	24	11
BCL10	25	0	2	2	0
TNFAIP3	29	3	9	8	0
NF-κBIA	0	3	5	20	0
NF-κBIE	0	2	0	18	0

or by enforced expression of downstream kinases such as IKK2 (canonical) and NIK (non-canonical). Although mutations in upstream signaling proteins are more commonly found in DLBCL and can drive strong NF-κB activity, these molecules also activate other pathways such as JNK and mTOR [11, 12]. On the other hand, enforced activation of IKK2 or NIK leads to specific NF-κB signaling only. Thus it has been shown in mature B cells that constitutively active IKK2, driving canonical NF-κB signaling, can substitute for non-canonical signaling driven by BAFF to promote survival and migration and that, following BCR activation, proliferation is increased [13]. Activation of NF-κB via the non-canonical pathway using enforced NIK expression also generates BAFF-

Table 4
Available Cre alleles specifically targeting B cells at the pro-B, mature, or germinal center stages

Cre allele	Site of insertion	B cell stage targeted	Efficiency[a]	Undesired targets
Mb1-cre [19] aka CD79a-cre	Endogenous *Cd79a* locus (leading to hemizygous WT Ig-α expression)	Pro-B cell	>95%	Germ cells, T cells [20]
Mb1-creERT2 [21] aka CD79a-creERT2	Endogenous *Cd79a* locus (leading to hemizygous WT Ig-α expression)	Pro-B cell	>90% after 2 mg tamoxifen for 5 days	Germ cells, T cells [20]
Mb1-MercreMer [22] aka CD79a-mercremer	Endogenous *Cd79a* locus (leading to hemizygous WT Ig-α expression)	Pro-B cell	High (>50%) but unquantified after 6 mg tamoxifen	Germ cells, T cells [20]
CD19-cre [23]	Endogenous *Cd19* locus (leading to hemizygous WT CD19 expression)	Pre-B cell, higher expression in mature B cells	55–95%	Possibly follicular dendritic cells; no data but some FDCs may express CD19
CD19-creERT2	LC-1 locus on chr.6C1 under control of CD19 promoter (preserving WT CD19 expression levels) [24] Or Endogenous *Cd19* locus (leading to hemizygous WT CD19 expression) [25]	Pre-B cell, higher expression in mature B cells. Only after administration of tamoxifen	27% (heterozygous cre) using 2.5 mg/day tamoxifen for 5 days 2% using 4 mg tamoxifen once	Possibly follicular dendritic cells; no data but some FDCs may express CD19
CD21-cre [26]	Unknown (BAC transgene)	T2 (CD21loIgMhi)—mature (CD21hiIgMlo)	65–80%	Germ cells, uterus, ovaries, testes, kidney, forebrain, heart, follicular dendritic cells [20]
CD23-cre [27]	Unknown (BAC transgene)	Splenic immature B cells, highest expression in mature B cells	4% in immature B cells rising to 20% in follicular B cells	None reported

(continued)

Table 4 (continued)

Cre allele	Site of insertion	B cell stage targeted	Efficiency[a]	Undesired targets
cg1-cre [28]	Endogenous *Ighg1* locus (impairing IgG1 expression from the targeted allele)	Germinal center B cells	>90%	Germ cells, B1 cells, other epithelia in cancer models
Cg1-creERT2 [29]	Endogenous *Ighg1*locus in 3′ UTR	Germinal center B cells	5% after 4 mg tanmoxifen on days 4, 5 and 6 after immunization	Memory cells (low frequency)
S1pr2-creERT2 [30]	Endogenous *S1pr2* locus (no data on effect on endogenous expression)	Germinal center B cells	>95% after 2 mg tamoxifen alternate days after immunization	No data. S1pr2 widely transcribed (cardiac myocytes, neurons, stem cells, germ cells, other hematopoietic)
AID-cre	Unknown (BAC transgene) [31] Or Endogenous *Aicda* locus [32]	Germinal center B cells	66–75%	Germ cells, testes, ovaries, Memory T cells [33]
AID-creERT2 [34]	Endogenous *Aicda* locus disrupting (leading to hemizygous AID expression)	Germinal center B cells. Only after administration of tamoxifen	20% after 10 mg tamoxifen on days 7 and 12 after immunization	Germ cells, testes, ovaries, Memory T cells [33]

[a]Recombination efficiency of Cre alleles varies with different reporter alleles. It is advisable to verify recombination frequency in your chosen experiment design

independent B cell hyperplasia with increased plasma cell formation [14]. However, these downstream activators of NF-κB signaling are not alone sufficient to drive lymphoma formation. Conversely, induction of *Myd88*^{L252P}, the mouse orthologue to the *MYD88*^{L265P} mutation found in human B cell lymphomas, can rarely lead to an indolent lymphoma [15] or (when expressed from the Rosa26 locus) to a lymphoplasmacytic proliferation resembling human Waldenström's macroglobulinemia [16]. Combining *Myd88*^{L252P} with BCL2 overexpression increases the rate of transformation but the resulting neoplasm displays plasmablastic features, including B220 downregulation and expression of CD138 [15]. Likewise, MYC and IKK2 co-activation drives malignant transformation but does not prevent plasmablastic differentiation, resulting in a plasmablastic lymphoma [17]. Conditional expression of a human *CARD11*^{L225LI} mutant was sufficient to drive an aggressive lymphoma in young mice (<10 days old) but these lymphomas express the plasma cell markers CD138, BLIMP1, and XBP1 in addition to the B cell marker B220 [18]. Thus NF-κB activation *per se* is not sufficient to induce a disease resembling DLBCL, and drives differentiation towards a plasmablast or plasma cell phenotype.

Combining NF-κB activation with partner oncogenic mutations can, by contrast, lead to a murine lymphoma that resembles human DLBCL. Loss of *Prdm1* (BLIMP1), found in approximately 25% of ABC-DLBCL, blocks B cell differentiation into plasmablasts, and, combined with IKK2 activation in mice, led to DLBCL that expressed IRF4 and other ABC-related genes [35]. BLIMP1 expression can be suppressed by BCL6, which is frequently expressed in ABC-DLBCL via a chromosomal translocation, and in mice enforced BCL6 expression together with conditional NIK activation also led to a DLBCL-like disease which predominantly has the expression profile of activated B cells [36]. Lymphomagenesis in these model systems occurs with a latency of over 1 year, but compound mutant mice with additional lymphoma-associated oncogenes or loss of tumor-suppressor genes display accelerated development of lymphoma modeling human DLBCL. Conditional inactivation of p53, combined with IKK2 activation and *Prdm1* deletion, generates DLBCL in mice with a latency of 6–10 months [37]. We have found that co-activation of MYC and IKK2 together with *Prdm1* deletion generates DLBCL in mice in 6–8 months (Maybury et al., unpublished).

These increasingly complex model systems can be used to study translational problems, including the identification and testing of therapeutic targets, and tracing the evolution of lymphoma during treatment and relapse. The p53/IKK2/*Prdm1* model demonstrated that p53 loss leads to PD-L1 downregulation and that checkpoint blockade with anti-PD-L1 augments the efficacy of anti-CD20 antibodies [37]. This protocol outlines the generation of NF-κB-driven mouse lymphomas, and methods for monitoring disease progression non-invasively by ultrasound, and invasively by partial splenectomy. These methods can be used to dynamically study the co-evolution of DLBCL and the microenvironment, and particularly the development of treatment resistance/relapse, in ways which are not possible from human diagnostic specimens.

2 Materials

2.1 Generating Conditional Mutations in B Cells

1. B cell-specific Cre recombinase allele.
2. cre-loxP-dependent NF-κB-activating allele(s).
3. Neubauer hemocytometer.
4. Polarized light microscope.
5. Imject alum.
6. Sterile phosphate buffered saline, LPS-free.
7. Sterile sheep red blood cells (SRBC): Shake defibrinated sheep blood and add 5–45 ml sterile PBS in a 50 ml falcon tube. Centrifuge at 1000 × *g* for 5 min. Aspirate supernatant. Resuspend the pellet in 45 ml PBS. Centrifuge at 1000 × *g* for 5 min. Aspirate supernatant. Resuspend the pellet in 4 ml PBS to form stock suspension. Take 10ul and make a 1:10^4 dilution. Place coverslip on hemocytometer and micropipette a small volume of the dilution to fill the counting chamber by capillary action. Examine using the polarized light microscope and count the red blood cells in the four large corner squares, take the mean and multiply by 10^8 to give the concentration in RBCs/ml of the stock suspension. Adjust the volume with PBS to give a stock suspension concentration of 5×10^9 ml^{-1}.
8. 2 mg/ml solution of NP-CGG (Chicken Gamma Globulin), hapten:protein ratio 20–29:1, in PBS.
9. 1 mg/ml NP-CGG emulsion: Shake the Imject alum bottle for 4 min. Add an equal volume of Imject alum dropwise to 2 mg/ml solution of NP-CGG in PBS, vortexing continuously for the next 30 min.
10. 40 mg/ml Tamoxifen in sunflower oil: Dissolve tamoxifen in sunflower oil at a concentration of 40 mg/ml, with warming and constant stirring for 30–60 min.

2.2 Detecting Lymphoma Development by Ultrasound

1. Isoflurane.
2. Oxygen-driven anesthetic delivery system including induction box, either using concentrator (Kent Scientific SomnoSuite or similar) or bottled oxygen.
3. Heat box.
4. Electric shaver.

5. Hair removal cream.
6. Viscotears eye gel.
7. Ultrasound gel.
8. Fujifilm VEVO 2100/3100 or similar setup for in vivo pre-clinical ultrasound imaging with heated stage and 3D motor stage for automated transducer scanning.
9. Fujifilm VevoLab software or similar for scan acquisition and analysis.
10. Linear ultrasound transducer.

2.3 Partial Splenectomy with Recovery

1. Meloxicam/custard mix: Mix equal volumes of 1.5 mg/ml Meloxicam oral suspension (for veterinary use) and egg custard baby food until homogenized.
2. Meloxicam/buprenorphine for injection: Mix 120μl of 5 mg/ml Meloxicam solution for injection, 20μl of 0.3 mg/ml of Buprenorphine for injection, and 160μl sterile saline for a total volume of 0.3 ml per mouse; scale up as required.
3. Surgical face mask.
4. Thermostat-controlled heat pad ×2 (one pre-op, one surgery).
5. Anesthetic delivery system via nose cone and induction box.
6. Viscotears gel.
7. Rectal thermometer.
8. Lubricant jelly.
9. Magnifying table lamp.
10. Glass bead sterilizer.
11. 0.1% Chlorhexidine solution.
12. GLAD Press'n Seal kitchen wrap (for drape).
13. High quality surgical instruments: Needle holder, scissors, McPherson forceps or other straight fine smooth-tipped forceps, fine-toothed forceps, ring forceps.
14. Battery powered cautery pen and batteries.
15. Single use scalpel.
16. Aluminum foil.
17. Sterile swabs.
18. Sutures: 5–0 vicryl curved reverse cutting (or another fine absorbable suture).
19. Sterile saline for injection.
20. 5% Lidocaine cream.
21. Temperature-controlled recovery rack.
22. Stitch cutter.

2.4 Obtaining Tumoral Specimens by Dissection

Dissection boards.

Forceps (McPherson or other straight fine smooth-tipped forceps).

Forceps (Non-toothed, Adson or similar).

Dissection scissors.

3 Method

3.1 Generating Conditional Mutations in B Cells

Select a Cre allele to most effectively address the research question. Different Cre alleles are active at different stages of B cell development and have different recombination efficiencies, toxicities, and off-target effects (*see* **Note 1**). Select Cre-loxP-dependent alleles considering the NF-κB pathway being targeted (*see* **Note 2**).

1. Cross the Cre-loxP-dependent allele(s) of choice with mice homozygous for a B cell-specific or germinal center-specific Cre allele (*see* **Note 3**). Experimental mice should be heterozygous for the Cre allele, as homozygosity increases Cre toxicity and may, depending on the allele, lead to loss of expression of the driver gene.
2. *Optional.* (*see* **Note 4**) Synchronizing recombination using cg1-Cre or AID-Cre can be achieved by immunizing mice. EITHER

   Inject 1×10^9 SRBCs in 200μl PBS via tail vein, or intraperitoneally.

   OR

   To identify antigen-specific B cells: inject 100μl NP-CGG/Alum emulsion (containing 100μg NP-CGG) intraperitoneally. Note: Do not inject intravenously due to risk of anaphylaxis.
3. *For creERT2 strains only.* Gavage 4 mg tamoxifen in 100μl sunflower oil once daily for 4 days (if using AID-CreERT2: start gavaging after immunization, at day +8 if the aim is to target the mature germinal center).

3.2 Detecting Lymphoma Development by Ultrasound

The latency of lymphoma development varies between experimental models by many months. Within individual models, latency may vary by up to 12 weeks (*see* **Note 5**). Monitoring animals can be undertaken by examination and palpation, by serial blood tests if a reliable blood marker is found, or by imaging. Reliable clinical signs vary depending on site of disease (*see* **Note 6**). Various imaging modalities can be used (ultrasound (US), CT, MRI, PET). Ultrasound is quick and involves less procedural burden on animals, and can accurately measure spleen volume but, due to overlying gas-containing structures, it cannot reliably image thoracic or abdominal lymph nodes. Note that NF-κB signaling is associated with extranodal disease [38] and this is difficult to detect non-invasively, unless a luminescent protein is incorporated in the disease model.

For small animal imaging, a high-frequency ultrasound system with an operating frequency of at least 20 MHz is recommended to obtain sufficient anatomical contrast and resolution for reliable measurements. Additionally, the system needs to be capable of acquiring a 3D image stack. This protocol is written for the Visual-Sonics Vevo 3100 ultrasound system, but it can easily be adapted to suit similar high-frequency ultrasound systems.

1. Start the ultrasound system, initialize the linear transducer, turn on heated stage set to 37 °C, and pre-heat the ultrasound gel and water to 37 °C in a heat box.
2. Set up a new study within the Vevo 3100 data management system, and a new series within it, using the "General Imaging" application and "Abdominal imaging" measurement package. Initialize the 3D motor. Place the linear transducer in the 3D motor stage.
3. Place mouse in induction box with isoflurane 5% for 1–3 min until unresponsive.
4. Apply viscotears gel to each eye and lie right side-down on heat pad with nosecone supplying 2–2.5% isoflurane (*see* **Note** 7).
5. Shave left flank to expose spleen, which can usually be seen beneath the skin.
6. Apply hair removal cream using cotton bud ensuring hair fully covered, wait 1–3 min.
7. Gently scrape off hair removal cream.
8. Moisten paper towels (using warm water) and wipe off residual cream. Dry thoroughly with paper towels.
9. Apply ultrasound gel liberally to left flank.
10. With Vevo 3100 set to B-mode, lower the transducer in the motor stage so that it is in contact with the gel and an image can be seen. The transducer should not press on the mouse. Manually scan (moving the heat pad+mouse underneath the fixed transducer) to determine the approximate size and locate midpoint of spleen. Orientate the animal on stage to allow gantry to scan longitudinally. Leave the heat pad + mouse positioned so the transducer is halfway down the spleen length.
11. Tap the 3D tab on the Vevo 3100 touch screen. In the 3D panel, slide the "Start" and "Stop" indicators to locate the spleen and center it relative to the transducer. Set the scan distance to the estimated spleen length in mm plus 10%. The available step size ranges from 0.04 to 0.50 mm depending on

resolution required. For splenic volume, a step size of 0.20 mm is sufficient. Tap "Scan" to acquire the selected volume (*see* **Note 8**).

12. Click Load 3D and scroll through the images to ensure the whole spleen has been imaged. If not, reposition the stage +mouse and rescan, adjusting the scan distance if required.
13. When satisfactory images are obtained, raise the transducer, clean the mouse, and return to cage. Observe the mouse until it has recovered from anesthetic.
14. The image segmentation and analysis to derive splenic volume measurements can be performed on the Vevo 3100 system or on a workstation running Vevo Lab software. Open the image data in the Vevo Lab software and load it into the 3D workspace.
15. On the "Measurement" panel, click "Parallel & Rotational Methods" and select "Closed" volume. Start the parallel segmentation.
16. Trace a contour around the spleen at regular intervals throughout the entire dataset and then click "Finish" to calculate the 3D volume in mm^3. Additionally, a wireframe volume rendering can be exported.

3.3 Partial Splenectomy with Recovery

Institutional procedures and veterinary guidance for recovery surgery should be followed and may deviate from the protocol below. There should be separate pre-operative and surgery areas, which may be in different rooms. Wear a fluid-resistant surgical mask during pre-operative preparation and surgery. The aim of perioperative care is to ensure that the mouse stays warm, well oxygenated, well hydrated, and to minimize pain and infection risk. The spleen is highly vascularized and the most common surgical difficulty with partial splenectomy is hemorrhage, particularly where the spleen is large due to tumor. In mice with advanced lymphoma, we have a surgical mortality rate of 20%.

1. Three and two days prior to surgery, supply egg custard (0.2 ml per mouse) in a petri-dish to acclimatize mice.
2. One day prior to surgery, single-house mice for surgery and supply 0.2 ml/mouse of Metacam/custard mix in a petri-dish.
3. One day prior to surgery, autoclave surgical instruments (needle holder, scissors, McPherson forceps, fine-toothed forceps, ring forceps) and aluminum foil 12 × 15 cm (1 sheet per mouse).
4. On the day of surgery, prepare surgical area as illustrated (Fig. 2) and wipe whole area with disinfectant. Secure nose cone perpendicular to heat pad to allow mouse to lie on right hand side. Place petri-dishes containing chlorhexidine 0.1% and sterile saline 0.9% on heat pad to warm.

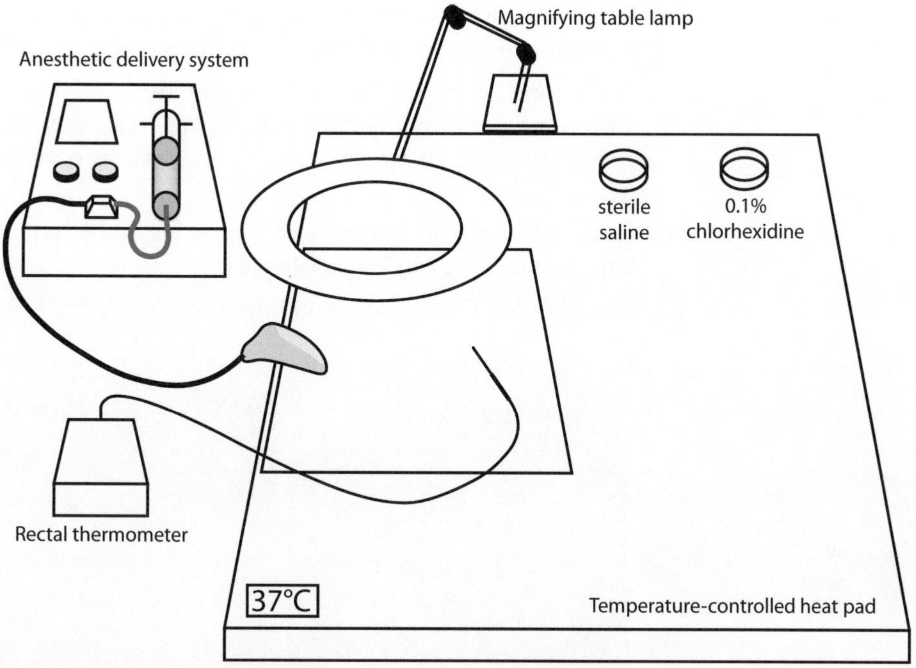

Fig. 2 Example layout of surgical area

5. Gather 1× per mouse (unopened): 5–0 vicryl suture, disposable sterile scalpel, 1 ml sterile syringe, two packs sterile small swabs, sterile gloves.

In pre-op area

6. Weigh mouse.
7. Place mouse in induction box with isoflurane 5% for 1–3 min until unresponsive.
8. Apply viscotears gel to each eye and lie right side-down on heat pad with nosecone supplying 2–2.5% isoflurane.
9. Scruff the back of neck and inject subcutaneously 5μl/g of the meloxicam/buprenorphine mixture.
10. Shave left flank, the spleen should be faintly visible through the skin.
11. Disinfect the shaved skin by wiping with swabs soaked in warmed chlorhexidine 0.1%, repeat three times.

In surgery area

12. Place mouse on surgical area heat pad with isoflurane 2–3% via nose cone. Titrate isoflurane concentration to respiratory rate 50–60 breaths/min (*see* **Note 9**).
13. Apply lubricant jelly to rectal thermometer, insert into rectum, and tape the wire to the tail to keep in place.

14. Disinfect the shaved skin by wiping with swabs soaked in warmed chlorhexidine 0.1%, repeat three times.
15. Apply long sheets of Press'n Seal, adhesive side-down, over mouse and surrounding operating area to create a sterile field, taking care not to touch the non-adhesive side within the sterile field.
16. Open surgical instruments and disposable sterile equipment onto sterile field. Put on sterile gloves.
17. Cut a hole in Press'n seal overlying the surgery site (Fig. 3a).

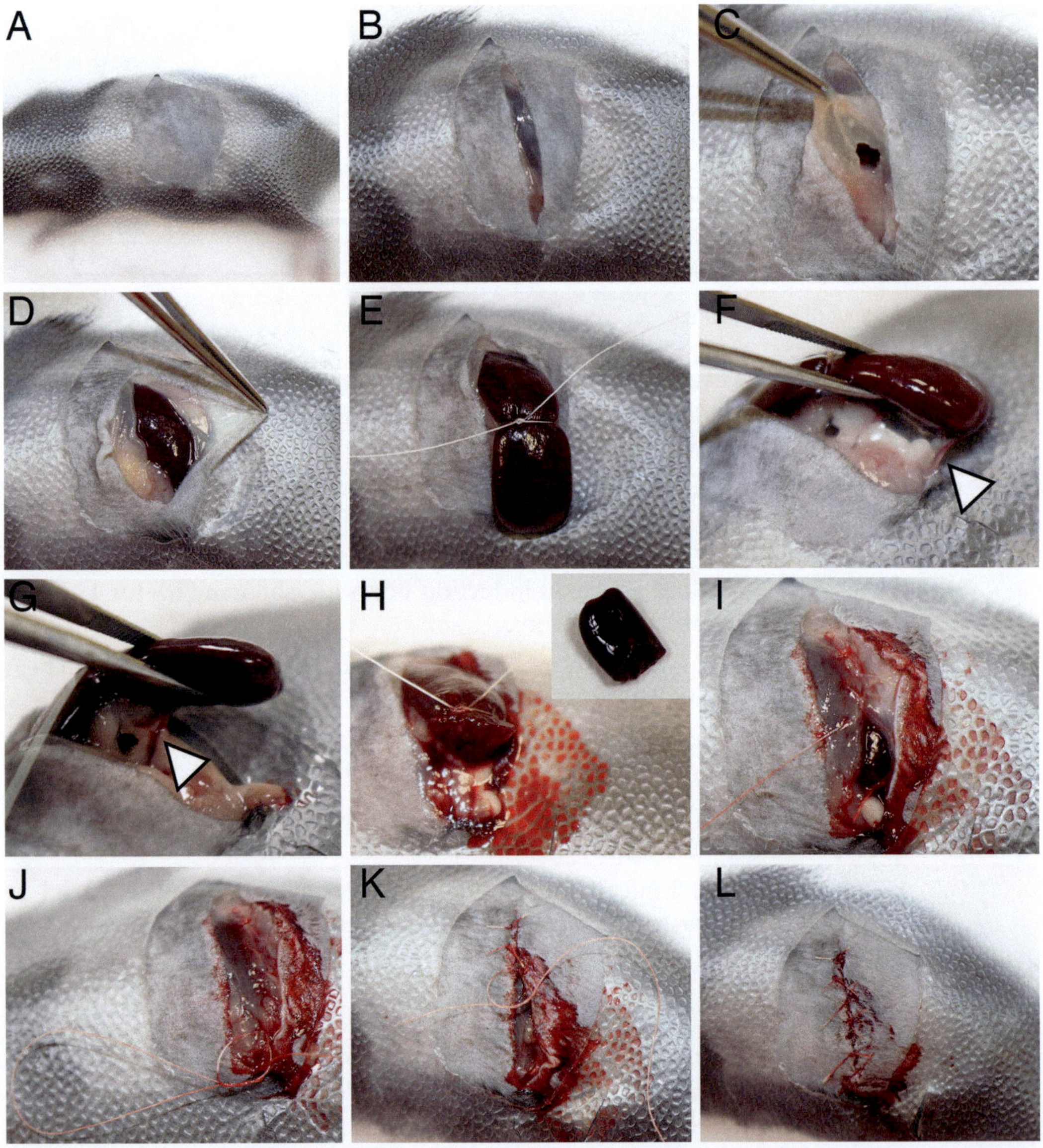

Fig. 3 Partial splenectomy. *See* Subheading 3.3 for descriptions. Arrowheads: *F:* lieno-pancreatic vessels. *G:* Lienal vessels

18. Pull skin surgery site taut, push down to tent the abdomen, and using the scalpel make a transverse incision overlying the spleen (*see* **Note 10**). Extend the cut, if necessary, with scissors (Fig. 3b).
19. Using toothed forceps, pinch abdominal wall, and make a small cut with scissors to breach the peritoneal cavity allowing air in (Fig. 3c), then extend the cut overlying the spleen (Fig. 3d).
20. Work ring forceps underneath the inferior pole of the spleen and externalize outside the abdominal cavity.
21. Using the McPherson forceps, feed the non-needle end of the suture beneath the middle of the spleen taking care to avoid catching blood vessels and pancreas, and tie the suture around the waist of the spleen, tight enough to indent the spleen but not to cut through the splenic capsule (*see* **Note 11**). Leave suture tails un-cut for now (Fig. 3e).
22. Gently reflect the inferior pole to expose the lieno-pancreatic (Fig. 3f arrowhead) and lienal (Fig. 3g arrowhead) vessels supplying the inferior pole (minimize tension to avoid avulsing the vessels).
23. Use ring forceps to hold the spleen pole up (Fig. 3f), allowing the proximal part of the lieno-pancreatic vessels to be clasped with McPherson forceps.
24. Use sterile aluminum foil to pick up and wrap the battery powered cautery pen. Cauterize the lieno-pancreatic vessels between the McPherson forceps and the spleen, ensuring hemostasis (*see* **Note 12**).
25. *Optional.* If the spleen suture has been placed proximal to the lineal vessels, they must also be cauterized. Again, clasp the proximal vessels and cauterize between the forceps and the spleen (Fig. 3g).
26. Holding the inferior pole of the spleen with ring forceps, use the scalpel to cut distal to the suture, freeing the distal spleen fragment (Fig. 3h). Store in chilled saline (if live) or formalin (if not).
27. Inspect the cut surface of the spleen for bleeding. If significant bleeding, use cautery cautiously, but avoid excess cautery otherwise the suture may come loose leading to uncontrolled bleeding.
28. When hemostasis is obtained, cut the waist suture tails and gently reinternalize the remaining spleen.
29. Use toothed and ring forceps to bring the muscle layer together. Suture using continuous sutures, taking care not to catch subcutaneous fat with the needle, closing most of the incision, allowing room for a syringe tip (Fig. 3i).

30. Administer 0.3–0.8 ml sterile saline (depending on estimated volume of blood loss) into peritoneal cavity through the remaining incision.
31. Finish closing muscle layer, tying the final knot using a loop and the end of the suture (Fig. 3j).
32. Close the skin layer using interrupted sutures, ensuring skin edges are accurately opposed (*see* **Note 13**) (Fig. 3k).
33. Clean the surgery site with a sterile swab soaked in sterile saline (Fig. 3l).
34. Apply lidocaine cream 5% generously to the wound line.
35. *Optional.* If multiple animals are being operated on, clean surgical instruments with sterile water and place in the hot bead sterilizer until the next animal is prepared.
36. Remove press'n seal wrap, turn off isoflurane and remove rectal thermometer.
37. Observe mouse on operating table for warmth and oxygen until starting to regain consciousness.
38. Return mouse to its cage and place cage in a recovery rack with temperature control (37 °C), containing a tray of water and humidity monitor. Ensure cage contains drinking water. Observe mouse until it self-rights and starts to move around.
39. Process the resected spleen fragment as required.
40. Return the cage to regular holding rack after 1–2 h.
41. Keep mouse singly housed for at least 48 hours and supply 0.2 ml/mouse of Metacam/custard mix in a petri-dish 24 and 48 h after surgery.
42. Inspect mouse daily for wound dehiscence, infection, or other adverse effects.
43. In the event of partial wound dehiscence, if detected early, it may be possible to re-anesthetize the mouse and re-suture the skin, provided there are no signs of infection.
44. Avoid scruffing mouse for 7 days to allow wound healing.
45. At 7–10 days after the procedure, use stitch cutter and forceps to remove any sutures still present.

3.4 Obtaining Tumoral Specimens by Dissection

In lymphoma models, there is often very mild but widespread lymphadenopathy. Truly tumoral lymph nodes are usually more isolated and many times larger than a physiological lymph node, and as such obvious at dissection. The photographs, which are from one mouse, illustrate how different anatomical sites are variably affected. Exhaustive lymph node dissection is time consuming, and the approach here aims to rapidly ascertain the extent of lymphoma while harvesting involved tissues. *See* Broeck et al. [39]

for an exhaustive anatomical description. Note some lymphomas disrupt the normal morphology and texture of lymph nodes, and may grow as structurally ill-defined and friable masses, particularly in the thorax (*see* **Note 14**).

1. Cull mice by cervical dislocation and exsanguination (*see* **Note 15**).
2. *Optional.* Consider perfusion with fixative (by cardiac puncture) under terminal anesthesia if only fixed specimens are required.
3. Wet fur with 70% ethanol and pin out on dissection board.
4. Make an anterior midline incision in skin, avoiding peritoneum, from groin to chin.
5. Use finger or round-ended scissors to blunt dissect skin off peritoneum, and pin back to reveal inguinal lymph nodes (Fig. 4a showing severe lymphadenopathy on left side only). Use forceps to tease apart subcutaneous fat to isolate inguinal lymph nodes.

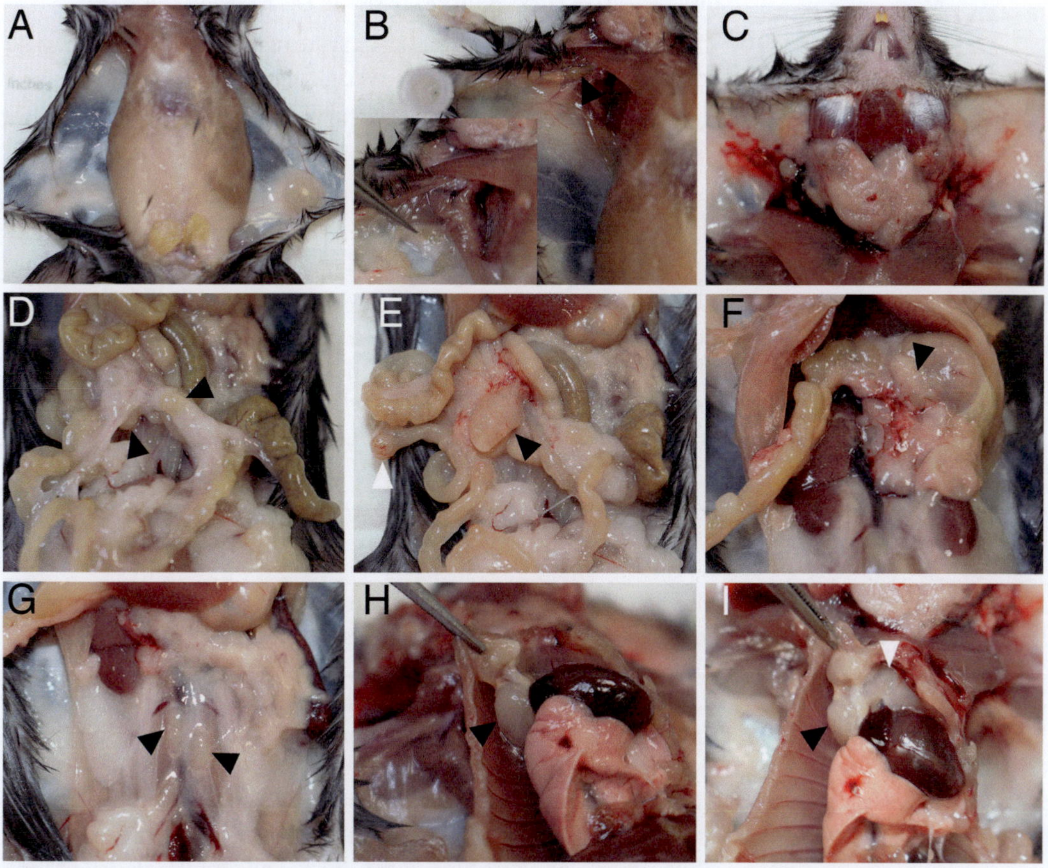

Fig. 4 Tumoral LN dissection. *See* Subheading 3.4 for description. Black arrowheads are lymph nodes. White arrowheads are enlarged Peyer's patch (4E) and thymus (4I)

6. Blunt dissect skin off thoracic wall to reveal axillary lymph node (Fig. 4b, arrowhead). Use forceps to tease apart posterior axillary wall to reveal accessory axillary lymph node (Fig. 4b inset).
7. Gently pull apart the neck skin either side of the incision. Cervical lymph nodes sit lateral to the thyroid beneath a thin connective tissue fascia. Use forceps to tease apart this fascia, uncovering the lymph nodes (Fig. 4c, fascia intact on right side, removed on left).
8. Make a midline incision in the peritoneum to the sternum and pin back. Milky ascites may be present. Carefully draw out the bowel starting with the caecum, observing the mesentery for enlarged lymph nodes. The jejunal lymph nodes are the most prominent in the mesentery (Fig. 4d, e black arrowheads showing grossly enlarged node).
9. Examine bowel to identify tumoral Peyer's patches (Fig. 4e white arrowhead) and examine the proximal mesentery for enlarged gastric and pancreaticoduodenal lymph nodes (Fig. 4f). Dilated bowel due to involved Peyer's patches or mucosa-associated lymphoid tissue (MALT) may be seen.
10. The omentum on the greater curvature of the stomach may be enlarged due to lymphoma arising from milky spots.
11. Remove the spleen, which may be uniformly enlarged or show discrete pale tumors.
12. Behind the bowel the para-aortic lymph nodes can be seen (Fig. 4g).
13. Make a cut in the diaphragm to allow air into the chest cavity. Use scissors to cut out the sternum and anterior rib cage, taking care to avoid thoracic contents, and cut out the diaphragm. Mediastinal lymph nodes are mostly posterior to the thymus (Fig. 4i, white arrowhead) and heart, but when enlarged they are easy to see (Fig. 4h, i).

4 Notes

1. The cg1-Cre in particular, when combined with certain oncogenic loxP-dependent alleles, can generate epithelial cancers (thyroid with R26-PI3K and R26-MYC [40]; stomach with R26-MYC (Calado et al., unpublished); T cell and lung with Kras [41]). This can be circumvented by generating bone marrow chimeras using RAG recipients and cg1-Cre donors. AID-Cre-driven Kras activation can generate papillomas and carcinomas [41], and activation of MYC using the AID-Cre allele gives rise to lung and liver cancers (Calado et al., unpublished).

2. In our experience, as the number of loxP-conditional alleles in the model increases, so does the proportion of lymphomas which have not recombined all the targeted alleles, which can be as high as 10–30% in a 3-allele model. This is likely to be dependent on the efficiency of the Cre recombinase allele. Each loxP-conditional allele should ideally include a marker protein such as GFP to positively identify cells in which recombination has taken place. This also facilitates the use of control strains with loxP-conditional marker proteins.
3. Ideally breeding stock should not carry both the Cre allele and loxP-dependent alleles as germline recombination can occur. If both are present in breeding mice, ensure genotyping strategies can detect germline recombination.
4. Immunization is not required for the generation of lymphomas in germinal center-specific Cre models (using cg1-Cre or AID-cre) due to spontaneous germinal center formation in mesenteric lymph nodes and Peyer's patches (and immunization does not accelerate this process). However, immunization is required to study a synchronized germinal center response, and the combination of immunization with tamoxifen should be used to generate lymphomas from the S1pr2-creERT2, AID-creERT2 and cg1-creERT2 models.
5. Variable latency makes studies of established lymphoma logistically challenging. Starting experiments in individual animals as they develop lymphoma ensures validity but introduces age as an uncontrolled variable. Conducting experiments with cohorts introduces variation in the stage of individual animals' lymphomagenesis which reduces validity of the model.
6. We have observed that mediastinal lymphadenopathy often causes reduced activity, slower scruff recovery, and tachypnea. Mesenteric lymphadenopathy or splenomegaly causes abdominal swelling but often mouse activity is normal. Involvement of Peyer's patches is common in NF-κB-driven models and often causes weight loss or abdominal swelling (due to bowel obstruction/ileus). Peripheral nodes can be seen as subcutaneous swellings.
7. Temperature monitoring is not necessary for the short time under anesthetic (<5 min).
8. 3D scanning on the Vevo 3100 can be synchronized with respiratory monitoring to avoid breathing artifacts in the spleen images, but this is cumbersome and makes little difference to estimates of spleen volume. Additionally, breathing artifacts can be removed from the dataset during post-acquisition image processing.

9. Extensive mediastinal disease increases the likelihood of respiratory arrest under anesthesia, so ensure isoflurane concentration is no higher than necessary to maintain anesthesia.
10. This initial incision requires practice on dead mice. Indeed, the whole procedure should be practiced several times prior to experiments on live animals. Speed of surgery is important especially in heavily diseased animals, as longer anesthetic times increase the chance of respiratory arrest.
11. Severe hemorrhage is the major danger with significantly enlarged spleens. Significant bleeding is often fatal.
12. We find half-used batteries are better as it reduces the heat of cautery enabling greater control.
13. Mice pick at sutures especially anterior sutures, so ensure some redundancy in the number of sutures and leave reasonably long threads, both for visualization purposes and to reduce their removal by animals.
14. *Ex vivo* fixation should be performed for at least 24 h in normal buffered formalin in >10× the volume of dissected tissue. Single cell suspensions require enzymatic digestion if accurate myeloid cell retrieval required. In tumoral spleens, the yield of viable cells/mg tissue is much lower than that of healthy tissue, for unclear reasons.
15. This culling maximizes viability of recovered cells and minimizes blood pooling in the spleen and liver.

Acknowledgments

This work was supported by The Francis Crick Institute which receives its core funding from Cancer Research UK (FC001057), the UK Medical Research Council (FC001057), the Wellcome Trust (FC001057); the MRC career development award MR/J008060/1, and Leukaemia UK (2019/PG/0001) to D.P. C.; by CRUK [C355/A26819] and FC AECC and AIRC under the Accelerator Award Program to J.F. and D.P.C.; B.M. was supported by a doctoral clinical fellowship from a CRUK Accelerator Award [C422/A26084].

References

1. Klein U, Heise N (2015) Unexpected functions of nuclear factor-κB during germinal center B-cell development: implications for lymphomagenesis. Curr Opin Hematol 22 (4):379–387
2. Taniguchi K, Karin M (2018) NF-κB, inflammation, immunity and cancer: coming of age. Nat Rev Immunol 18:309–324
3. Alizadeh AA, Eisen MB, Davis RE et al (2000) Distinct types of diffuse large B-cell lymphoma identified by gene expression profiling. Nature 403:503
4. Compagno M, Lim WK, Grunn A et al (2009) Mutations of multiple genes cause deregulation of NF-kappaB in diffuse large B-cell lymphoma. Nature 459:717–721

5. Chapuy B, Stewart C, Dunford AJ et al (2018) Molecular subtypes of diffuse large B cell lymphoma are associated with distinct pathogenic mechanisms and outcomes. Nat Med 24:679–690
6. Lacy SE, Barrans SL, Beer P et al (2020) Targeted sequencing in DLBCL, molecular subtypes, and outcomes: a Haematological Malignancy Research Network report. Blood 135(20):1759–1771
7. Davis RE, Brown KD, Siebenlist U et al (2001) Constitutive nuclear factor kappaB activity is required for survival of activated B cell-like diffuse large B cell lymphoma cells. J Exp Med 194:1861–1874
8. Ngo VN, Davis RE, Lamy L et al (2006) A loss-of-function RNA interference screen for molecular targets in cancer. Nature 441:106–110
9. Reddy A, Zhang J, Davis NS et al (2017) Genetic and functional drivers of diffuse large B cell lymphoma. Cell 171:481–494.e15
10. Schmitz R, Wright GW, Huang DW et al (2018) Genetics and pathogenesis of diffuse large B-cell lymphoma. N Engl J Med 378:1396–1407
11. Jeelall YS, Wang JQ, Law H-D et al (2012) Human lymphoma mutations reveal CARD11 as the switch between self-antigen-induced B cell death or proliferation and autoantibody production. J Exp Med 209:1907–1917
12. Phelan JD, Young RM, Webster DE et al (2018) A multiprotein supercomplex controlling oncogenic signalling in lymphoma. Nature 560:387–391
13. Sasaki Y, Derudder E, Hobeika E et al (2006) Canonical NF-κB activity, dispensable for B cell development, replaces BAFF-receptor signals and promotes B cell proliferation upon activation. Immunity 24:729–739
14. Sasaki Y, Calado DP, Derudder E et al (2008) NIK overexpression amplifies, whereas ablation of its TRAF3-binding domain replaces BAFF: BAFF-R-mediated survival signals in B cells. Proc Natl Acad Sci 105:10883–10888
15. Knittel G, Liedgens P, Korovkina D et al (2016) B-cell-specific conditional expression of Myd88p.L252P leads to the development of diffuse large B-cell lymphoma in mice. Blood 127:2732–2741
16. Ouk C, Roland L, Saintamand A et al (2019) B-cell enforced expression of the mouse ortholog of MYD88L265P is responsible for Waldenstrom-like B-cell lymphoma. bioRxiv 2019:794024. https://doi.org/10.1101/794024
17. Barbosa R, Xu AQ, D'Andrea D et al (2020) Co-activation of NF-κB and MYC renders cancer cells addicted to IL6 for survival and phenotypic stability. https://doi.org/10.1101/2020.04.12.038414
18. Knies N, Alankus B, Weilemann A et al (2015) Lymphomagenic CARD11/BCL10/MALT1 signaling drives malignant B-cell proliferation via cooperative NF-κB and JNK activation. Proc Natl Acad Sci U S A 112:E7230–E7238
19. Hobeika E, Thiemann S, Storch B et al (2006) Testing gene function early in the B cell lineage in mb1-cre mice. Proc Natl Acad Sci U S A 103:13789–13794
20. Schmidt-Supprian M, Rajewsky K (2007) Vagaries of conditional gene targeting. Nat Immunol 8:665–668
21. Schweighoffer E, Vanes L, Nys J et al (2013) The BAFF receptor transduces survival signals by co-opting the B cell receptor signaling pathway. Immunity 38:475–488
22. Hobeika E, Dautzenberg M, Levit-Zerdoun E et al (2018) Conditional selection of B cells in mice with an inducible B cell development. Front Immunol 9:1806
23. Rickert RC, Roes J, Rajewsky K (1997) B lymphocyte-specific, Cre-mediated mutagenesis in mice. Nucleic Acids Res 25:1317–1318
24. Boross P, Breukel C, van Loo PF et al (2009) Highly B lymphocyte-specific tamoxifen inducible transgene expression of CreERT2 by using the LC-1 locus BAC vector. Genesis 47:729–735
25. Yasuda T, Wirtz T, Zhang B et al (2013) Studying Epstein–Barr virus pathologies and immune surveillance by reconstructing EBV infection in mice. Cold Spring Harb Symp Quant Biol 78:259–263
26. Kraus M, Alimzhanov MB, Rajewsky N et al (2004) Survival of resting mature B lymphocytes depends on BCR signaling via the Igalpha/beta heterodimer. Cell 117:787–800
27. Kwon K, Hutter C, Sun Q et al (2008) Instructive role of the transcription factor E2A in early B lymphopoiesis and germinal center B cell development. Immunity 28:751–762
28. Casola S, Cattoretti G, Uyttersprot N et al (2006) Tracking germinal center B cells expressing germ-line immunoglobulin gamma1 transcripts by conditional gene targeting. Proc Natl Acad Sci U S A 103:7396–7401
29. Weber T, Seagal J, Wingler W et al (2019) A novel allele for inducible Cre expression in germinal center B cells. Eur J Immunol 49:192–194

30. Shinnakasu R, Inoue T, Kometani K et al (2016) Regulated selection of germinal-center cells into the memory B cell compartment. Nat Immunol 17:861–869
31. Crouch EE, Li Z, Takizawa M et al (2007) Regulation of AID expression in the immune response. J Exp Med 204:1145–1156
32. Robbiani DF, Bothmer A, Callen E et al (2008) AID is required for the chromosomal breaks in c-myc that lead to c-myc/IgH translocations. Cell 135:1028–1038
33. Qin H, Suzuki K, Nakata M et al (2011) Activation-induced cytidine deaminase expression in CD4+ T cells is associated with a unique IL-10-producing subset that increases with age. PLoS One 6:e29141
34. Dogan I, Bertocci B, Vilmont V et al (2009) Multiple layers of B cell memory with different effector functions. Nat Immunol 10:1292–1299
35. Calado DP, Zhang B, Srinivasan L et al (2010) Constitutive canonical NF-κB activation cooperates with disruption of BLIMP1 in the pathogenesis of activated B cell-like diffuse large cell lymphoma. Cancer Cell 18:580–589
36. Zhang B, Calado DP, Wang Z et al (2015) An oncogenic role for alternative NF-κB signaling in DLBCL revealed upon deregulated BCL6 expression. Cell Rep 11:715–726
37. Pascual M, Mena-Varas M, Robles EF et al (2019) PD-1/PD-L1 immune checkpoint and p53 loss facilitate tumor progression in activated B cell diffuse large B-cell lymphomas. Blood 133(22):2401–2412
38. Rovira J, Karube K, Valera A et al (2016) MYD88 L265P mutations, but no other variants, identify a subpopulation of DLBCL patients of activated b-cell origin, extranodal involvement, and poor outcome. Clin Cancer Res 22:2755–2764
39. Van den Broeck W, Derore A, Simoens P (2006) Anatomy and nomenclature of murine lymph nodes: fescriptive study and nomenclatory standardization in BALB/cAnNCrl mice. J Immunol Methods 312:12–19
40. Sander S, Calado DP, Srinivasan L et al (2012) Synergy between PI3K Signaling and MYC in Burkitt lymphomagenesis. Cancer Cell 22:167–179
41. Mullins CD, Su MY, Hucthagowder V et al (2013) Germinal center B-cells resist transformation by kras independently of tumor suppressor arf. PLoS One 8:e67941

Chapter 21

The Screening of Combinatorial Peptide Libraries for Targeting Key Molecules or Protein–Protein Interactions in the NF-κB Pathway

Laura Tornatore, Daria Capece, Annamaria Sandomenico, Daniela Verzella, Davide Vecchiotti, Francesca Zazzeroni, Menotti Ruvo, and Guido Franzoso

Abstract

Peptides are emerging as an increasingly dependable class of therapeutics in the treatment of cancer and metabolic and cardiovascular diseases, which are all areas of high interest to the pharmaceutical industry. The global market for peptide therapeutics was valued at about 25 billion USD in 2018 and is estimated to reach 57.2 billion USD by the end of 2027. Here, we describe a method for the screening and deconvolution of combinatorial peptide libraries to discover compounds that target discrete signaling components of the NF-κB pathway. Recently, we used this approach to specifically disrupt the interaction between the JNK-activating kinase, MKK7, and the NF-κB-regulated antiapoptotic factor, GADD45β, in multiple myeloma (MM). We showed that the GADD45β/MKK7 complex is a functionally critical survival module downstream of NF-κB in MM cells and as such provides an attractive therapeutic target to selectively inhibit NF-κB antiapoptotic signaling in cancer cells. By integrating the library screening and deconvolution methods described here with a rational chemical optimization strategy, we developed the first-in-class GADD45β/MKK7 inhibitor, DTP3 (a D-tripeptide), which is now being trialed in MM and diffuse large B-cell lymphoma (DLBCL) patients. The same drug discovery approach may be generally applied to therapeutically target other key components of the NF-κB pathway in cancers beyond MM and DLBCL, as well as in non-malignant NF-κB-driven diseases.

Key words Combinatorial chemistry, Peptides, Drug discovery, Targeted therapy, Cancer, Apoptosis, NF-κB, GADD45β, MKK7

1 Introduction

The screening of combinatorial peptide libraries has proven to be a powerful tool to identify peptides or peptidomimetics that bind to a wide range of molecular targets, such as protein kinases, phosphatases, cell surface receptors, and proteases [1, 2]. The solid-phase peptide synthesis (SPPS) method, first conceived by Nobel laureate

Guido Franzoso and Francesca Zazzeroni (eds.), *NF-κB Transcription Factors: Methods and Protocols*, Methods in Molecular Biology, vol. 2366, https://doi.org/10.1007/978-1-0716-1669-7_21,

Robert Bruce Merrifield, has had an enormous impact on the development of the peptide synthesis technology and its many applications in drug discovery. Indeed, SPPS is widely considered to be faster, more efficient, and less expensive than any solution-phase peptide synthesis methodology [3, 4]. The introduction of one-bead-one-compound (OBOC) combinatorial methods has subsequently broadened the application of peptide library synthesis and screening technologies, leading to the development of simplified procedures that substantially reduce the cost and time for the identification of new hits [4–7]. Optimized synthetic procedures enable the preparation of millions of different custom-made peptides or peptidomimetics in the time frame of only a few days and in the format of combinatorial mixtures that can be iteratively analyzed in a time- and cost-efficient manner. As such, the peptide library technology provides a powerful ready-to-use tool for drug discovery, laboratory-scale drug screening, preclinical drug target validation, and other pharmaceutical applications [8].

Peptides represent a unique class of pharmaceutical compounds, in that they share molecular features with both proteins and small molecules, being at the same time therapeutically and biochemically different from one another and yet containing common structural elements. Due to the low cost of production and the generally high affinity of peptides for their molecular targets, an increasingly large number of bioactive peptide are entering the drug discovery and clinical development pipelines as candidate therapeutics in diverse disease indications, such as metabolic diseases, cardiovascular diseases, and cancer [9]. As of 2018, more than 60 peptide drugs (excluding insulins and analogous small proteins) were clinically approved in the USA, Europe, and Japan. An additional 150 peptides were concurrently in active clinical development, while a further 260 were assessed in human trials, but failed to reach approval [9, 10]. According to recent reports, the global market value of peptide therapeutics surpassed 25 billion USD in 2018 and is forecasted to expand at a Compound Annual Growth Rate (CAGR) of 7.9% between 2019 and 2027 [9, 11].

The use of combinatorial peptide libraries provides an effective means to identify molecules directed against novel therapeutic targets and then rapidly progress them through the drug development pipeline by affording the off-the-shelf synthetic chemistry for subsequent hit optimization and drug candidate selection. Indeed, when combined with robust screening assays and in silico molecular design approaches [4], this platform can provide bioactive compounds that present attractive drug-like characteristics for clinical development [12–14]. To streamline the screening process, we introduced the concept of "simplified library" and a bioinformatic approach for the design of focused sets of peptides that efficiently bind to the protein target [4, 5]. In brief, a simplified library is constructed using a reduced, non-redundant set of amino acids (typically 10 or 12, instead of 20 or more), each accounting for specific chemical and physical properties (i.e., hydrophilicity, hydrophobicity, charge, size, and aromaticity) [15–17]. This reduces the library complexity and facilitates the peptide synthesis and screening process. Simplified libraries of tri- or tetrapeptides are generally more homogeneous than those of larger peptides, due to a higher synthetic efficiency and the presence of fewer impurities.

Over the last decade, combinatorial library approaches similar to the one described here have been successfully applied to drug discovery across many fields of medicine [18], resulting in the identification of novel inhibitors of Janus kinase 2 (JAK2), acyl-peptide hydrolase (APEH) inhibitors, antioxidant peptides, and other molecules [19–21].

Here, we describe the methodology for the screening and iterative deconvolution of a simplified combinatorial library of 20,736 *L*-tetrapeptides. We recently used this approach to isolate compounds capable of disrupting the complex formed by the NF-κB-regulated antiapoptotic factor, GADD45β, and the JNK-activating kinase, MKK7, which we had previously identified as a cancer-selective and functionally critical survival module downstream of NF-κB and novel therapeutic target in multiple myeloma (MM) (Figs. 1 and 2) [12, 22–27]. The combinatorial *L*-tetrapeptide library for this screen was synthesized by using the pre-mix approach [4, 5], which remains the preferred method of combinatorial chemistry for production of random peptide libraries [12, 20], despite the development in the early 1990s of several alternative techniques [28, 29]. Following iterative deconvolution of the library by ELISA competition assays, we conducted a secondary screen and optimization of the resulting hits, yielding two structurally related acetylated (Ac) *L*-tetrapeptides, namely Ac-LTP1 and Ac-LTP2. Notably, both *L*-tetrapeptides disrupted the GADD45β/MKK7 complex, in vitro, with half-maximum inhibitory concentration (IC_{50}) values in the sub-nM range (Fig. 2) [12], and retained in vitro potency upon synthesis in the *D* configuration, a strategy that can be used in some cases to induce resistance to proteolysis [30, 31]. Indeed, the *D*-enantiomers of Ac-LTP1 and Ac-LTP2 (termed Ac-DTP1 and Ac-DTP2, respectively) displayed no loss of activity in vitro and, unlike their *L*-counterparts, were stable in human serum at 37 °C (Fig. 2) [12].

To improve the bioavailability of the D-peptides, while retaining specific pharmacological activity towards the GADD45β/MKK7 complex, we applied a chemical optimization strategy based on structure–activity relationship (SAR) and pharmacophore analysis [12]. By combining these two methods, we developed DTP3, a *D*-tripeptide of 525 Da of molecular weight (MW) (Fig. 2) [12], which retained all the main characteristics of the parental *D*-tetrapeptides in terms of bioactivity and specificity, including sub-nM activity and stability in vitro, while exhibiting an

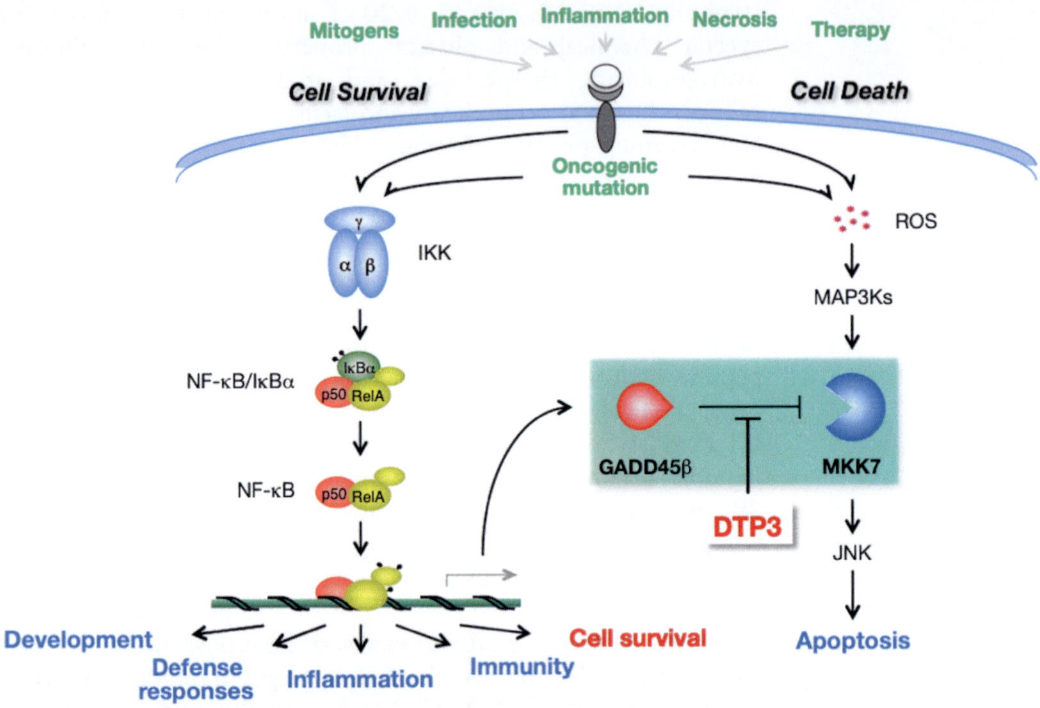

Fig. 1 Schematic representation of the GADD45β/MKK7-targeting therapeutic strategy. Depicted is the NF-κB-dependent, GADD45β/MKK7-mediated survival mechanism suppressing apoptotic JNK signaling in MM cells. Also illustrated is the first-in-class GADD45β/MKK7 inhibitor, DTP3, developed by integrating combinatorial chemistry and a chemical optimization strategy. DTP3 represents a therapeutic candidate with a novel mode of action that selectively blocks NF-κB survival signaling in malignant cells of MM and potentially other NF-κB-driven cancers

improved pharmacokinetic profile. We showed that, by disrupting the GADD45β/MKK7 complex, DTP3 specifically kills MM cells, ex vivo and in vivo, ablating MM xenografts in mice, with no apparent adverse effects, nor toxicity to healthy cells (Fig. 2) [12–14, 32]. Additional studies in rodent and non-rodent species demonstrated that DTP3 combines on-target-selective pharmacology, therapeutic anticancer efficacy, favorable drug-like properties, long plasma half-life, and good bioavailability, with no target-organs of toxicity and no adverse effects that preclude its clinical development in oncology [12–14, 32, 33]. Accordingly, a Phase I/IIa trial of DTP3 is ongoing in patients with MM and diffuse large B-cell lymphoma (DLBCL).

Future studies will determine the clinical safety and therapeutic efficacy of DTP3 in patients with MM and/or DLBCL [34]. Notwithstanding, the basic principle of targeting a pathogenetically critical axis of the NF-κB pathway, rather than NF-κB globally, could provide an effective strategy to therapeutically block NF-κB signaling in a disease-specific manner in NF-κB-driven cancers beyond MM and DLBCL and, potentially, non-malignant NF-κB-driven diseases.

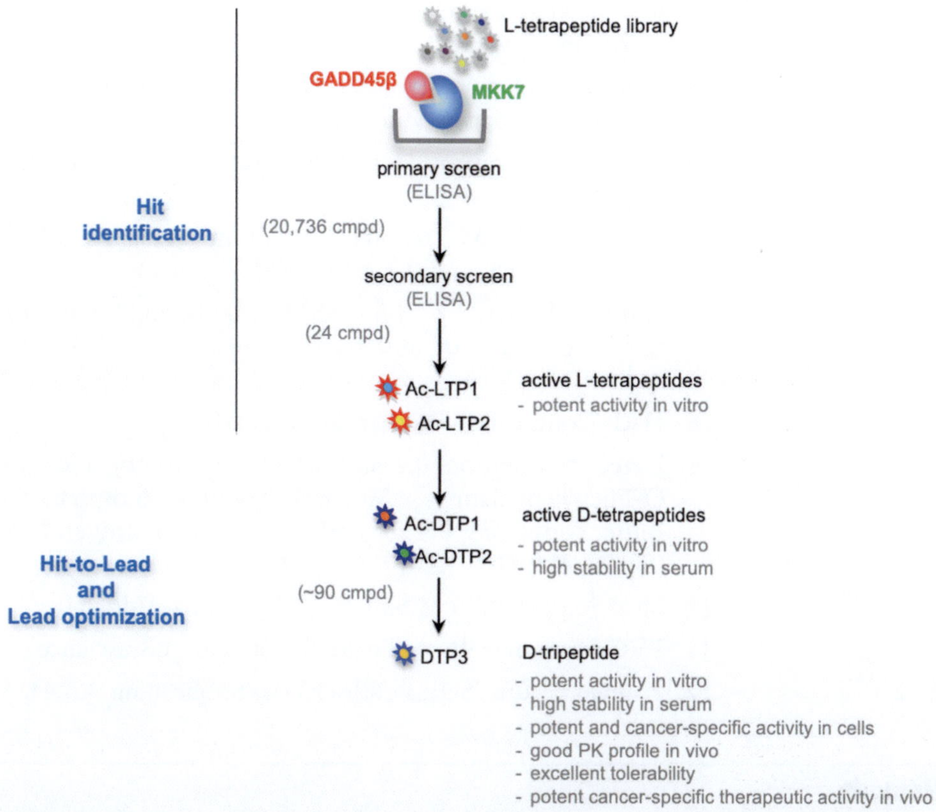

Fig. 2 Summary of the drug discovery strategy which led to the discovery of DTP3. *Cmpd* compounds

2 Materials

2.1 Protein Biotinylation

1. Buffer A, pH 7.0–7.5: 100 mM NaCl, 1 mM DTT, 1 mM EDTA, 0.1 mM PMSF.
2. His_6-Tagged Human (h)GADD45β (His_6-hGADD45β).
3. Working Solution: 2.3 mg of EZ Link NHS-LC-Biotin (Pierce) dissolved in 1 mL of 100% cold Ethanol.
4. Stop Solution: 100 mM Glycine Solution or Tris Solution.
5. Buffer E: 20 mM Tris, pH 7.5, 150 mM NaCl, 1 mM DTT, 0.1 mM EDTA and 0.1 mM PMSF.
6. Dimethylsulfoxide (DMSO).
7. HisTrap Columns (GE Healthcare).
8. Amicon Ultra-4 Centrifugal Filter.

2.2 ELISA Competition Assay

1. Biotin-Labeled His_6-hGADD45β.
2. Glutathione S-Transferase (GST)-hMKK7.

3. GST protein (expressed and purified using the same conditions as for GST-hMKK7).
4. Blocking Solution: Non-fat Dry Milk (NFDM Biorad), 1% Phosphate Buffered Saline (PBS) 1×.
5. Individual Tetrapeptides and Mixtures of Tetrapeptides that are to be tested.
6. Buffer E: 20 mM Tris, pH 7.5, 150 mM NaCl, 1 mM DTT, 0.1 mM EDTA, and 0.1 mM PMSF.
7. Washing Buffer: 1× PBS (modified by having a final concentration of NaCl of 500 mM), 0.04% TWEEN 20 (Sigma-Aldrich).
8. HRP-Conjugated Streptavidin (Sigma).
9. Detection Solution: 0.4 mg/mL of the chromogenic substrate O-Phenylenediamine, 50 mM Sodium Phosphate-Citrate Buffer containing 0.4 mg/mL of Urea in Hydrogen Peroxide (Sigma-Aldrich).
10. Stop Solution: 2.5 M Sulfuric Acid (Sigma-Aldrich).
11. PVC Microtiter 96-well plate flat bottom (eBioscience).
12. Human Serum (Sigma-Aldrich) (*see* Subheading 3.3.4).

3 Methods

Carry out the procedures at room temperature unless otherwise specified.

3.1 Tetrapeptide Library Design and Synthesis

1. Synthesize the Combinatorial Tetrapeptide Library using the solid-phase method, by applying conventional 9-Fluorenylmethoxycarbonyl (Fmoc)/*tert*-Butyl (*t*Bu) chemistry and the pre-mix approach to achieve sequence randomization, as previously described [12, 20, 21] (*see* **Notes 1** and **2**). For synthesis of the Simplified *L*-Tetrapeptide library, use a selection of 12 commercially available *L*-amino acids (GL-Biochem; Novabiochem) that occupy different chemical spaces (*see* Table 1) [12, 20] (*see* **Notes 1** and **2**).
2. Analyze the 12 Tetrapeptide mixtures to assess the identity and distribution of the peptide components by using Liquid Chromatography-Mass Spectrometry (LC-MS) [12] (*see* **Note 3**).
3. Lyophilize and then dissolve the tetrapeptide mixtures in DMSO at a concentration of 10 mg/mL (stock solution) and store them in aliquots at −20 °C until they are used [12, 20].

Table 1
Properties of the "amino acidic building blocks" used for library design

Selected amino acid	Side chain	Side chain properties		
Glutamine	$-CH_2CH_2CONH_2$	Polar	Neutral	Amide
Serine	$-CH_2OH$	Polar	Neutral	Hydrophobic
Arginine	$-(CH_2)_3NHC(NH)NH_2$	Polar	Basic (strongly)	
Alanine	$-CH_3$	Non polar	Neutral	Hydrophobic
Tyrosine	$-CH_2-C_6H_4OH$	Polar	Neutral	Aromatic
Proline	$-CH_2CH_2CH_2$	Non polar	Neutral	Hydrophobic
Methionine	$-CH_2CH_2SCH_3$	Non polar	Neutral	Hydrophobic
Cysteine	$-CH_2SH$	Non polar	Neutral	–
Phenylalanine	$-CH_2C_6H_5$	Non polar	Neutral	Aromatic
Leucine	$-CH_2CH(CH_3)_2$	Non polar	Neutral	Aliphatic
Histidine	$-CH_2-C_3H_3N_2$	Polar	Basic (weakly)	Aromatic
Aspartic acid	$-CH_2COOH$	Polar	Acidic	–

Related to Fig. 2

3.2 Protein Purification and Biotinylation

1. Purify the human GST-hMKK7 and His$_6$-hGADD45β recombinant proteins using HisTrap Columns (GE Healthcare), as described previously [35] (*see* **Note 4**).
2. Dialyze the fractions of His$_6$-hGADD45β purified using HisTrap Columns against Buffer A at 4 °C for 16 h.
3. Concentrate the His$_6$-hGADD45β protein using the Amicon Ultra-4 Centrifugal Filter to obtain a final protein concentration of 1 mg/mL.
4. Add 50 μL of Working Solution to a 1-mL aliquot of the His$_6$-hGADD5β protein concentrated at 1 mg/mL in ice.
5. Incubate the His$_6$-hGADD5β protein mixture for 1 h in ice and then stop the reaction by adding 50 μL of Stop Solution.
6. Dialyze the Biotin-Labeled His$_6$-hGADD5β protein against Buffer A for 16 h at 4 °C.
7. Prepare aliquots of the Biotin-Labeled His$_6$-hGADD5β protein in ice.
8. Store the aliquots of Biotin-Labeled His$_6$-hGADD45β at −80 °C (*see* **Note 5**).
9. Measure the concentration of the Biotin-Labeled His$_6$-hGADD45β protein using the Bradford method.
10. Analyze the purity of Biotin-Labeled His$_6$-hGADD45β protein preparation by using LC-MS (*see* **Note 6**).

3.3 Iterative Deconvolution of the Combinatorial Tetrapeptide Library

3.3.1 ELISA Competition Assay

1. Coat the wells of a PVC Microtiter Plate with GST-hMKK7 at the concentration of 42 nM in Buffer E, pH 7.5 (*see* **Note 7**).
2. Cover the plate with adhesive plastic and incubate it in the dark overnight at 4 °C.
3. Wash the plate three times with Washing Solution. The solutions and washes are removed by flicking the plate over a sink. The remaining drops are removed by gently patting the plate on a paper towel.
4. Block the remaining protein-binding sites on the coated wells by adding 200 μL of Blocking Solution per well.
5. Cover the plate with adhesive plastic and incubate it for at least 1–2 h in the dark at 37 °C.
6. Wash the plate three times with Washing Solution. The solutions and washes are removed by flicking the plate over a sink. The remaining drops are removed by gently patting the plate on a paper towel.
7. Add 100 μL of soluble Biotin-Labeled His_6-hGADD45β at the concentration of 21 nM and 100 μL of the competing Tetrapeptide Mixtures or Single Tetrapeptides at the nominal concentration of 42 nM to each well (*see* **Notes 7** and **8**).
8. Incubate the plate(s) in the dark for 1 h at 37 °C.
9. Repeat **step 3**.
10. Add 100 μL of 1 μg/mL HRP-Conjugated Streptavidin to each well.
11. Incubate the plate(s) in the dark for 1 h at 37 °C.
12. Repeat **step 3**.
13. Add 100 μL of Detection Solution to each well.
14. Add 50 μL of Stop Solution to each well.
15. Read the plate at 490 nm using an ELISA Spectrophotometer.
16. Analyze the data using Excel and GraphPad software.
17. Express the Optical Density (OD) values as percentages of inhibition of the His_6-hGADD45β binding to GST-hMKK7, relative to the binding measured in the absence of a competitor (*see* **Note 9**).
18. Identify the most active Tetrapeptide(s) as the one yielding the maximal percentage of inhibition value and use it as template for the synthesis of a Second-Generation Library.

3.3.2 Synthesizing the Second-Generation Library

1. Design 24 Second-Generation *L*-Tetrapeptides using as template(s) the most active peptide(s) identified in the Combinatorial Library screen (Subheading 3.3.1) and including the *L*-amino acids which were excluded from the synthesis of the initial Combinatorial *L*-Tetrapeptides Library, as described previously [12].

2. Synthesize the 24 Second-Generation *L*-Tetrapeptides by using the solid-phase Fmoc/tBu method and replacing the N-terminal Fmoc-(βAla)$_2$ Tag with an Acetyl (Ac) group. Purify.
3. Analyze the 24 Second-Generation *L*-Tetrapeptides by LC-MS [12] (*see* **Note 3**).
4. Lyophilize and then dissolve the Peptides Mixtures in DMSO at the concentration of 10 mg/mL (Stock Solution) and store them in aliquots at −20 °C until they are used [12, 20].

3.3.3 Examining the Dose-Dependent Effects

1. Carry out the ELISA GADD45β/MKK7 competition assay using the conditions described in Subheading 3.3.1 (**steps 1–6**).
2. Add 100 μL of soluble Biotin-Labeled His$_6$-hGADD45β at the concentration of 21 nM and 100 μL of the competing Tetrapeptides at increasing concentrations, ranging from 0.001 to 21 nM (*see* **Note 9**).
3. Carry out the ELISA GADD45β/MKK7 competition assay using the conditions described in Subheading 3.3.1 (**steps 8–17**) (*see* **Note 9**).

3.3.4 Synthesizing D-Enantiomers

1. Synthesize the *D*-Enantiomers of the most active *L*-Tetrapeptides identified from deconvolution of the Combinatorial Library, including those identified from the screen of the Second-Generation *L*-Tetrapeptide pool (*see* **Note 10**).
2. Purify and characterize the *D*-Tetrapeptides from **step 1** using LC-MS [12] (*see* **Note 3**).
3. Lyophilize and then dissolve the *D*-Tetrapeptides in DMSO at a concentration of 10 mg/mL (Stock Solution) and store them in aliquots at −20 °C until they are used [12, 20].

3.3.5 Examining the Tetrapeptide Stability in Biological Fluids

1. Incubate an aliquot of the *L*- and *D*-Tetrapeptides at the final concentration of 10 μM in human serum for 48 h at 37 °C.
2. In parallel, incubate a separate aliquot of the *L*- and *D*-Tetrapeptides at the concentration of 10 μM in Buffer E for 48 h at 37 °C.
3. Carry out the ELISA GADD45β/MKK7 competition assay using the conditions described in Subheading 3.3.1 (**steps 1–6**).
4. Add 50 μL of soluble Biotin-Labeled His$_6$-hGADD45β at the concentration of 42 nM to each well, and then add 50 μL of the Tetrapeptide solutions incubated with or without human serum, ranging from 0.002 to 42 nM (*see* **Note 7**).

5. Carry out the ELISA GADD45β/MKK7 competition assay using the conditions described in Subheading 3.3.1 (**steps 8–17**).

3.4 Undertaking the Hit-To-Lead and Lead Optimization of the Selected D-Tetrapeptides

1. Use an appropriate chemical optimization strategy based on structure–activity relationship (SAR) and pharmacophore analysis to optimize the cellular uptake and in vivo bioavailability of the tetrapeptides isolated from Subheading 3.3.5, as previously described [12].
2. Test the resulting compounds in ELISA GADD45β/MKK7 competition assays, as described in Subheading 3.3.3, and select those showing the strongest activity.
3. Test the stability of the compounds selected in **step 2** in biological fluids, as described in Subheading 3.3.5 (*see* **Note 11**).

4 Notes

1. The Combinatorial Tetrapeptide Library reported here is commonly referred to as "Simplified Combinatorial Peptide Library," because it utilizes a limited number of non-redundant amino acids for the assembly of the peptides. Typically, only 12 of the 20 natural amino acids are used for peptide synthesis based on their diversity in lipophilicity, MW, charge, polarity, and aromaticity, as previously described [12, 20]. This approach simplifies both the synthesis and deconvolution of the library. For the Combinatorial Library reported here, we selected the following amino acids: Asp (D), representing acidic amino acids; Arg (R) and His (H), representing basic amino acids, with His also representing aromatic residues; Gln (Q) and S-Acetamidomethyl(Acm)-Cys (referred to as C), representing amidated side chains; Phe (F) and Tyr (Y), representing aromatic amino acids; Ser (S), representing amino acids containing hydroxyl groups; Leu (L) and Met (M), representing bulky hydrophobic amino acids; Ala (A), representing small hydrophobic amino acids; Pro (P), producing a bent in the peptide conformation (*see* Table 1) [12].
2. The Combinatorial *L*-Tetrapeptide library described here is synthesized using the pre-mix method (*see* Fig. 2). In the adopted design, the library consists of 20,736 *L*-Tetrapeptides, representing all possible combinations of the twelve amino acids used at each of the four positions (i.e., 12^4 peptides in total), as previously described [12]. In this arrangement, the solid support is divided into 12 equal portions (12 being the number of amino acids used at each step), and each portion is then coupled to a specific Fmoc-derivatized amino acid (>99%

of purity) [36]. Amino acids are pre-activated by one of the previously described methods [37].

The Combinatorial Library used for the identification of GADD45β/MKK7-targeting agents [12] had the general formula "Fmoc-(βAla)$_2$-X_1-X_2-X_3-X_4-$CONH_2$" and contained all possible combinations of the 12 *L*-amino acids listed in (Table 1), at each position from X_1 to X_4. The *L*-Tetrapeptides were joined to an N-terminal β-Ala2 spacer linked to a Fmoc Tag. This was introduced to provide a large hydrophobic and fluorescent group for assisting in various types of screening assays. The two N-terminal β-Ala residues were introduced as spacers between the peptides and the Fmoc group. An amide ($CONH_2$) group was present at the C-termini of the peptides. Other combinatorial peptide libraries, similar to the one described here, were used for different screens in other studies [12, 19–21].

3. The identity and purity of each peptide should be verified by LC-MS. Peptide spectra can be acquired in the positive mode between 200 and 2000 *m/z*, using approximately 50 ng of peptide, at a flow rate of 0.20 mL/min by applying a gradient from 5% to 70% of Solvent B (CH_3CN, 0.05% TFA) over Solvent A (H_2O, 0.08% TFA), within a time period of 100 min. Where necessary, the gradient can be changed to elute compounds within a time frame suitable for LC-MS and diode array analyses. Peptide identities are verified on the basis of the extracted ion chromatogram for each species.
4. The full-length hGADD45β protein fused to an N-terminal poly-histidine (His_6) tag is purified by affinity chromatography using HisTrap Columns (GE Healthcare), as previously described [12, 35]. The full-length hMKK7 protein fused to an N-terminal GST tag (GST-MKK7) is purified by using GST-Trap, as previously described [12, 35], and then used for ELISA competition assays. The GST protein is similarly purified by GST-Trap and used as control in the assays, as appropriate.
5. Aliquots of Biotin-Labeled His_6-hGADD45β cannot be re-frozen or kept at 4 °C.
6. The presence of a single biotin molecule increases the MW of His_6-hGADD45β by approximately 339 Da, and so the derivatization of the protein can be monitored by LC-MS.
7. Routinely run duplicate or triplicate experiments and set up 96-well plates that are coated with GST (*see* **step 1**) or Biotin-Labeled His_6-hGADD45β bound to GST-hMKK7 in the absence of competitor (*see* **step 7**) as blank and positive control, respectively. Use at least one or two unrelated peptides as negative controls.

8. Test the purity of the Biotin-Labeled His_6-hGADD45β protein by running a dose-dependent response in MKK7-binding ELISA assays. Carry out the ELISA assay as described in Subheading 3.2 (**step 10**), but in **step 7** add Biotin-Labeled His_6-hGADD45β to the wells at increasing concentrations, ranging from 5.2 to 168 nM, and incubate in the dark for 1 h at 37 °C.
9. The peptides that can disrupt the His_6-hGADD45β/GST-hMKK7 interaction are considered active compounds. IC_{50} values are defined as the mean concentrations of the compounds producing a 50% inhibition of His_6-hGADD45β binding to GST-hMKK7, relative to the binding measured in the absence of competitors. For calculating the IC_{50} values, it is recommended to test at least seven concentrations of the peptides being investigated.
10. The synthesis of *D*-Enantiomers of the *L*-Tetrapeptides is a strategy that can be used in some cases to render tetrapeptides resistant to proteolysis [30, 31].
11. The most active compounds must be subsequently investigated in cell-based assays to assess their cellular uptake and biological activities, as previously described [4, 12, 21]. The pharmacokinetic profiles of the most active compound(s) should be thoroughly investigated, in vitro and in vivo, before proceeding with any experimentation in animal models.

Acknowledgments

This work was supported in part by Medical Research Council (MRC) Biomedical Catalyst grant MR/L005069/1, Bloodwise project grant 15003, and NIHR Imperial Biomedical Research Centre (BRC) Push for Impact Award WIIS_P81135 to G.F. Infrastructure support for this research was provided in part by the NIHR Imperial BRC. The authors would like to acknowledge and thank Mr. Cristian Ng for his assistance with the literature review.

Conflict of Interest: *G.F., L.T., and M.R. are named inventors on patents relating to this research.*

References

1. Bozovičar K, Bratkovič T (2019) Evolving a peptide: library platforms and diversification strategies. Int J Mol Sci 21(1):215
2. Bononi FC, Luyt LG (2015) Synthesis and cell-based screening of one-bead-one-compound peptide libraries. Methods Mol Biol 1248:223–237
3. Merrifield RB (1963) Solid phase peptide synthesis. 1. Synthesis of a tetrapeptide. J Am Chem Soc 85:2149–2154
4. Sandomenico A, Caporale A, Doti N, Doti N, Cross S, Cruciani G et al (2020) Synthetic peptide libraries: from random mixtures to

in vivo testing. Curr Med Chem 27 (6):997–1016

5. Marasco D, Perretta G, Sabatella M, Ruvo M (2008) Past and future perspectives of synthetic peptide libraries. Curr Protein Pept Sci 9(5):447–467
6. Lam KS, Lebl M, Krchnák V (1997) The "one-bead-one-compound" combinatorial library method. Chem Rev 97:411–448
7. Coin I, Beyermann M, Bienert M (2007) Solid-phase peptide synthesis: from standard procedures to the synthesis of difficult sequences. Nat Protoc 2:3247–3256
8. Hughes JP, Rees S, Kalindjian SB, Philpott KL (2011) Principles of early drug discovery. Br J Pharmacol 162(6):1239–1249
9. Lee AC, Harris JL, Khanna KK, Hong JH (2019) A comprehensive review on current advances in peptide drug development and design. Int J Mol Sci 20(10):2383
10. Lau JL, Dunn MK (2018) Therapeutic peptides: historical perspectives, current development trends, and future directions. Bioorg Med Chem 26(10):2700–2707
11. Peptide therapeutics market – global industry analysis, size, share, growth, trends, and forecast 2019–2027. https://www.transparencymarketresearch.com/peptide-therapeutics-market.html. Rep Id: TMRGL418. Published: 2020-02-28
12. Tornatore L, Sandomenico A, Raimondo D, Low C, Rocci A, Capece D et al (2014) Cancer-selective targeting of the NF-κB survival pathway with GADD45β/MKK7 inhibitors. Cancer Cell 26:495–508
13. Tornatore L, Capece D, D'Andrea D, Begalli F, Verzella D, Bennett J et al (2019) Preclinical toxicology and safety pharmacology of the first-in-class GADD45β/MKK7 inhibitor and clinical candidate, DTP3. Toxicol Rep 6:369–379
14. Tornatore L, Capece D, D'Andrea D, Begalli F, Verzella D, Bennett J et al (2018) Clinical proof of concept for a safe and effective NF-κB-targeting strategy in multiple myeloma. Br J Haematol 185(3):588–592
15. Houghten RA, Pinilla C, Blondelle SE, Appel JR, Dooley CT, Cuervo JH (1991) Generation and use of synthetic peptide combinatorial libraries for basic research and drug discovery. Nature 354(6348):84–86
16. Lonardo E, Parish CL, Ponticelli S, Marasco D, Ribeiro D, Ruvo M et al (2010) A small synthetic cripto blocking peptide improves neural induction, dopaminergic differentiation, and functional integration of mouse embryonic stem cells in a rat model of Parkinson's disease. Stem Cells 28(8):1326–1337
17. Ponticelli S, Marasco D, Tarallo V, Albuquerque RJC, Mitola S, Takeda A et al (2008) Modulation of angiogenesis by a tetrameric tripeptide that antagonizes vascular endothelial growth factor receptor 1. J Biol Chem 283 (49):34250–34259
18. Liu R, Li X, Lam KS (2017) Combinatorial chemistry in drug discovery. Curr Opin Chem Biol 38:117–126
19. Doti N, Scognamiglio PL, Madonna S, Scarponi C, Ruvo M, Perretta G et al (2012) New mimetic peptides of the kinase-inhibitory region (KIR) of SOCS1 through focused peptide libraries. Biochem J 443(1):231–240
20. Sandomenico A, Russo A, Palmieri G, Bergamo P, Gogliettino M, Falcigno L, Ruvo M (2012) Small peptide inhibitors of acetyl-peptide hydrolase having an uncommon mechanism of inhibition and a stable bent conformation. J Med Chem 55:2102–2111
21. Sandomenico A, Severino V, Apone F, De Lucia A, Caporale A, Doti N et al (2017) Trifluoroacetylated tyrosine-rich D-tetrapeptides have potent antioxidant activity. Peptides 89:50–59
22. De Smaele E, Zazzeroni F, Papa S, Nguyen DU, Jin R, Jones J et al (2001) Induction of gadd45β by NF-κB downregulates pro-apoptotic JNK signalling. Nature 414:308–313
23. Papa S, Zazzeroni F, Bubici C, Jayawardena S, Alvarez K, Matsuda S et al (2004) GADD45β mediates the NF-κB suppression of JNK signalling by targeting MKK7/JNKK2. Nat Cell Biol 6:146–153
24. Papa S, Monti SM, Vitale RM, Bubici C, Jayawardena S, Alvarez K et al (2007) Insights into the structural basis of the GADD45-β-mediated inactivation of the JNK kinase, MKK7/JNKK2. J Biol Chem 282:19029–19041
25. Begalli F, Bennett J, Capece D, Verzella D, D'Andrea D, Tornatore L et al (2017) Unlocking the NF-κB conundrum: embracing complexity to achieve specificity. Biomedicine 5 (3):50
26. Bennett J, Capece D, Begalli F, Verzella D, D'Andrea D, Tornatore L et al (2018) NF-κB in the crosshairs: rethinking an old riddle. Int J Biochem Cell Biol 95:108–112
27. Capece D, D'Andrea D, Verzella D, Tornatore L, Begalli F, Bennett J et al (2018) Turning an old GADDget into a troublemaker. Cell Death Differ 25(4):640–642

28. Furka A, Sebestyén F, Asgedom M, Dibó G (1991) General method for rapid synthesis of multicomponent peptide mixtures. Int J Pept Protein Res 37(6):487–493
29. Lam K, Salmon S, Hersh E, Hruby VJ, Kazmierski WM, Knappet RJ (1991) A new type of synthetic peptide library for identifying ligand-binding activity. Nature 354:82–84
30. Zhou N, Luo Z, Luo J, Fan X, Cayabyab M, Hiraoka M et al (2002) Exploring the stereochemistry of CXCR4-peptide recognition and inhibiting HIV-1 entry with D-peptides derived from chemokines. J Biol Chem 277:17476–17485
31. Nickl CK, Raidas SK, Zhao H, Sausbier M, Ruth P, Tegge W, Brayden JE, Dostmann WR (2010) (D)-Amino acid analogues of DT-2 as highly selective and superior inhibitors of cGMP-dependent protein kinase alpha. Biochim Biophys Acta 1804:524–532
32. Rega C, Russo R, Focà A, Sandomenico A, Iaccarino E, Raimondo D et al (2018) Probing the interaction interface of the GADD45β/MKK7 and MKK7/DTP3 complexes by chemical cross-linking mass spectrometry. Int J Biol Macromol 114:114–123
33. Karin M (2014) Whipping NF-κB to submission via GADD45 and MKK7. Cancer Cell 26 (4):447–449
34. Verzella D, Bennett J, Fischietti M, Thotakura AK, Recordati C, Pasqualini F et al (2018) GADD45β loss ablates innate immunosuppression in cancer. Cancer Res 78:1275–1292
35. Tornatore L, Marasco D, Dathan N, Vitale RM, Benedetti E, Papa S et al (2008) Gadd45 beta forms a homodimeric complex that binds tightly to MKK7. J Mol Biol 378(1):97–111
36. Behrendt R, White P, Offer J (2016) Advances in Fmoc solid-phase peptide synthesis. J Pept Sci 22:4–27
37. Caporale A, Doti N, Monti A, Sandomenico A, Ruvo M (2018) Automatic procedures for the synthesis of difficult peptides using oxyma as activating reagent: a comparative study on the use of bases and on different deprotection and agitation conditions. Peptides 102:38–46

Index

Guido Franzoso and Francesca Zazzeroni (eds.), *NF-κB Transcription Factors: Methods and Protocols*, Methods in Molecular Biology, vol. 2366, https://doi.org/10.1007/978-1-0716-1669-7,

T

U

W

MIX
Papier aus verantwortungsvollen Quellen
Paper from responsible sources
FSC® C105338

If you have any concerns about our products,
you can contact us on
ProductSafety@springernature.com

In case Publisher is established outside the EU,
the EU authorized representative is:
Springer Nature Customer Service Center GmbH
Europaplatz 3, 69115 Heidelberg, Germany

Printed by Libri Plureos GmbH
in Hamburg, Germany